STUDENT STUDY GUIDE

DIGITAL SYSTEMS

Principles and Applications

sixth edition

STUDENT STUDY GUIDE

Frank J. Ambrosio

DIGITAL SYSTEMS

Principles and Applications

sixth edition

Ronald J. Tocci
Neal Widmer

Prentice Hall
Englewood Cliffs, NJ 07632

Production Editor: *Cathy O'Connell*
Prepress/Manufacturing Buyer: *Ilene Sanford*
Supplements Acquisitions Editor: *Judy Casillo*

A Simon & Schuster Company
Englewood Cliffs, New Jersey 07632

Printed in the United States of America

10 9 8 7 6 5

ISBN 0-13-302050-9

PRENTICE-HALL INTERNATIONAL (UK) LIMITED, *London*
PRENTICE-HALL OF AUSTRALIA PTY. LIMITED, *Sydney*
PRENTICE-HALL CANADA INC., *Toronto*
PRENTICE-HALL HISPANOAMERICANA, S.A., *Mexico*
PRENTICE-HALL OF INDIA PRIVATE LIMITED, *New Delhi*
PRENTICE-HALL OF JAPAN, INC., *Tokyo*
SIMON & SCHUSTER ASIA PTE. LTD., *Singapore*
EDITORA PRENTICE-HALL DO BRASIL, LTDA., *Rio de Janeiro*

CONTENTS

SOLUTIONS

HOW TO USE THE STUDY GUIDE

The *Student Study Guide* should be used in conjunction with the 6th edition of the DIGITAL SYSTEMS **Principles and Applications** text book by **Ronald J. Tocci.** The user of this *Student Study Guide* should attempt to acquire a reasonable understanding of the material by first reading the text book. Should the student find that he/she has difficulties in comprehending certain parts of a chapter, then consultation of the *Student Study Guide* might be all that is needed to bring a full and clear understanding of the topic in question. Some students may find that a full understanding of the text material is achieved just by reading the text and doing the problems at the end of each chapter. For those students, the *Student Study Guide* will provide additional problems for greater proficiency and mastery of the subject matter. In addition, this Study Guide will provide a valuable source of detailed information for review purposes and for examination preparation.

The glossary at the beginning of each chapter contains all of the new terms and their definition introduced in that particular chapter. In brackets, at the end of each definition, the text book chapter and corresponding section reference is provided in case that a more in-depth explanation of the new term is needed.

The exercises, and the end-of-chapter-quiz, will thoroughly review the material covered in each chapter of the text book. New circuit diagrams as well as existing ones from the text book will be used throughout. Wherever appropriate, troubleshooting case studies will be used to review and supplement the exercises. These troubleshooting case studies will have their particular problem number preceded by ⊕ for readily identification. More difficult and challenging problems will be preceded with 💡. Problems whose answer require a timing diagram are identified with ⌛. In any case, the step-by-step solution for each individual problem is given in the second half (page 234) of the *Student Study Guide.*

The student should work out each solution in the space provided after each problem. Some problems require only a written answer, while others require diagrams, calculations, timing diagrams, etc. The Figures used in the *Solutions* part of this Study Guide, have a *'P'* preceding their number as to reference them to their respective Problem number (i.e. Figure P10.14, means that, this figure is part of the solution of problem 10.14). Figures without a preceding 'P' are used only in the *Problems* section of the Study Guide.

ACKNOWLEDGMENTS

My sincere thanks and appreciation go to **Professor Ronald J. Tocci** for his unselfish contributions to this Study Guide. His constant willingness to answer a question or make an insightful suggestion made the writing of this Study Guide a more pleasant and enjoyable experience.

Moreover, I am grateful for the helpful comments provided by users of the fifth edition. I am also thankful for all the suggestions made by Jerry Hartman of ITT Technical Institute. Many of his suggestions have been incorporated into this edition. I hope that the improvements that have been made in this edition will provide students, instructors, and all other users with a more effective tool for learning about Digital Systems.

And finally, a special word of gratitude goes to **Dr. Barbara J. Agor**, for her patience, guidance, and total dedication to my education during my first years in the United States of America.

*To my parents **José** and **Georgina Ambrósio**, whose efforts and unselfish sacrifices made my education possible.*

*And, to my wife **Ana** and son **Filip**, for their patience and understanding during the writing of this Study Guide.*

1 INTRODUCTORY CONCEPTS

Objectives

Upon completion of this chapter, you will be able to:

- Distinguish between analog and digital representations.
- Name the advantages, disadvantages, and major differences among analog, digital, and hybrid systems.
- Understand the need for analog-to-digital converters (ADCs) and digital-to-analog converters (DACs).
- Convert between decimal and binary numbers.
- Identify typical digital signals.
- Cite several integrated-circuit fabrication technologies.
- Identify a timing diagram.
- State the differences between parallel and serial transmission.
- Describe the property of memory.
- Describe the major parts of a digital computer and understand their functions.

Glossary of key terms covered in this chapter:

- **Analog Representation** - Representation of a quantity that varies over a continuous range of values. [sec.1.1]
- **Analog System** - Combination of devices designed to manipulate physical quantities that are represented in analog form. [sec.1.2]
- **Analog-to-Digital converter** - Circuit that converts an Analog quantity to a corresponding Digital quantity. [sec.1.2]
- **Arithmetic Unit** - Part of the Computer dedicated to all of the arithmetic and logical operations. [sec.1.8]
- **Binary Digit** - Bit. [sec.1.3]
- **Binary Point** - A mark which separates the integer from the fractional portion of a binary quantity. [sec.1.3]
- **Binary System** - Number system in which there are only two possible digit values, 0 and 1. [sec.1.3]
- **Bit** - A digit in the Binary System. [sec.1.3]
- **CMOS (Complementary Metal Oxide Semiconductor)** - Integrated circuit technology which uses MOSFETS as the principal circuit element. [sec.1.5]
- **Control Unit** - This unit provides decoding of program instructions and the necessary timing and control signals for the execution of such instructions. [sec.1.8]
- **Decimal System** - Number system which uses ten different digits or symbols to represent a quantity. [sec.1.3]
- **Digital Integrated Circuits** - Self-containing circuits which have been made by using one of several integrated-circuit fabrication technologies. [sec.1.5]
- **Digital Representation** - Representation of a quantity that varies in discreet steps over a range of values. [sec.1.1]
- **Digital-to-Analog converter** - Circuit that converts a Digital quantity to a corresponding Analog quantity. [sec.1.2]

- **Digital Computer** - A system of hardware that performs arithmetic and logic operations, manipulates data and makes decisions. [sec.1.8]
- **Digital System** - Combination of devices designed to manipulate physical quantities that are represented in digital form. [sec.1.2]
- **Flip-Flop** - A memory device capable of storing a logic level. [sec.1.7]
- **Hybrid System** - A system that employs both analog and digital techniques. [sec.1.2]
- **Input Unit** - This unit facilitates the feeding of information into the computer's memory unit. [sec.1.8]
- **Least Significant Bit (LSB)** - The rightmost bit (smallest weight) of a binary expressed quantity. [sec.1.3]
- **Least Significant Digit (LSD)** - The digit that carries the least weight in a particular number. [sec.1.3]
- **Logic Circuits** - Any circuit that behaves according to a set of logic rules. [sec.1.5]
- **Mainframe** - (See maxicomputer.)
- **Maxicomputer** - The largest computers available, used to maintain and update large quantities of data and information. [sec.1.8]
- **Memory** - The ability of a circuit's output to remain at one state even after the input condition which caused that state is removed. [sec.1.7]
- **Memory Unit** - This unit stores instructions and data received from the Input unit, as well as results from the Arithmetic Logic Unit. [sec.1.8]
- **Microcomputer** - The newest member of the computer family consisting of microprocessor chip, memory chips and I/O interface chips. In some cases all of the aforementioned are in one single IC. [sec.1.8]
- **Minicomputers** - Computers that are generally faster and possess more capabilities than Microcomputers. [sec.1.8]
- **Most Significant Bit (MSB)** - The leftmost binary bit (largest weight) of a binary expressed quantity. [sec.1.3]
- **Most Significant Digit (MSD)** - The digit that carries the most weight in a particular number. [sec.1.3]

- **NMOS (N-channel Metal Oxide Semiconductor)** - Integrated-Circuit Technology which uses N-Channel MOSFETS as the principal circuit element. [sec.1.5]

- **Output Unit** - This unit receives data from the memory unit and presents it to the operator. [sec.1.8]

- **Parallel Transmission** - The simultaneous transfer of binary information from one place to another. [sec.1.6]

- **Positional-value system** - A system in which the value of a digit is dependent on its relative position. [sec.1.3]

- **Program** - A sequence of binary-coded instructions designed to accomplish a particular task by a computer. [sec.1.2/1.8]

- **Serial Transmission** - The transfer of binary information from one place to another a bit at a time. [sec.1.6]

- **Supercomputers** - Computers with the greatest operating speed and computational power. [sec.1.8]

- **Timing Diagram** - A depiction of logic levels as it relates to time. [sec.1.6]

- **Toggle** - The process of changing from one binary state to the other. [sec.1.3]

- **TTL (Transistor Transistor Logic)** - Integrated-Circuit Technolog which uses the bipolar transistor as the principal circuit element. [sec.1.5]

Problems

SECTION 1.1 *Numerical Representations*

1.1 Complete the following statements.

(a) __________ is the continuous representation of a quantity, while __________ is the discrete representation of a quantity.

(b) Systems in which both analog and digital quantities are manipulated are called __________.

SECTION 1.3 *Digital Number Systems*

1.2 Convert the following binary numbers to their equivalent decimal values:

(a) 10110_2 = ________ $_{10}$ (b) 11101_2 = ________ $_{10}$
(c) 11011110110010_2 = ________ $_{10}$ (d) 1101.1101_2 = ________ $_{10}$
(e) 0.111011101_2 = ________ $_{10}$

1.3 What is the largest number that can be represented using 12 bits?

1.4 How many binary bits are needed to represent 1015_{10}?

1.5 A certain digital circuit is supposed to count in a descending sequence from 15_{10} to 0. The binary counting sequence below is observed. What counts are missing from the expected correct sequence?

1111_2, 1101_2, 1100_2, 1011_2, 1001_2, 1000_2, 0111_2, 0110_2, 0101_2, 0011_2, 0010_2, 0001_2, 0000_2.

SECTION 1.6 *Parallel and Serial Transmission*

1.6 Binary number 00110_2 is being transmitted serially from circuit A to circuit B of Figure 1.1(a). Draw the associated timing diagram for the data as it flows from circuit A to circuit B. (The first bit to be transmitted is the LSB.)

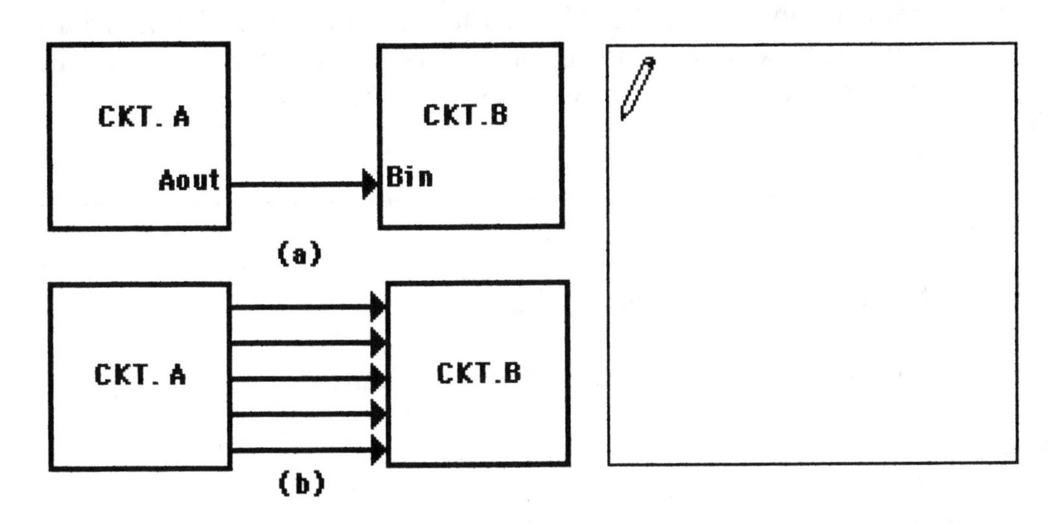

Figure 1.1 (a) Serial Transmission; (b) Parallel Transmission

1.7 If each bit in problem 1.6 takes 10ms to transmit, how long will it take for the complete transmission?

1.8 Generally speaking, how much faster is the transmission performed by the circuit of Figure 1.1(b) compared with that of Figure 1.1(a)?

1.9 Repeat problem 1.6 if the MSB is the first bit to be transmitted.

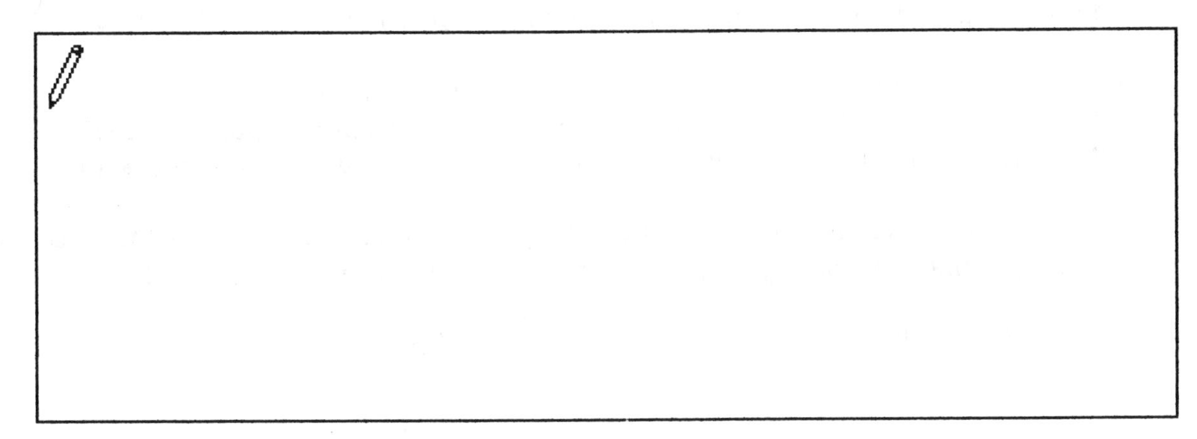

1.10 Binary number 1100011010_2 has to be transmitted from circuit X to circuit Y. How many lines between the transmitter and the receiver are required if a parallel transmission is to be performed.

____

SECTION 1.8 *Digital Computers*

1.11 What are the five major units of a computer?

1. ____________ 2. ____________ 3. ____________
4. ____________ 5. ____________

1.12 What term is often used to denote computers with the greatest operating speed and computational power?

____

TEST 1

1. Digital representations of numerical values may best be described as having characteristics:

   (a) that vary in constant and direct proportion to the quantities they represent.
   (b) that vary constantly over a continuous range of values.
   (c) that are difficult to interpret because they are continuously changing.
   (d) that vary in discrete steps in proportion to the values they represent.

2. Which of the following represents the largest number that can be obtained in the decimal number system when the MSD positional value is 10^5?

   (a) $10{,}000_{10}$ (b) $9{,}999_{10}$ (c) $100{,}000_{10}$ (d) $99{,}999_{10}$

3. What is the largest count that can be obtained with six binary bits?

(a) 31_{10} (b) 32_{10} (c) 63_{10} (d) 64_{10}

4. What is the largest number that can be obtained with six binary bits?

(a) 31_{10} (b) 32_{10} (c) 63_{10} (d) 64_{10}

5. How many binary bits are needed to represent 185_{10}?

(a) 5 (b) 6 (c) 7 (d) 8

6. Serial data transmission is faster than parallel data transmission.

(a) True (b) False

7. If the LSB of a binary number is a 0, the equivalent decimal number is even.

(a) True (b) False

8. The video display terminal of a computer system could be called:

(a) an input unit (b) an output unit (c) the memory (d) the control unit

9. When counting in a binary sequence which of the binary bits toggle for each successive count?

(a) MSB (b) ADC (c) LSB (d) DAC

10. Latches and flip-flops are examples of:

(a) elements that store digital information
(b) memory elements
(c) digital circuits that have memory
(d) all of the above

2 NUMBER SYSTEMS AND CODES

Objectives

Upon completion of this chapter, you will be able to:

- Use two different methods to perform decimal-to binary conversions.
- Cite several advantages of the octal and hexadecimal number systems.
- Convert from the hexadecimal or octal number system to either the decimal or binary number system.
- Express decimal numbers using the BCD code.
- Understand the difference between the BCD code and the straight binary code.
- Cite the major differences between the Gray code and the binary code.
- Understand the need for alphanumeric codes, especially the ASCII code.
- Describe the parity method for error detection.
- Determine parity (odd or even) of digital data.

Glossary of key terms covered in this chapter:

- **Alphanumeric Codes** - Codes that represent numbers, letters, punctuation marks and special characters. [sec.2.8]
- **ASCII Code** - A 7-bit alphanumeric code used by most computer manufacturers. [sec.2.8]
- **Binary-Coded-Decimal Code (BCD Code)** - Four-bit code used to represent each digit of a decimal number. [sec.2.5]
- **Electrical Noise** - Spurious fluctuations in voltage and/or current in any electronic system. [sec.2.9]
- **Encoding** - When a group of symbols is used to represent numbers, letters or words. [sec.2.5]
- **Even Parity** - Total number of 1s (including the parity bit) that are contained in the code group is an even number. [sec.2.9]
- **Gray Code** - Code which never has more than one bit changing when going from one step to the next. [sec.2.7]
- **Hexadecimal Number System** - Number system which has a base of sixteen. Digits 0 through 9 plus letters A through F are used to express a hexadecimal number. [sec.2.4]
- **Minimum-Change Codes** - Codes in which only one bit in the code group changes when going from one step to the next. [sec.2.7]
- **Octal Number System** - Number system which has a base of eight. In this number system digits from 0 to 7 are used to express an octal number. [sec.2.3]
- **Odd Parity** - Total number of 1s (including the parity bit) that are contained in the code group is an odd number. [sec.2.9]
- **Parity Bit** - Additional bit that is attached to each code group before being transferred from one location to another. [sec.2.9]
- **Parity Method** - Scheme used for error detection during the transmission of data. [sec.2.9]
- **Repeated Division** - Method used to convert a decimal number to its equivalent binary-system representation. [sec.2.2]

- **Straight Binary Coding** - Decimal number represented by its equivalent binary number. [sec.2.5]

- **Unweighted Code** - Code in which bit positions in the code groups do not have any specific weight assigned to them. [sec.2.7]

Problems

SECTION 2.1 *Binary-to-Decimal Conversions*

2.1 Convert the following binary numbers to their decimal equivalent.

(a) 1101_2 (b) 1010_2 (c) 1100011010_2 (d) 1000001101_2 (e) 11101_2

SECTION 2.2 *Decimal-to-Binary Conversions*

2.2 Convert the following decimal numbers to their binary equivalent.

(a) 51_{10} (b) 13_{10} (c) 137_{10} (d) 567_{10}

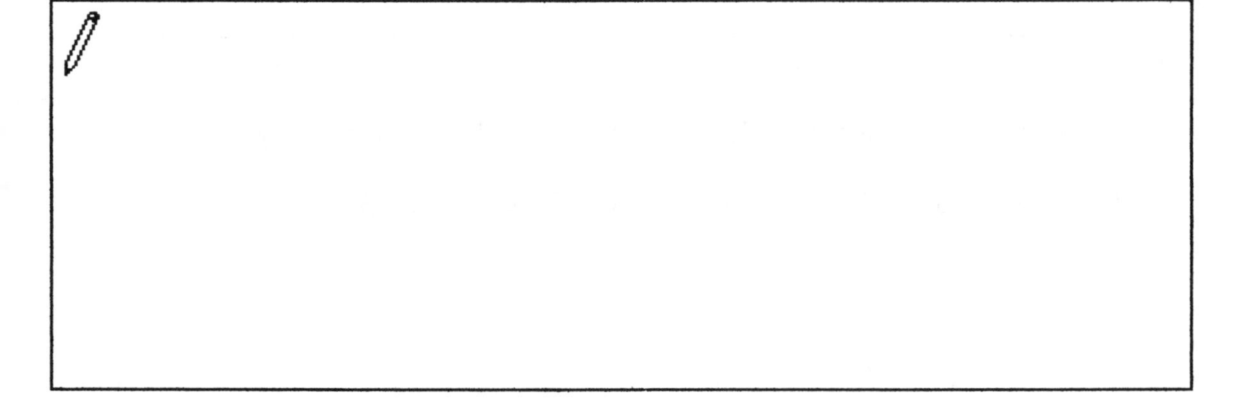

2.3 How many bits are required to express the decimal number 1023_{10}?

2.4 What is the highest decimal number that can be represented by a 7-bit binary number?

SECTION 2.3 *Octal Number System*

2.5 Convert the following octal numbers to their decimal equivalent.

(a) 217_8 (b) 55_8 (c) 5076_8 (d) 511_8 (e) 100_8 (f) 898_8

2.6 Convert the following decimal numbers to their octal equivalent.

(a) 323_{10} (b) 123_{10} (c) 898_{10} (d) 32536_{10} (e) 245_{10}

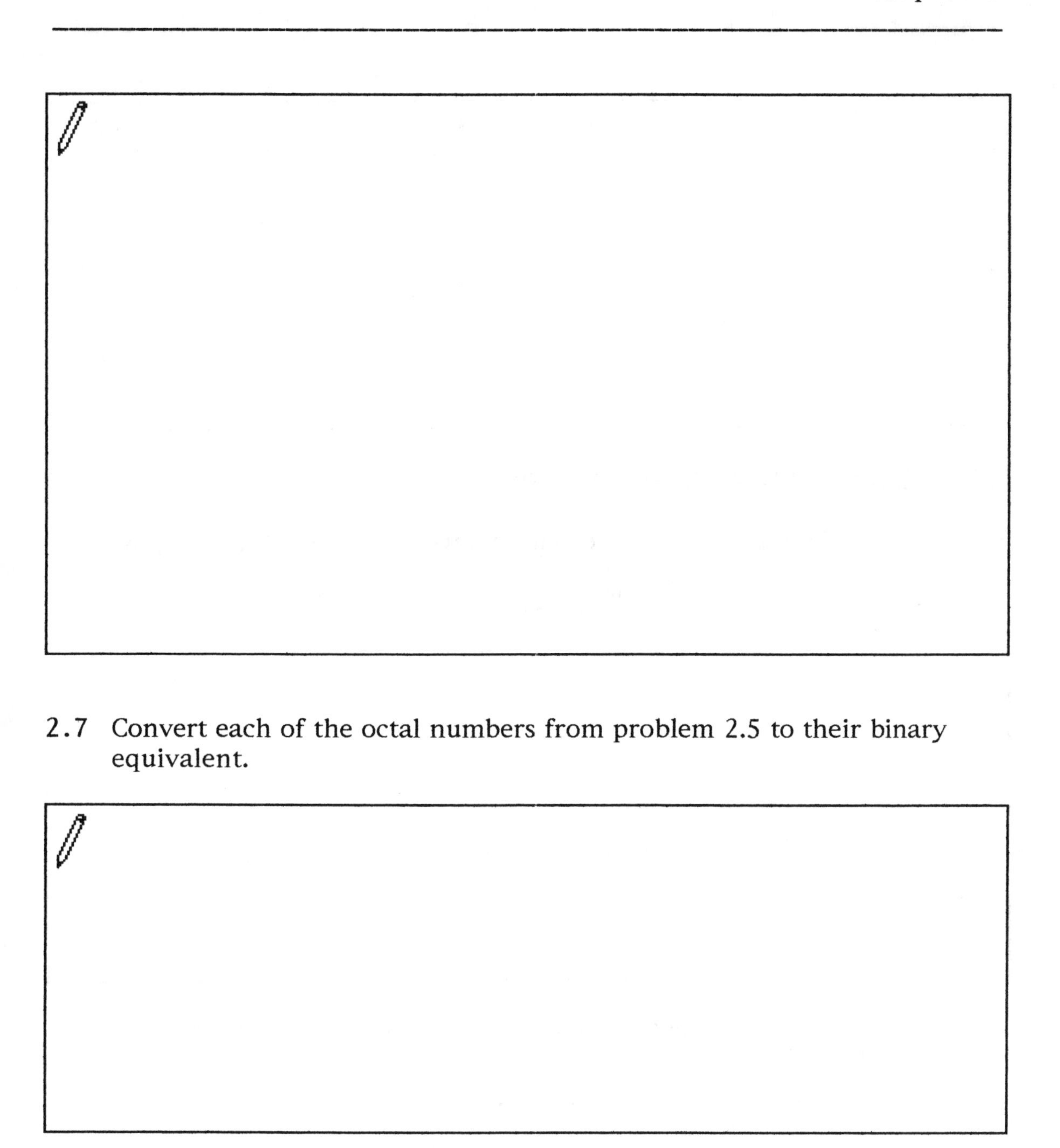

2.7 Convert each of the octal numbers from problem 2.5 to their binary equivalent.

2.8 Convert each of the binary numbers from problem 2.1 to their octal equivalent.

SECTION 2.4 *Hexadecimal Number System*

2.9 Convert the following hexadecimal numbers to their decimal equivalent.

(a) FF_{16} (b) $AD3_{16}$ (c) 589_{16} (d) $3AFD_{16}$ (e) $FEED_{16}$

2.10 Convert each of the decimal numbers from problem 2.6 to their hexadecimal equivalent.

2.11 Convert each of the hexadecimal numbers from problem 2.9 to their binary equivalent.

2.12 Convert each of the binary numbers from problem 2.1 to their hexadecimal equivalent.

2.13 Perform the following conversions between base-5 and decimal.

(a) 1230_5 (b) 777_{10}

2.14 Perform the following conversions between base-4 and decimal.

(a) 333_4 (b) 777_{10}

2.15 Complete the following counting sequence using the hexadecimal system.

96_{16}, 97_{16}, 98_{16}, ...AF_{16}.

SECTION 2.5 *BCD Code*

2.16 Encode the following decimal numbers into BCD.

(a) 63_{10} (b) 105_{10} (c) 757_{10} (d) 999_{10} (e) 36543_{10}

2.17 Convert the following BCD numbers to their decimal equivalent.

(a) 0111010000101001_{BCD} (b) 101000100111_{BCD} (c) 010100111001_{BCD}

SECTION 2.6 *Putting it All Together*

2.18 What is the binary, octal, hexadecimal, and BCD equivalent of 1000_{10}?

SECTION 2.7 *Gray Code*

2.19 Convert each of the following decimal numbers to their Gray code equivalents:

(a) 1_{10} (b) 2_{10} (c) 3_{10} (d) 7_{10} (e) 13_{10}

SECTION 2.8 *Alphanumerical Codes*

2.20 Encode the following messages into ASCII code.

(a) FLIP-FLOP (b) PRINT (5+10/2)

SECTION 2.9 *Parity Method For Error Detection*

2.21 Add the required even parity bit to each of the following binary quantities.

(a) 00110001_2 (b) 100001_2

2.22 A transmitter is going to send messages (a) and (b) of problem 2.20 to a receiver. Redo problem 2.20 if the receiver checks for odd parity.

2.23 For problem 2.22 message (a), the following hexadecimal codes are received and stored in the computer's memory. What is the probable cause for the malfunction? How could it be eliminated?

50_{16}, CF_{16}, CC_{16}, $C6_{16}$, $2D_{16}$, 50_{16}, $C9_{16}$, CC_{16}, $C6_{16}$

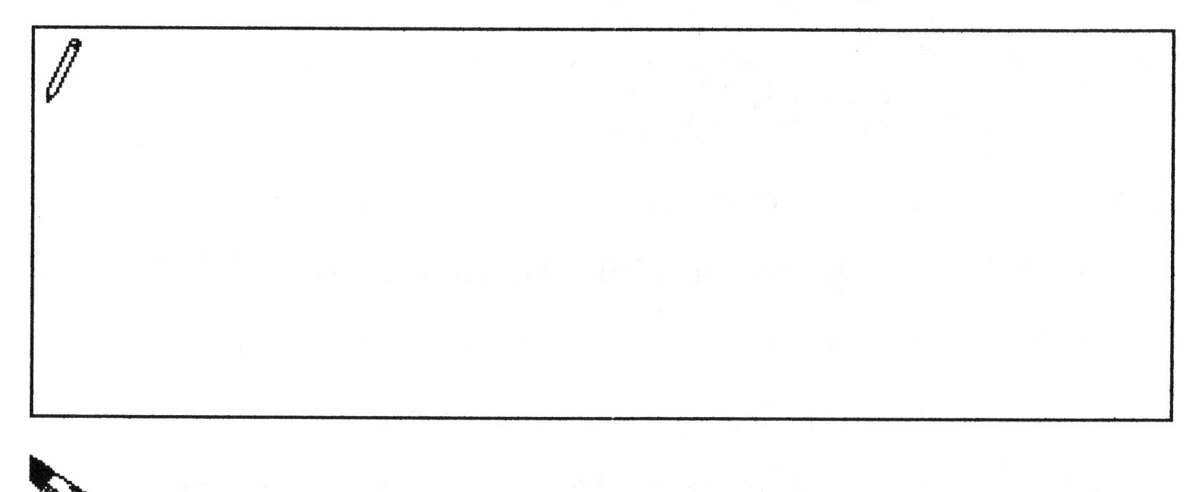

TEST 2

1. Convert 11001101_2 to its decimal equivalent.

   (a) 515_{10} (b) 205_{10} (c) 221_{10} (d) 207_{10}

2. Convert 1050_{10} to binary.

   (a) 10001100100_2 (b) 11011100101_2 (c) 10000011010_2

3. Which of the following numbers represent the largest decimal number?

   (a) 65_{10} (b) FA_{16} (c) 77_8 (d) 111111_2

4. Which of the decimal numbers is equivalent to $3E23_{16}$?

   (a) $16,807_{10}$ (b) $15,907_{10}$ (c) $22,564_{10}$ (d) None of the above.

5. Which of the following numbers is not equivalent to the other three?

   (a)125_{10} (b) 1111101_2 (c) 175_8 (d) $7F_{16}$

6. Translate the Hexadecimal counting $0FF_{16}$-104_{16} to an Octal equivalent sequence.

   (a) 377_8, 378_8, 379_8, 380_8, 381_8, 382_8.
   (b) 777_8, 1000_8, 1001_8, 1002_8, 1003_8, 1004_8.
   (c) 100_8, 101_8, 102_8, 103_8, 104_8, 105_8.
   (d) 377_8, 400_8, 401_8, 402_8, 403_8, 404_8.

7. Which of the following represents the binary equivalent of 321_4?

   (a) 0111101_2 (b) 0111001_2 (c) 0011001_2 (d) 111011_2

8. Which of the following represents the decimal equivalent of 321_4?

   (a) 56_{10} (b) 58_{10} (c) 128_{10} (d) None of the above

9. Number 9 does not exist in the number system whose base is nine.

   (a) True (b) False

10. When converting from decimal to octal using the repeated division method, the last remainder becomes the MSD.

    (a) True (b) False

3 LOGIC GATES AND BOOLEAN ALGEBRA

Objectives

Upon completion of this chapter, you will be able to:

- Analyze the INVERTER circuit.
- Describe the operation of and construct the truth tables for the AND, NAND, OR, and NOR gates.
- Draw timing diagrams for the various logic-circuit gates.
- Write the Boolean expression for the logic gates and combinations of logic gates.
- Implement logic circuits using basic AND, OR, and NOT gates.
- Simplify complex logic circuits by applying the various Boolean algebra laws and rules.
- Simplify intricate Boolean equations by applying DeMorgan's theorems.
- Use either of the universal gates (NAND or NOR) to implement the circuit represented by a Boolean expression.
- Explain the advantages of constructing a logic circuit diagram using the alternate gate symbols, versus the standard logic-gate symbols.
- Describe the concept of active-LOW and active-HIGH logic signals.
- Draw and interpret logic circuits that use the new IEEE/ANSI standard symbols.

Glossary of key terms covered in this chapter:

- **Active logic levels** - When an input or output line of a logic circuit symbol has a bubble, that line is Active-Low. On the other hand, if it doesn't have a bubble, then that line is Active-High. [sec.3.13]

- **AND gate** - The digital circuit which implements the AND operation. The output of this circuit is high (logic level 1), only if all of its inputs are high. [sec.3.4]

- **AND operation** - A Boolean Algebra operation in which the symbol ( ) is used to indicate the ANDing of two or more logic variables. The result of the AND operation will be high (logic level 1), only if all variables are high. [sec.3.1/3.4]

- **Asserted** - Term used to describe the state of a logic signal. The term "Asserted" is synonymous with "Active." [sec.3.14]

- **Associative Laws** - Laws which state that the manner in which variables in an AND expression or OR expression are grouped together does not affect the final result. [sec.3.10]

- **Bi-State Signals** - Signals that have both an active-LOW and an active-HIGH state (i.e. $RD/\overline{WR}$). [3.14]

- **Boolean Algebra** - Algebraic process used as a tool in the design and analysis of digital systems. In Boolean Algebra only two values are possible, "0" and "1." [sec.3.1]

- **Boolean Theorems** - Rules which can be applied to Boolean Algebra in order to simplify logic expressions. [sec.3.10]

- **Bubbles** - Small circles on the input or output lines of logic circuit symbols which represent inversion of that particular signal. If a bubble is present, that input or output is said to be Active-Low. [sec.3.13]

- **Commutative laws** - Laws which state that the order in which two variables are ORed or ANDed is unimportant. [sec.3.10]

- **Complementation** - See NOT operation.

- **DeMorgan's Theorems** - The first theorem states that the complement of a sum (OR-operation) equals the product (AND-operation) of the complements. The second theorem states that the complement of a product (AND-operation) equals the sum (OR-operation) of the complements. [sec.3.11]

- **Dependency Notation** - A method used to pictorially represent the relationship between inputs and outputs of logic circuits. This method employs the usage of qualifying symbols embedded near the top center or geometric center of a symbol element. [sec.3.15]

- **Distributive Law** - Law which states that an expression can be expanded by multiplying term by term. This law also states that if we have a sum of two or more terms, each of which contains a common variable, the common variable can be factored out. [sec.3.10]

- **IEEE/ANSI** - Institute of Electrical and Electronics Engineers/American National Standards Institute. [sec.3.15]

- **INVERTER** - Also referred to as the NOT circuit, this logic circuit implements the NOT operation. An INVERTER has only one input, and its output logic level is always the opposite of this input's logic level. [sec.3.5]

- **Inversion** - See NOT operation.

- **Logic Level** - State of a voltage variable. The range of such voltage is expressed either by a "1" (High) or a "0" (Low). [sec.3.1]

- **NAND gate** - Logic circuit which operates like an AND gate followed by an INVERTER. The output of a NAND gate is low (logic level 0), only if all inputs are high (logic level 1). [sec.3.9]

- **NOR gate** - Logic circuit which operates like an OR gate followed by an INVERTER. The output of a NOR gate is low (logic level 0), when any or all inputs are high (logic level 1). [sec.3.9]

- **NOT circuit** - see INVERTER. [sec.3.5]

- **NOT operation** - A Boolean Algebra operation in which the overbar ($\bar{\ }$), or the prime (') symbol is used to indicate the Inversion of one or more logic variables. The result of a NOT operation is always the complement of the expression being NOTed. [sec.3.1/3.5]

- **OR gate** - The digital circuit which implements the OR operation. The output of this circuit is high (logic level 1), if any or all of its inputs are high. [sec.3.3]

- **OR operation** - A Boolean Algebra operation in which the symbol (+) is used to indicate the ORing of two or more logic variables. The result of the OR operation will be high (logic level 1), if one or more variables is high. [sec.3.1/3.3]

- **Truth Table** - A logic table which depicts a circuit's output response to the various combinations of the logic levels at its inputs. [sec.3.2]

- **Unasserted** - Term used to describe the state of a logic signal. The term "Unasserted" is synonymous with "Inactive." [sec.3.14]

Problems

SECTION 3.1 *Boolean Constants and Variables*

3.1 Complete the following statements.

(a) Logical Addition, can also be referred to as _____ Addition.
(b) Inversion or logical complementation is often called the _____ operation.
(c) The ______ operation can also be referred to as the Logical Multiplication.

SECTIONS 3.2-3.5 *Truth Tables/OR, AND and NOT Operations*

3.2 (a) Apply waveforms A and B of Figure 3.1(a) to a two input OR gate and determine the resulting output waveform X.

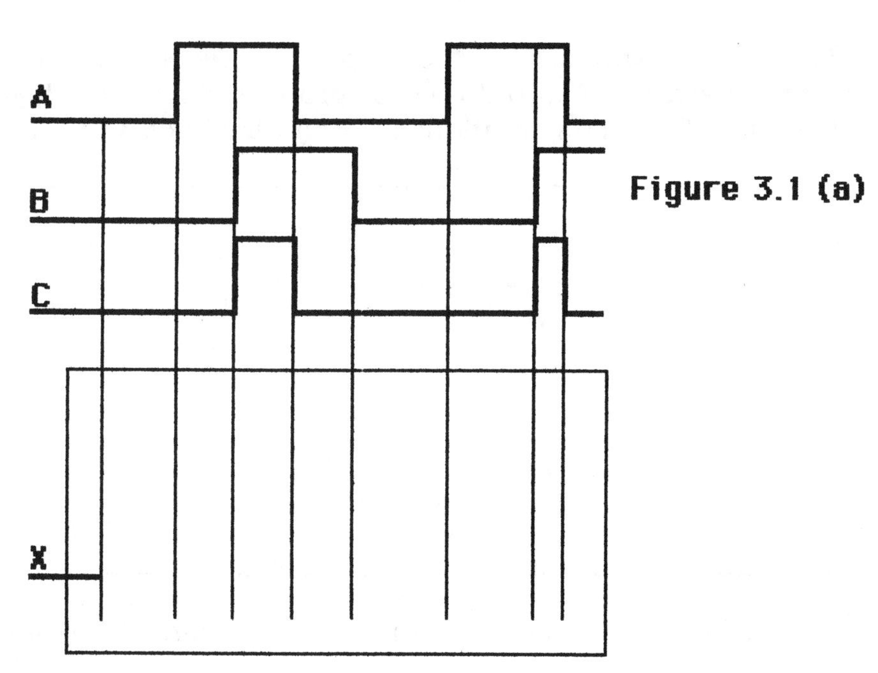

Figure 3.1 (a)

3.2 (b) Invert waveforms A and B of Figure 3.1(b) and draw the output waveform X for a 2-input AND gate.

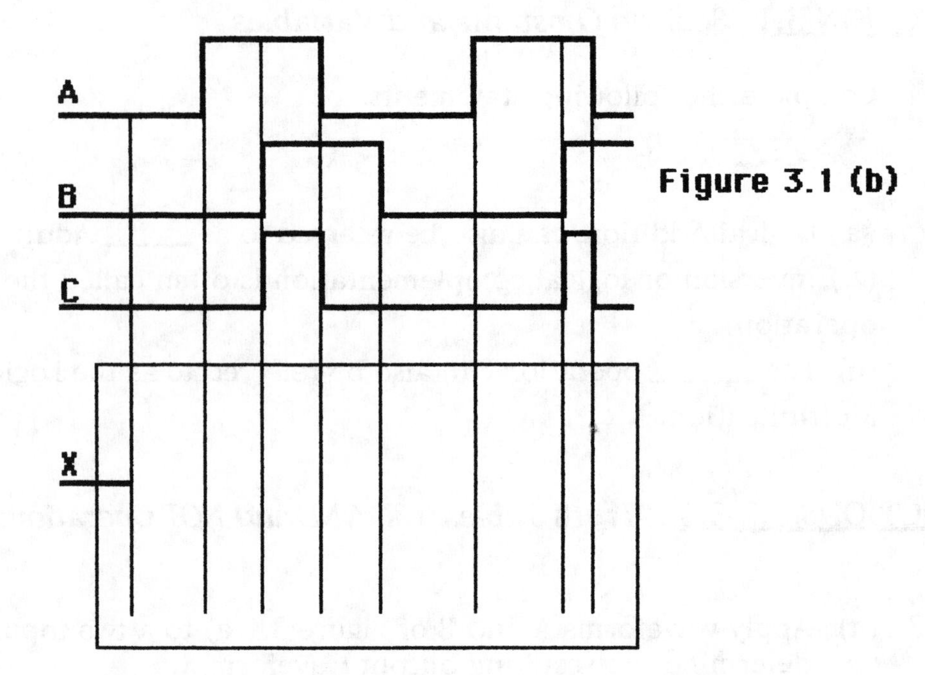

Figure 3.1 (b)

3.3 The Boolean expression at the output of a certain logic circuit is $X = \overline{A} + B$. Apply waveforms A and B used in problem 3.2 to that logic circuit, and determine for what conditions of A and B will output X be LOW?

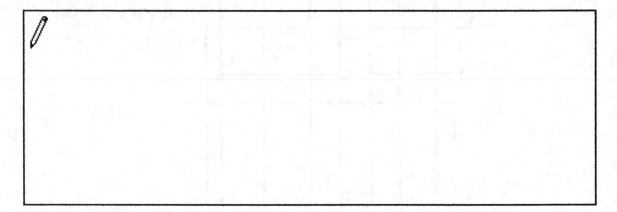

3.4 Determine whether the following statements are True or False:

(a) The output of a 3-input AND gate is HIGH when one or all of its inputs is HIGH. [True] , [False]

(b) The output of a 3-input OR gate is LOW when one or all of its inputs is LOW. [True] , [False]
(c) If the inputs of a 2-input AND gate are tied together, then the output waveform will follow the input signal. [True] , [False]

SECTIONS 3.6-3.7 *Describing Logic Circuits Algebraically/ Evaluating Logic-Circuit Outputs*

3.5 For the circuit of Figure 3.2 insert INVERTERS at the outputs of the OR and AND gates respectively. Determine the new output expression for X.

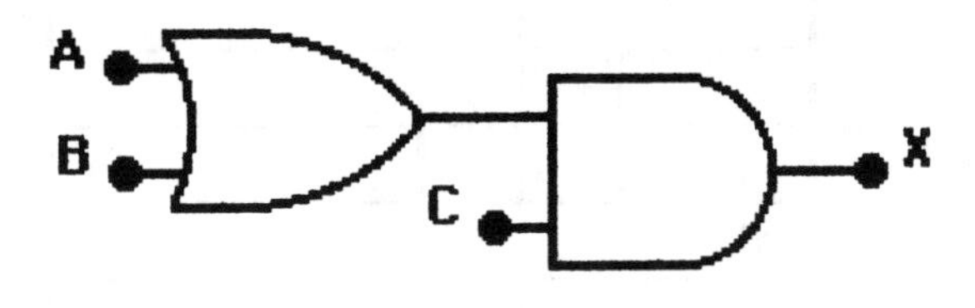

Figure 3.2

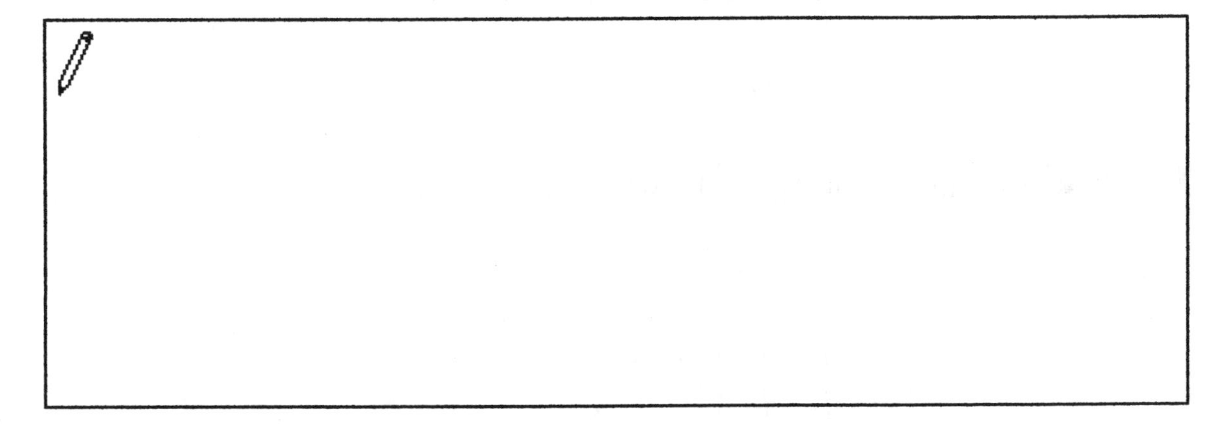

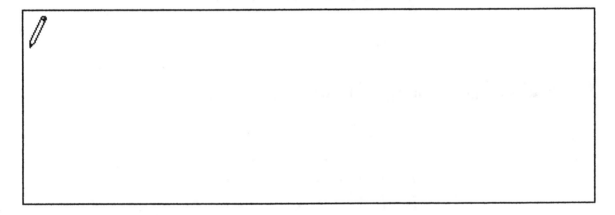

3.6 A circuit's output expression is $Y=\overline{(ABC)+D}$. Fill in the truth table for that circuit.

A	B	C	D	$Y=\overline{(ABC)+D}$

3.7 Evaluate output Y of Figure 3.3 when A=0, B=1 and C=1.

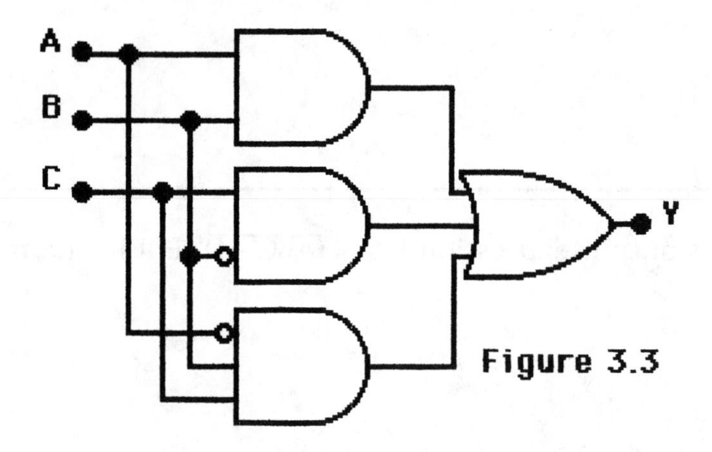

Figure 3.3

SECTIONS 3.8-3.9 *Implementing Circuits From Boolean Expressions/NOR gates and NAND Gates*

3.8 Draw the circuit diagrams that implement the following expressions.

(a) $X = \overline{(\overline{A+B})\bullet(\overline{B}\,C)}$ (b) $X = \overline{\overline{ABC}\bullet(A+D)}$

SECTIONS 3.10-3.11 *Boolean Theorems/DeMorgan's Theorems*

3.9 Simplify the following expressions.

(a) $X = \overline{A}\,\overline{B}\,\overline{C} + \overline{B}\,\overline{C}\,\overline{C}$ (b) $X = \overline{A}BC + A\overline{B}C + \overline{A}\,\overline{B}C$

(c) $X = \overline{KL(\overline{M+N})KL}$ (d) $X = \overline{\overline{AB}(A+B)}$

SECTION 3.12 *Universality of NAND Gates and NOR Gates*

3.10 How many 2-input NOR gates are required to implement the Boolean expression (A + B)(C + D)?

3.11 Implement expression $X = ((A + \overline{B}) \cdot (B + C) \cdot B)$ using one 7400 IC package.

3.12 Show how $X = A + B + \overline{C}$ can be implemented with one 2-input NOR and one 2-input NAND gate.

SECTIONS 3.13-3.15 *Alternate Logic-Gate Representation/ Which Gate Representation to Use/ New IEEE Standard Logic Symbols.*

3.13 Using the Alternate Logic-Gate Representation, modify the circuit of Figure 3.4 if output Z is to be active LOW.

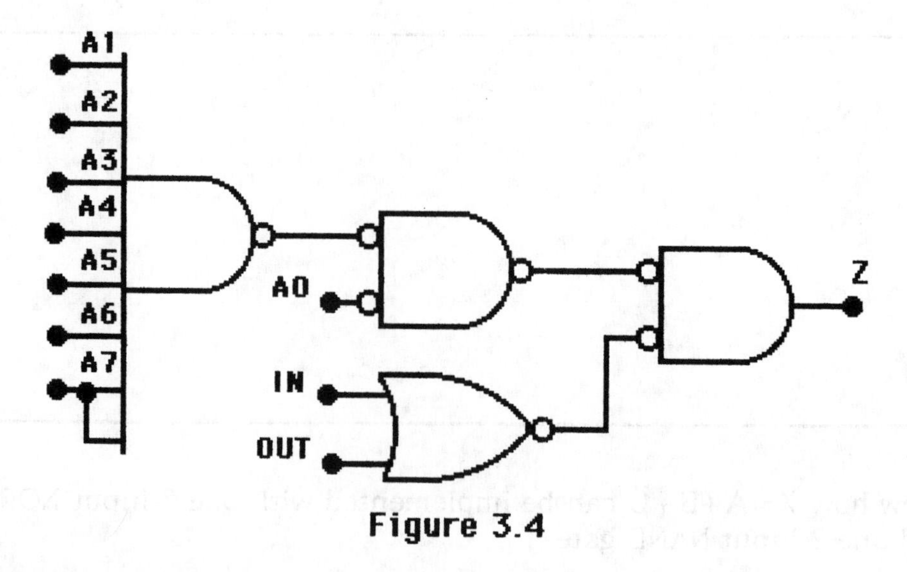

Figure 3.4

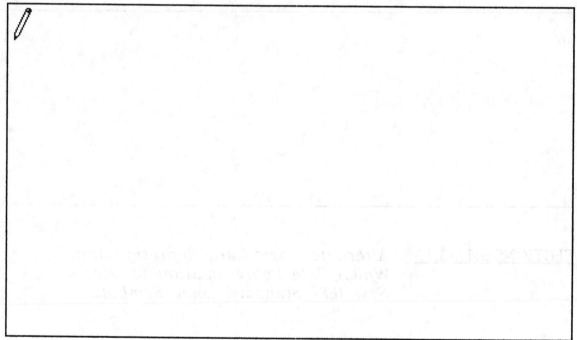

3.14 A logic LOW from output X of circuit of Figure 3.5 is needed to turn on an LED. Redraw the circuit so that it more accurately represents the circuit's operation.

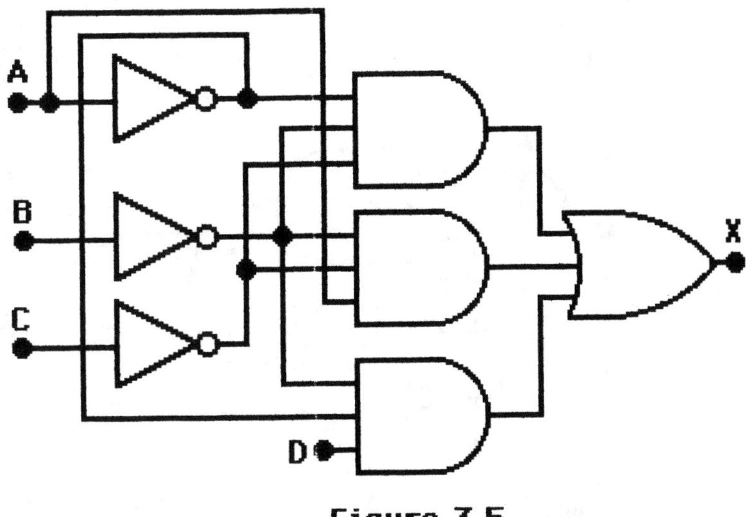

Figure 3.5

3.15 Under what conditions is the output $\overline{\text{LIGHT}}$ unasserted in Figure 3.6?

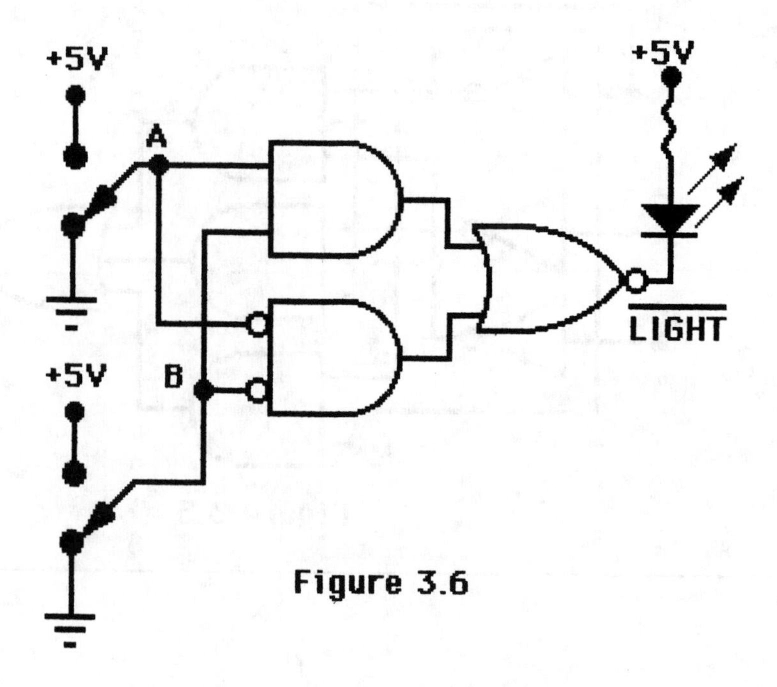

Figure 3.6

3.16 Using the dependency notation symbology, redraw the simplified circuit of problem 3.11.

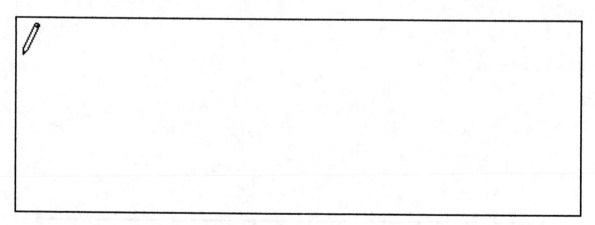

3.17 Implement the Boolean expression $Z=(AB+\overline{C})F+\overline{C}+D+E$ by using dependency notation symbology.

3.18 (a) By using Boolean algebra simplify expression Z of problem 3.17.

(b) What is the minimum amount of logic gates required to implement the simplified expression?

(c) Name each one of the required circuits.

TEST 3

1. The output of this gate is LOW, only if all inputs are HIGH.

   (a) AND **(b)** NAND **(c)** OR **(d)** NOR

2. A 2-input NAND gate with its inputs tied together will function as what type of logic circuit?

   (a) AND **(b)** INVERTER **(c)** OR **(d)** NOR

3. Three signals (A, B, and C) are ORed together. Their product is then ANDed with signals K, L, and Q. Which of the following expressions indicates the correct order of operations?

   (a) A+B+CKLQ **(b)** A+B+C(KLQ) **(c)** A+(B+CKLQ) **(d)** (A+B+C)KLQ

4. Given the Boolean expression $X=[(\overline{A+B})+(\overline{CD+\bar{F}+\bar{G}})+(E\bar{F}+G+\bar{H})]$ and the values A=0, B=1, C=1, D=0, E=1 F=1, G=0, H=0, the value of X is:

   (a) 0 **(b)** 1 **(c)** Undetermined

5. Which of the following Boolean expressions represents the DeMorganized version of the expression $X=\overline{(\overline{A+B})+(\bar{C}+\bar{D})}$?

   (a) (A+B)+(C+D) **(b)** (AB)+(CD) **(c)** (A+B)(CD) **(d)** (AB)(C+D)

6. Which of the following Boolean expressions represents a three input OR gate?

   (a) $\bar{A}\cdot\bar{B}\cdot\bar{C}$ **(b)** $\overline{\bar{A}+\bar{B}+\bar{C}}$ **(c)** $\overline{A\cdot B\cdot C}$ **(d)** $\overline{\bar{A}\cdot\bar{B}\cdot\bar{C}}$

7. Suppose that a Boolean expression represents the output of a logic circuit and that it contains four variables (four inputs). How many different input combinations must be listed in this circuit's truth table?

   (a) 4_{10} (b) 8_{10} (c) 16_{10} (d) 32_{10}

8. The term "Unasserted" is synonymous to "Active."

   (a) [TRUE] (b) [FALSE]

9. The Boolean expression $X = \overline{\overline{\overline{A + B + C}}}$ represents the output of a three input NOR gate fed to an INVERTER whose output is also fed to another INVERTER.

   (a) [TRUE] (b) [FALSE]

10. You would need a minimum of four universal two-input NAND gates to perform the logical operation of a two-input OR gate.

    (a) [TRUE] (b) [FALSE]

4 COMBINATIONAL LOGIC CIRCUITS

Objectives

Upon completion of this chapter, you will be able to:

- Convert a logic expression into a sum-of-products expression.
- Perform the necessary steps to derive a sum-of-products expression in order to design a combinational logic circuit in its simplest form.
- Use the Karnaugh map as a tool to simplify and design logic circuits.
- Explain the operation of both exclusive-OR and exclusive-NOR circuits.
- Design logic circuits with and without the help of a truth table.
- Identify and understand inhibit circuits.
- Cite the basic characteristics of digital ICs.
- Understand the inherent operative differences between TTL and CMOS.
- Use the basic troubleshooting rules of digital systems.
- Deduce from measured results the faults of malfunctioning combinational logic circuits.

Glossary of key terms covered in this chapter:

- **Bipolar ICs** - Integrated Digital Circuits in which NPN and PNP transistors are the main circuit elements. [sec.4.9]
- **CMOS** - Complementary metal-oxide semiconductor. A logic family which belongs to the category of unipolar digital ICs. [sec.4.9]
- **Combinational logic circuits** - Circuits made up of combinations of logic gates. [sec.4.4]
- **Current Tracer** - Digital troubleshooting tool which senses the changing magnetic field around a conductor. [sec.4.13]
- **DIP** - "Dual-in-line package." The most common type of IC packages. [sec.4.9]
- **"Don't Care"** - An input condition of a logic variable in which neither a 1 (HIGH) nor a 0 (LOW) will affect the final output result. [sec.4.5]
- **Exclusive-NOR circuit** - A two input logic circuit which produces a HIGH output only when the inputs are equal. [sec.4.6]
- **Exclusive-OR circuit** - A two input logic circuit which produces a HIGH output only when the inputs are different. [sec.4.6]
- **Floating inputs** - Input signals which are left disconnected in a logic circuit. [sec.4.9]
- **Indeterminate** - Whenever a logic voltage level of a particular logic family falls out of the required range of voltages for either a logic 0 or logic 1. [sec.4.9.]
- **Inhibit circuits** - Logic circuits (gates) that control the passage of an input signal through to the output. [sec.4.8]
- **Karnaugh map** - A two-dimensional form of a truth table used to simplify a Sum-of-products expression. [sec.4.5]
- **Logic probe** - Digital troubleshooting tool which senses the logic level at a particular point in a circuit. [sec.4.10]

- **Looping** - When adjacent squares in a Karnaugh map containing 1s are combined for purpose of simplification of a sum-of products expression. [sec.4.5]
- **LSI** - Large-Scale Integration. [sec.4.9]
- **MSI** - Medium-Scale Integration. [sec.4.9]
- **Octets** - A group of eight 1s that are adjacent to each other within a Karnaugh map. [sec.4.5]
- **Pairs** - A group of two 1s that are adjacent to each other within a Karnaugh map. [sec.4.5]
- **Parity Checker** - A logic circuit that checks transmitted data for the proper parity, and produces an error output (E) when a single-bit error has occurred. [sec.4.5]
- **Parity Generator** - A logic circuit that generates an even and an odd parity bit. [sec.4.5]
- **Quads** - A group of four 1s that are adjacent to each other within a Karnaugh map. [sec.4.5]
- **Solder bridges** - Splashes of solder that short two or more points together. [sec.4.12]
- **SSI** - Small-Scale Integration. [sec.4.9]
- **Substrate** - A piece of semiconductor material which is part of the building block of any digital IC. [sec.4.9]
- **Sum-of-products form** - Logic expression consisting of two or more AND terms (products) that are ORed together. [sec.4.1]
- **TTL** - Transistor-transistor logic. Logic family which belongs to the category of bipolar digital ICs. [sec.4.9]
- **ULSI** - Ultralarge-Scale Integration. [sec.4.9]
- **Unipolar ICs** - Integrated Digital Circuits where unipolar Field Effect Transistors (MOSFETs) are the main circuit elements. [sec.4.9]
- **VLSI** - Very Large-Scale Integration. [sec.4.9]

Problems

SECTIONS 4.1-4.3 *Sum-of-Products Form/Simplifying Logic Circuits/Algebraic Simplification*

4.1 Simplify the following expressions using Boolean algebra.

(a) $X = ABC + A\bar{B}C + \bar{A}BC + (\overline{A+B})C$

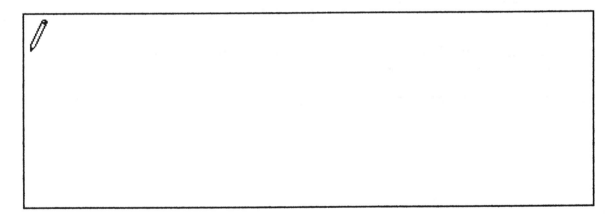

(b) $K = \bar{X}\bar{Y}Z + \bar{X}YZ + X\bar{Y}\bar{Z} + X\bar{Y}Z + (\overline{\bar{X}+\bar{Y}+\bar{Z}})$

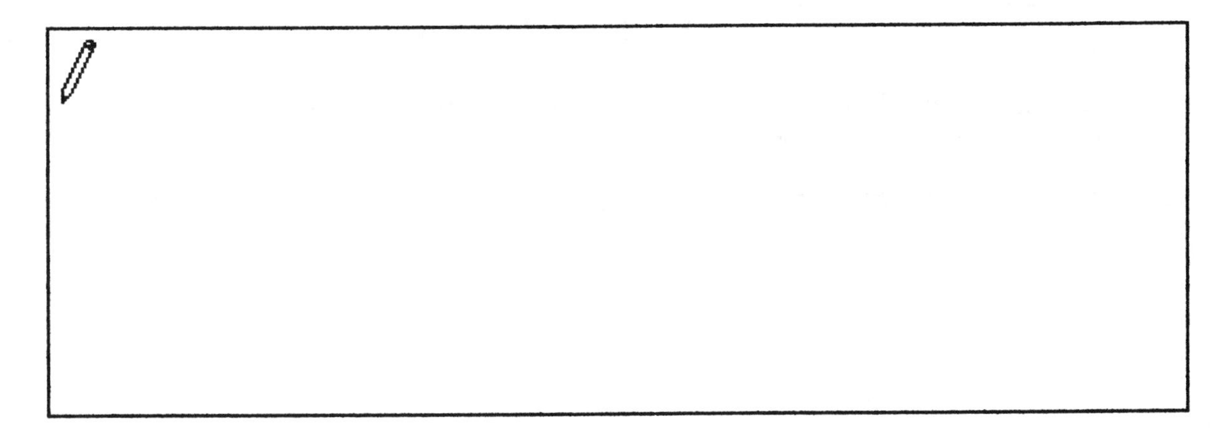

(c) $W = (K+N+M)(K+\bar{N}+M) + (\bar{K}+N+M)(\bar{K}+\bar{N}+M)$

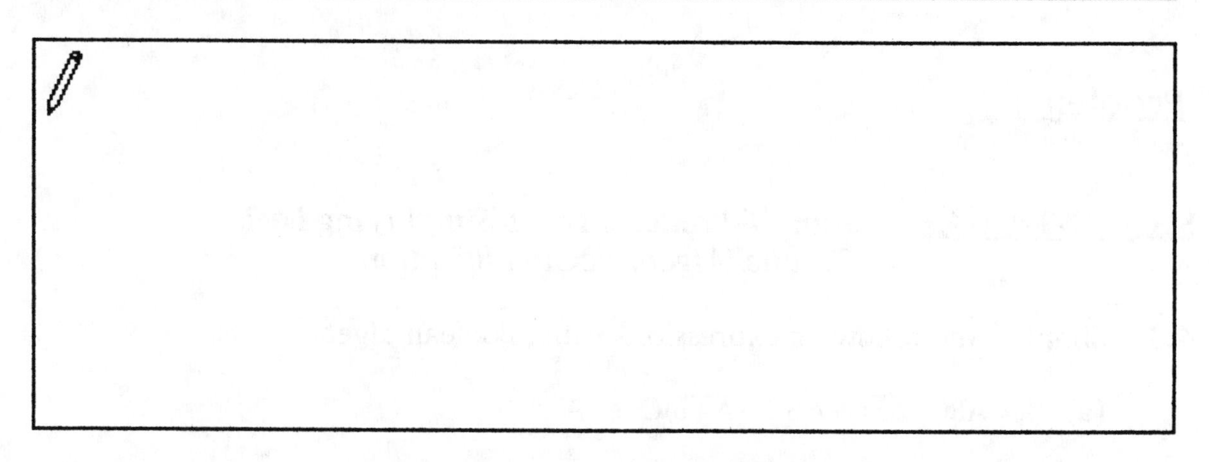

4.2 Refer to Figure 4.1 . Let DRIV=A; $\overline{BELTP} = \overline{BELTD} = B$; PASS=C; IGN=E; $\overline{ALARM} = X$.

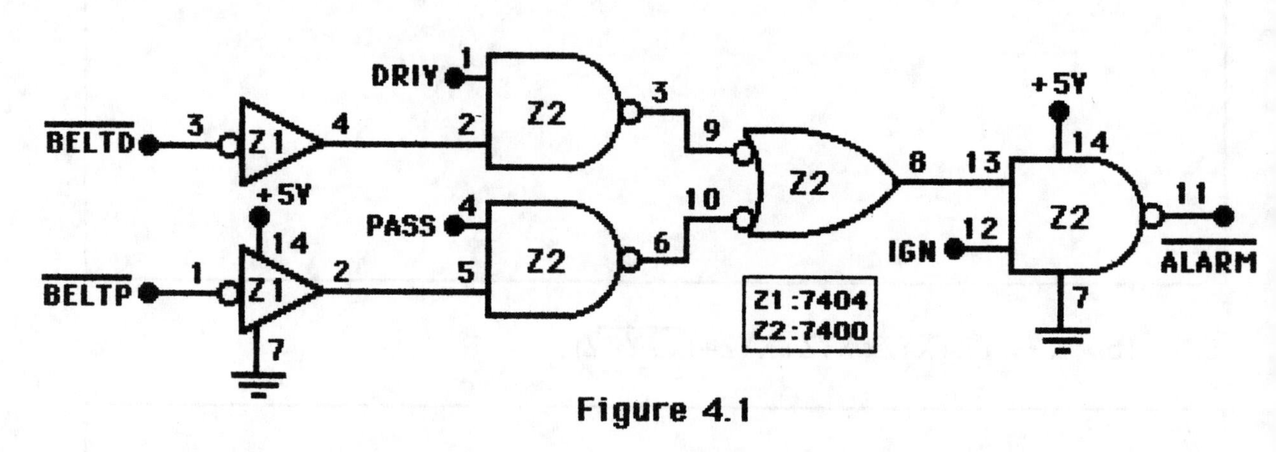

Figure 4.1

(a) Determine the Boolean expression for the output X using the newly assigned variables (Don't simplify it, yet!)

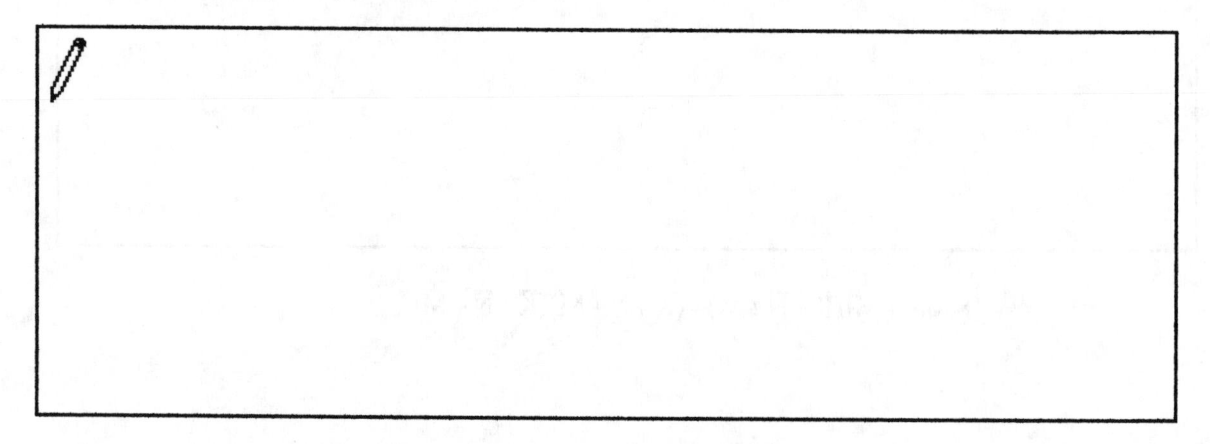

(b) Now, simplify the expression of step (a).

(c) Implement the simplified expression using only 3-input NOR gates.

SECTION 4.4 *Designing Combinational Logic Circuits*

4.3 Design a logic circuit whose output is HIGH when inputs A and B are LOW, or when inputs C and D are HIGH.

(Step 1: Set up the truth table)

(Step 2: Write the AND term for each case where the output is a 1.)

(Step 3: Write the sum-of-products expression for the output.)

(Step 4: Simplify the output expression using Boolean algebra.)

(Step 5: Implement the circuit for the final expression.)

4.4 A certain logic circuit in the NASA Shuttle is supposed to turn ON a green indicator in the astronauts' cabin to indicate that the pressure in the external fuel tanks has reached the desired and necessary level for lift-off. The desired pressure is set by the on-board computer and is encoded in a 3-bit binary number. A pressure transducer monitors the fuel tanks and relays its information to the on-board computer via a 3-bit binary number. The weight of the LSB of this number is equivalent to 100 p.s.i. Design the logic circuit necessary to turn on the green indicator.

(Use the step-by-step procedure used in problem 4.3.)

4.5 Design the necessary logic for the situation in Problem 4.4, if a red indicator is to turn ON when the pressure inside the tanks rises above 600 p.s.i.

4.6 Design a circuit whose output X is HIGH only when inputs A and B are HIGH but A and C are different.

4.7 Design a circuit whose output Z is HIGH when inputs C and D are LOW and inputs A or B are HIGH.

(Use the step-by-step procedure used in problem 4.3.)

4.8 A burglar alarm is required to sound off whenever an intrusion takes place. The alarm should sound off when the entrance door (E) or the window (W) is opened. It should also sound off when the entrance door and the window are closed, but a motion detector (M) detects any movement inside of the house. And finally, the alarm must go off if a panic button (P) is depressed, regardless of the status of the other sensors in the house. Design the necessary logic circuit for the desired burglar alarm.

4.9 A late-model automobile has an 8-bit microcomputer that is used to check certain sensors in the car before the ignition can be turned on. Each sensor has an associated name and its corresponding μcomputer address, they are as follows:

(I) Doors Closed and Locked (DCL)____ $FCXX_{16}$.
(II) Seat Belts Buckled-up (SBB) ______ $FDXX_{16}$.
(III) Electrical System Check (ESC) ____ $FEXX_{16}$.
(IV) Ignition ON ($\overline{ION}$)____________ $FFXX_{16}$.

The μcomputer sends the proper logic HIGH signal to each of the sensors (I thru IV) and monitors each response. If all the responses are appropriate then a final logic LOW signal is sent to start the engine (IV). Naturally, all of this is checked in microseconds as the ignition key is turned on to start the car.
Design the necessary logic that will decode the address bus of the 8-bit μcomputer and whose outputs control each of the sensors used.

SECTIONS 4.5-4.6 *The Karnaugh map Method/Exclusive-OR and Exclusive-NOR Circuits*

4.10 Simplify the following S-of-P expressions using Karnaugh mapping.

(a) $X = \overline{A}\overline{B}\overline{C}\overline{D} + \overline{A}\overline{B}C\overline{D} + A\overline{B}C\overline{D} + \overline{A}B\overline{C}D + \overline{A}BCD + A\overline{B}\overline{C}\overline{D}$

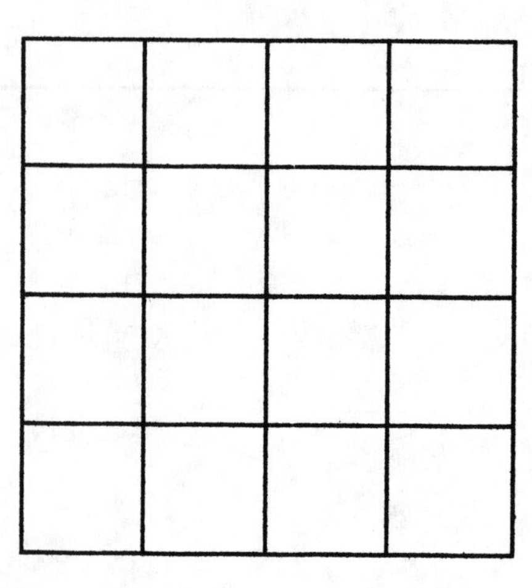

X=__________________

(b) $Z = \overline{K}\overline{N}M + \overline{K}\overline{N}\overline{M} + \overline{K}N\overline{M}$

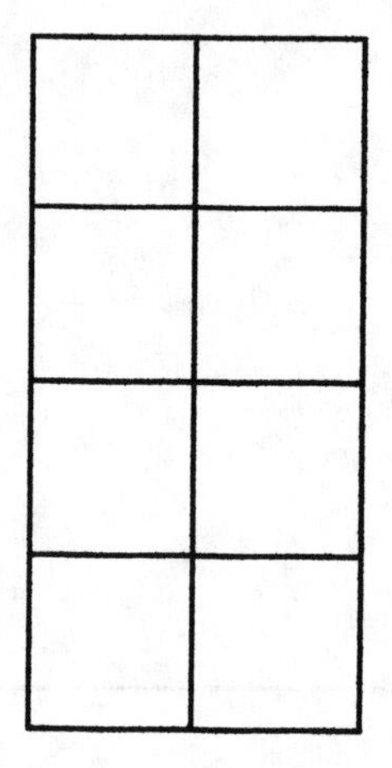

Z=__________________

(c) $Y = \overline{P}Q\overline{R}S + \overline{P}QRS + PQ\overline{R}S + PQRS$

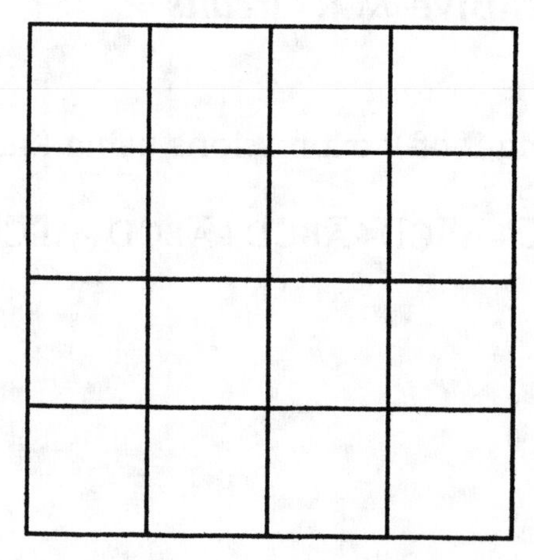

Y=__________________

4.11 Write and implement the S-of-P expression for the following statements using Karnaugh maps:

(a) A,B,C and D are the inputs of a counter. Z is the output and is to be HIGH for counts 0,1,2,3,7 and 15.

(Step 1: Set up the truth table.)
(Step 2: Write the AND term for each case where the output is 1.)
(Step 3: Write the sum-of-products expression for the output.)
(Step 4: Simplify the output expression using the Karnaugh map.)

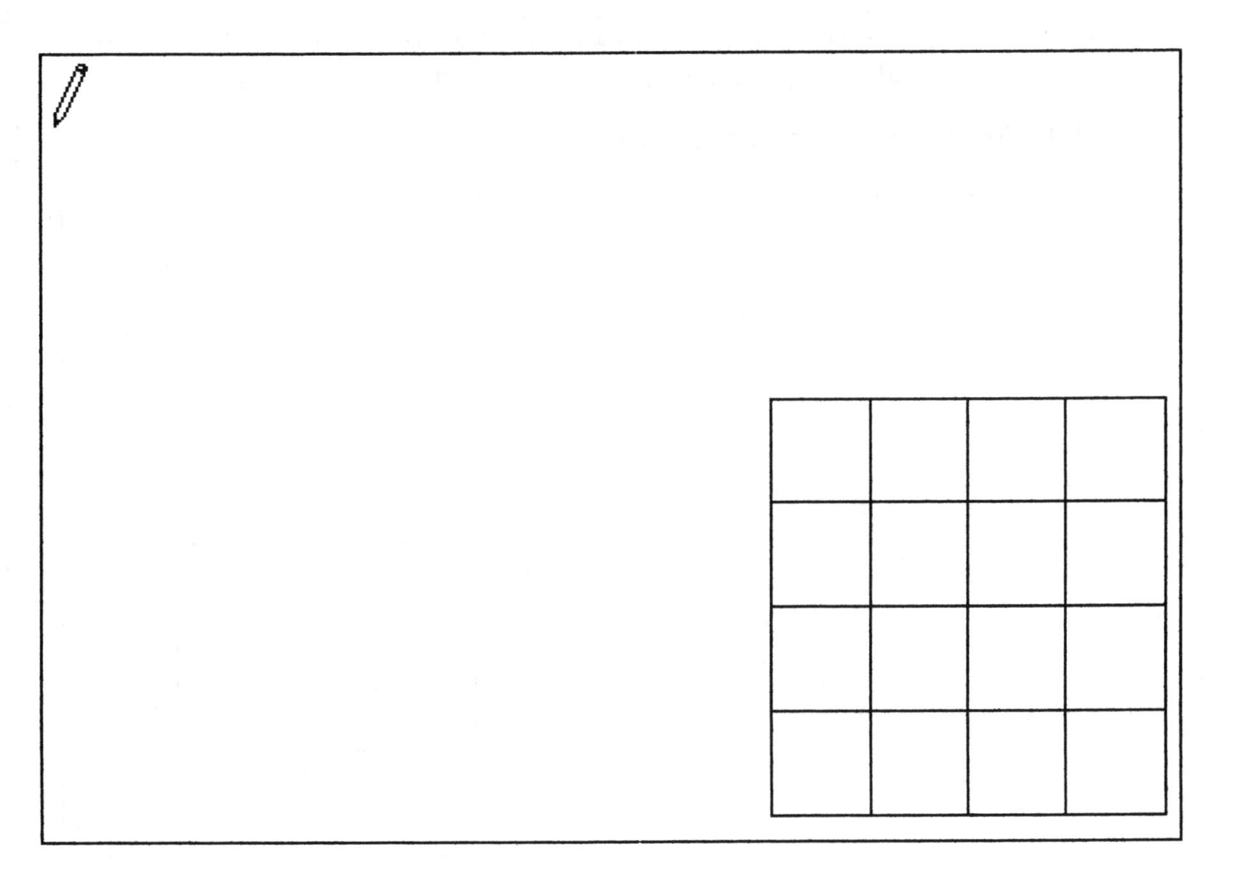

(Step 5: Implement the circuit for the final expression.)

(b) The output Z of a circuit whose inputs are K,L,M and N, is to be HIGH whenever all of the inputs are equal, or when K is equal to M.

(Follow steps 1-5 of pervious problem)

SECTIONS 4.7 *Parity Generator and Checker*

4.12 Design a logic circuit that has three data inputs (A, B, C) and two outputs (Y and Z). Output Y will be HIGH if there is an even number of 1s in the input data. Output Z will be HIGH if there is an odd number of 1s in the data input.

4.13 Design the Parity Checker Circuit for the Parity Generator circuit of problem 4.12.

(This circuit should check for both ODD and EVEN parity.)

SECTIONS 4.9 *Basic Characteristics of Digital ICs*

4.14 Complete the following statements by filling in the blanks.

(a) Bipolar transistors are used by the _____ logic family, while the _____ logic family uses P-channel and N-channel MOSFETs.
(b) If a TTL input is left unconnected, that input acts like a logic _____.
(c) If a _____ input is left floating, the IC may become overheated and eventually destroy itself.
(d) If an IC has fewer than 12 gates built on its substrate, then it's considered ________ _________ Integration.

However, if it has over 10,000 gates built on its substrate, then the IC falls into the category of ______ ______ ______ Integration.

SECTIONS 4.10 *Troubleshooting Digital Systems*

⌖4.15 A technician has just wire-wrapped the circuit of Figure 4.2(a). Determine what are the most probable causes for the malfunction demonstrated by the logic probe indications in the table of Figure 4.2(b).

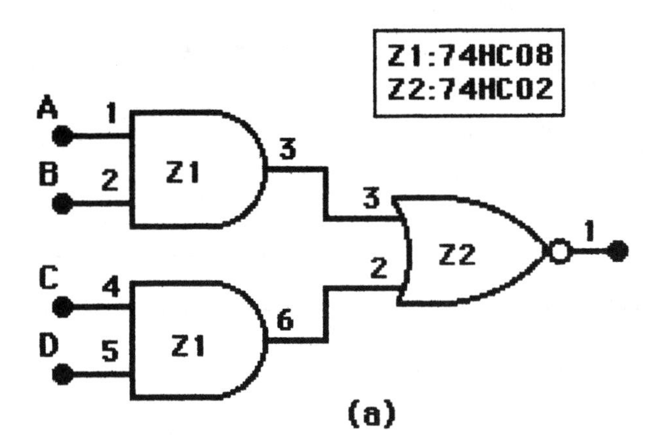

Pin	Condition
Z1-1	Pulsing
Z1-2	HIGH
Z1-3	Pulsing
Z1-4	HIGH
Z1-5	HIGH
Z1-6	LOW
Z2-3	Pulsing
Z2-2	LOW
Z2-1	Pulsing

(b)

Figure 4.2

⌖4.16 Refer to the TTL circuit of Figure 4.3(a). The logic probe conditions in the table of Figure 4.3(b) are observed. Determine the most probable cause for the malfunctioning.

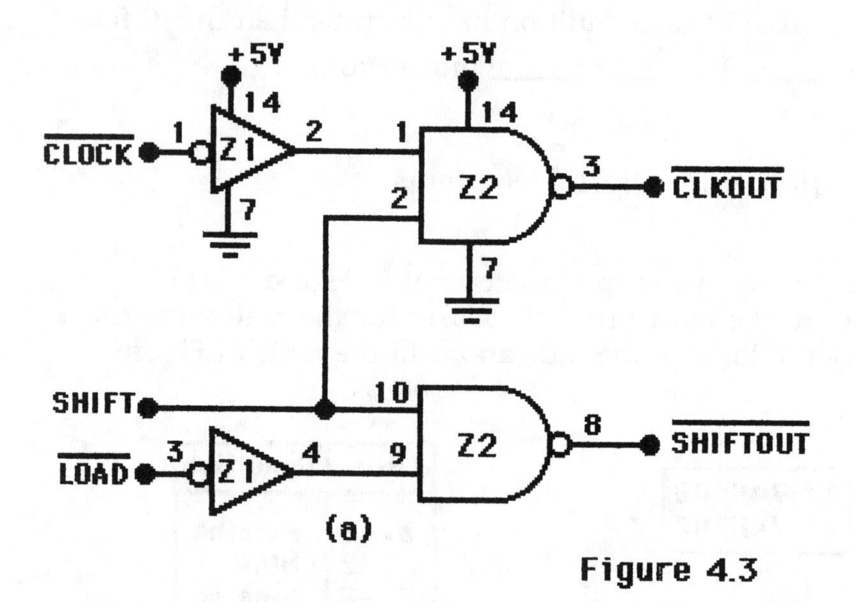

Pin	Condition
Z1-1	Pulsing
Z1-2	Pulsing
Z1-3	LOW
Z1-4	HIGH
Z2-1	Pulsing
Z2-2	Indeterminate
Z2-3	Pulsing
Z2-9	HIGH
Z2-10	LOW
Z2-8	HIGH

(b)

Figure 4.3

4.17 Refer to the TTL circuit of Figure 4.4(a). A technician constructs the circuit, plugs the ICs into the IC sockets, and then tests the circuit. The logic probe conditions listed in the table of Figure 4.4(b) are viewed. Determine the most probable cause for the observed circuit malfunction.

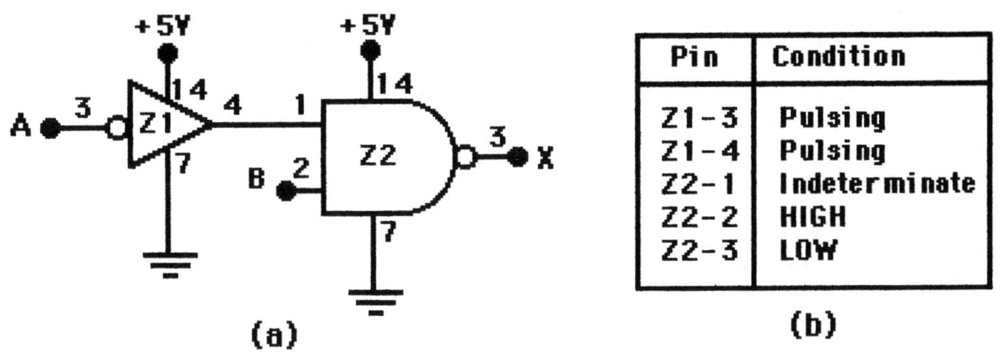

Pin	Condition
Z1-3	Pulsing
Z1-4	Pulsing
Z2-1	Indeterminate
Z2-2	HIGH
Z2-3	LOW

Figure 4.4

4.18 Refer to circuit of Figure 4.5. Output X should be HIGH only when the output from the BCD counter is equal to decimal counts 2, 3 or 9. The technician designed the "Logic Circuit" and upon testing it, he notices that the BCD counter is working properly, but output X is always HIGH. State 5 possible causes for the malfunction.

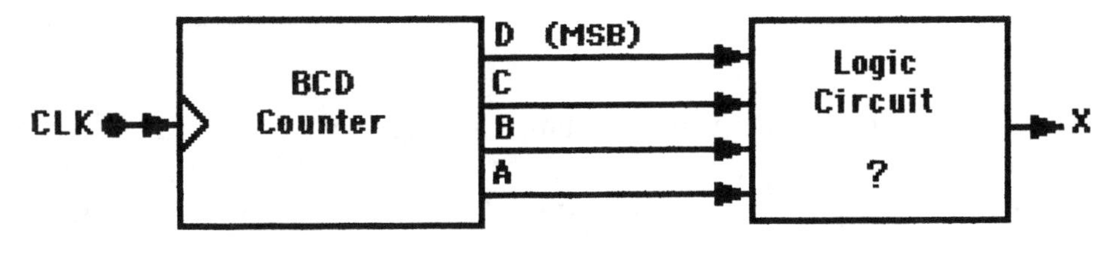

Figure 4.5

1. ______________________________
2. ______________________________
3. ______________________________
4. ______________________________
5. ______________________________

4.19 The logic probe conditions listed in Figure 4.6(b) are observed for the TTL circuit of Figure 4.6(a). Consider the following list of possible problems. For each one indicate 'yes' or 'no'. Explain your reasoning for each 'no' response.

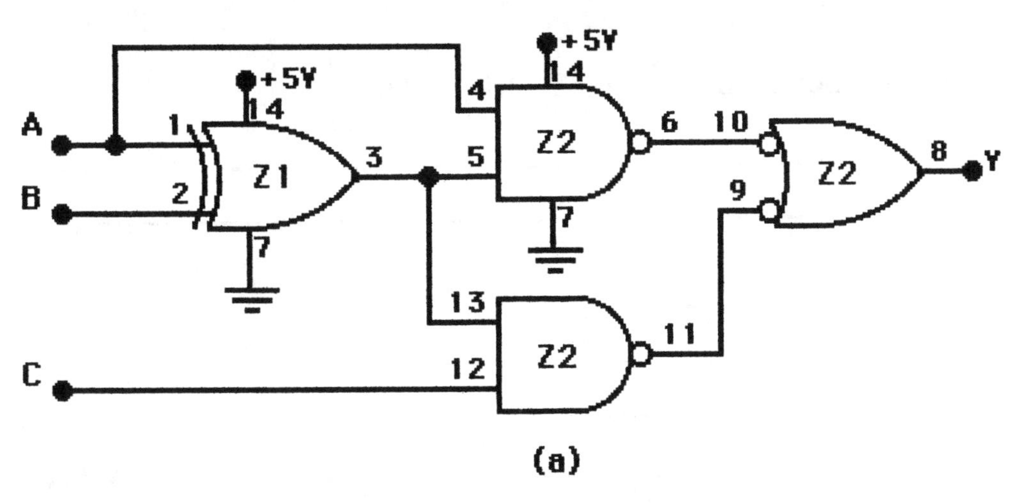

Pin	Condition
Z1-1	LOW
Z1-2	HIGH
Z1-3	HIGH
Z2-5	HIGH
Z2-4	LOW
Z2-6	Indeterminate
Z2-13	HIGH
Z2-12	Pulsing
Z2-11	Indeterminate
Z2-10	Indeterminate
Z2-9	Indeterminate
Z2-8	Indeterminate

(b)

Figure 4.6

(a) Integrated circuit Z2 has pin 13 broken.

(b) Integrated circuit Z2 has pin 13 internally shorted to ground.

(c) Connection from Z2-9 to Z2-11 is broken.

(d) Z2-9 is shorted to Z2-10.

(e) Integrated circuit Z2 doesn't have Vcc applied to it.

4.20 A technician builds the circuit of Figure 4.6 (a) using CMOS ICs. She uses an exclusive OR gate (Z1) and three 3-input NAND gates (Z2). During the testing of the circuit, she finds that the outputs of the NAND gates are undetermined and at unpredictable logic levels. She also notices that IC Z2 is getting excessively warm. What is the most probable cause for the malfunction?

TEST 4

1. Simplify the expression $Y=\bar{W}\bar{F}Z+\bar{W}FZ+W\bar{F}Z$.

   (a) $Y=Z(\bar{W}+\bar{F})$ (b) $Y=Z(\bar{W}\cdot\bar{F})$ (c) $Y=Z(\overline{W+F})$ (d) $Y=\bar{Z}(\overline{W\cdot F})$

2. The Boolean expression $X=\overline{AB+\bar{C}}+A\bar{C}+B$ can be implemented using one three input:

   (a) OR gate (b) AND gate (c) NAND gate (d) None of the above

3. Which of the following represent a sum-of-products expression?

   (a) $X=\overline{\bar{A}B\bar{C}}+ABC+\bar{A}BC$ (b) $X=\bar{A}\bar{K}L+AKL+(\overline{A+\bar{K}+\bar{L}})$
   (c) $X=\bar{A}BC+A\bar{B}C+\bar{A}B\bar{C}$ (d) None of the above

4. How many variables are eliminated when a quad is formed in a Karnaugh map?

   (a) One (b) Two (c) Three (d) Four

5. The logic gate that produces a LOW output only when the inputs are equal is the:

   (a) Exclusive NOR (b) Exclusive OR (c) NAND (d) NOR

6. Which two input gate acts as an inverter when one of its inputs is permanently tied to ground?

   (a) Exclusive NOR (b) Exclusive OR (c) NAND (d) NOR

7. What are the minimum amount of gates need to implement the expression $X=AB+\bar{A}\bar{B}$?

   (a) 5 (b) 4 (c) 3 (d) 1

8. A four variable truth table yields an output X which is HIGH for fifteen conditions and LOW for one condition. The easiest and fastest way to implement the circuit whose output is X, is to:

   (a) first write the S-of-P expression for all cases where X=1, then simplify the X expression using Boolean algebra, and then draw the circuit.

   (b) first write the S-of-P expression for all cases where X=1, then simplify the X expression using the Karnaugh map, and then draw the circuit.

   (c) first write the S-of-P expression for the case where X=0, then invert the expression using DeMorgan's theorems, simplify the expression using Boolean algebra, and then draw the circuit.

   (d) first write the S-of-P expression for the case where X=0, then invert the expression using DeMorgan's theorems, simplify the expression using the Karnaugh map, and then draw the circuit.

9. The four corner squares of a four variable Karnaugh map form a quad.

   (a) [TRUE] (b) [FALSE]

10. A parity generator circuit has to add an even parity bit to every word that a transmitter sends. If the transmitter sends the binary word 11001110_2, then the parity generator must add a 1 to the word.

   (a) [TRUE] (b) [FALSE]

5 FLIP-FLOPS AND RELATED DEVICES

Objectives

Upon completion of this chapter, you will be able to:

- Construct and analyze the operation of a latch Flip-Flop made from NAND or NOR gates.
- Debounce a mechanical switch by using a latch circuit.
- Describe the difference between synchronous and asynchronous systems.
- Understand several types of edge-triggered Flip-Flops, such as the J-K, D-type, and S-C.
- Analyze and apply the various Flip-Flop timing parameters specified by the manufacturers.
- Describe a pulse-steering and an edge-detector circuit.
- Understand the major differences between parallel and serial data transfers.
- Draw the output timing waveforms of several types of Flip-Flops in response to a set of input signals.
- Analyze the various IEEE/ANSI Flip-Flop symbols.
- Use state transition diagrams to describe counter operation.
- Cite various Flip-Flop applications.
- Use Flip-Flops in synchronous circuits.
- Connect shift registers as data transfer circuits.
- Employ Flip-Flops as frequency-division and counting circuits.
- Understand the typical characteristics of Schmitt triggers.
- Apply two different types of one-shots in circuit design.
- Design a free-running oscillator using a 555 timer.
- Recognize and predict the effects of clock skew on synchronous circuits.
- Troubleshoot various types of Flip-Flop circuits.

Glossary of key terms covered in this chapter:

- **Astable Multivibrator** - Digital circuit which oscillates between two unstable output states. [sec.5.23]
- **Asynchronous Active Pulse Width** - The minimum time duration that a DC SET or DC CLEAR input has to be kept in its active state in order to reliably SET or CLEAR the Flip-Flop. [sec.5.11]
- **Asynchronous Inputs** - Flip-Flop inputs which can affect the operation of the Flip-Flop independently of the synchronous and Clock inputs. [sec.5.9]
- **Asynchronous Systems** - Systems in which outputs can change states any time one or more of the inputs change. [sec.5.4]
- **Asynchronous Transfer** - Data transfer performed without the aid of the Clock. [sec.5.17]
- **Binary Counter** - A group of Flip-Flops connected in a special arrangement in which the states of the Flip-Flops represent the binary number equivalent to the number of pulses that have occurred at the input of the counter. [sec.5.19]
- **Bistable Multivibrator** - Name which is sometimes used to describe a Flip-Flop. [sec.5.23]
- **Cleared State** - The Q=0 state of a Flip-Flop. [sec.5.1]
- **Clock** - A digital signal in the form of a rectangular pulse train or a squarewave. [sec.5.4]
- **Clock Pulse HIGH** [$t_{W(H)}$] - The minimum time duration that a Clock signal must remain HIGH before going LOW. [sec.5.11]
- **Clock Pulse LOW** [$t_{W(L)}$] - The minimum time duration that a Clock signal must remain LOW before going HIGH. [sec.5.11]
- **Clock Skew** - When because of propagation delays a Clock signal arrives at the Clock inputs of different Flip-Flops at different times. [sec.5.24]
- **Clock Transition Times** - Duration given by the manufacturer of a particular IC for the rise and fall times of the Clock signal transitions used by that IC. [sec.5.11]

- **Clocked D Flip-Flop** - Type of Flip-Flop where the D (Data) input is the synchronous input. [sec.5.7]
- **Clocked Flip-Flops** - Flip-Flops which have a Clock input. [sec.5.4]
- **Clocked J-K Flip-Flop** - Type of Flip-Flop where the inputs J and K are the synchronous inputs. [sec.5.6]
- **Clocked S-C Flip-Flop** - Type of Flip-Flop where the inputs SET and CLEAR are the synchronous inputs. [sec.5.5]
- **Common-Control Block** - Symbol used by the IEEE/ANSI standard to describe when one or more inputs are common to more than one of the circuits in an IC. [sec.5.10]
- **Contact Bounce** - Random voltage transitions produced by operating a mechanical switch. [sec.5.1]
- **Control Inputs** - Control input signals synchronized with the active Clock transition determine the output state of a Flip-Flop. [sec.5.4]
- **D Latch** - Circuit which contains a NAND gate latch and two steering NAND gates. [sec.5.8]
- **Data Lock-Out** - Feature on some Master/Slave Flip-Flops by which the Master Flip-Flop is disable a short time after the positive-going Clock transition. [sec.5.13]
- **DC CLEAR** - Asynchronous Flip-Flop input used to clear Q immediately to 0. [sec.5.9]
- **DC SET** - Asynchronous Flip-Flop input used to set Q immediately to 1. [sec.5.9]
- **Edge-Detector** - Circuit which produces a narrow positive spike that occurs coincident with the active transition of a Clock input pulse. [sec.5.5]
- **Edge-Triggered** - The manner in which a Flip-Flop is activated by a signal transition. It may be either a positive or negative edge-triggered Flip-Flop. [sec.5.4]
- **555 Timer** - A TTL-compatible IC which can wired to operate in several different modes such as a One-Shot and an Astable Multivibrator. [sec.5.23]
- **Free-Running Multivibrator** - See Astable Multivibrator. [sec.5.23]

- **Hold Time** (t_H) - Time interval immediately following the active transition of the Clock signal during which the control input has to be maintained at the proper level. [sec.5.4/5.11]
- **Jam Transfer** - See Asynchronous Transfer. [sec.5.17]
- **Latch** - A term synonymous with Flip-Flop. [sec. 5.1]
- **Master/Slave Flip-Flops** - Flip-Flops which have as their internal structure two Flip-Flops - a Master and a Slave. [sec.5.13]
- **Maximum Clocking Frequency** (f_{MAX}) - The highest frequency that may be applied to the Clock input of a Flip-Flop and still have trigger reliably. [sec.5.11]
- **MOD Number** - The number of different states that a counter can sequence through. [sec.5.19]
- **Monostable Multivibrator** - See One-Shot. [sec.5.21]
- **NAND Gate Latch** - A Flip-Flop constructed from two NAND gates. [sec.5.1]
- **Negative-Going Threshold** (V_{T-}) - A voltage level inherent to a Schmitt Trigger circuit which if dropped below will cause the output to change. [sec.5.20]
- **Negative-Going Transition** (NGT) - When a Clock signal changes from a logic 1 to a logic 0. [sec.5.4]
- **Noise** - Spurious voltage fluctuations that may be present in the environment and cause digital circuits to malfunction. [sec.5.24]
- **Nonretriggerable One-Shot** - Type of One-Shot which will not respond to a trigger input signal while in its quasi-state. [sec.5.21]
- **NOR Gate Latch** - A Flip-Flop constructed from two NOR gates. [sec.5.2]
- **One-Shot** - A circuit which belongs to the Flip-Flop family but which has only one stable state (normally Q=0). [sec.5.21]
- **Override Inputs** - Synonymous with Asynchronous inputs. [sec.5.9]
- **Parallel Data Transfer** - Operation by which the contents of a register are transferred simultaneously to another register. [sec.5.17]

- **Positive-Going Threshold** (V_{T+}) - A voltage level inherent to a Schmitt Trigger circuit which if exceeded will cause the output to change. [sec.5.20]
- **Positive-Going Transition (PGT)** - When a Clock signal changes from a logic 0 to a logic 1. [sec.5.4]
- **Postponed Output Indicator** - Symbol used by the IEEE/ANSI standard to show the fact that the effects of the control inputs of a Flip-Flop do not reach the output Q until the Clock returns to its initial level. [sec.5.13]
- PRESET - Term synonymous with DC SET. [sec.5.9]
- **Propagation Delays** (t_{PLH}/t_{PHL}) - Delay from the time a signal is applied to the time when the output makes its change. [sec.5.11]
- **Pulse-Triggered Flip-Flop** - Name which is sometimes used in reference to a Master Slave Flip-Flop. [sec.5.13]
- **Quasi-Stable** - State in which a One-Shot stays momentarily (normally Q=1) before returning to its normal state (normally Q=0). [sec.5.21]
- **Registers** - A group of Flip-Flops capable of storing data. [sec.5.17]
- RESET - Term synonymous with DC CLEAR. [sec.5.9]
- **Reset State** - The Q=0 state of a Flip-Flop. [sec.5.1]
- **Retriggerable One-Shot** - Type of One-Shot which will respond to a trigger input signal while in its quasi-state. [sec.5.21]
- **Schmitt-Trigger** - Digital circuit which accepts a slow-changing input signal and produces a rapid oscillation-free transition at the output. [sec.5.20]
- **Serial Data Transfer** - When data are transferred from one place to another one bit at a time. [sec.5.17/5.18]
- **Set State** - The Q=1 state of a Flip-Flop. [sec.5.1]
- **Set-Up Time** (t_S) - The time interval immediately preceding the active transition of the Clock signal during which the control input has to be maintained at the proper level. [sec.5.4/5.11]
- **Shift Register** - Digital circuit which accepts binary data from some input source and then shifts these data through a chain of Flip-Flops one bit at a time. [sec.5.18]

- **Synchronous Inputs** - See Control Inputs. [sec.5.9]
- **Synchronous Systems** - Systems in which the output can change states only when a Clock occurs. [sec.5.4]
- **Synchronous Transfer** - Data transfer performed by using the synchronous and the Clock inputs of a Flip-Flop. [sec.5.17]
- **Toggle Mode** - When a Flip-Flop changes states for each Clock pulse. [sec.5.6]
- **Transparent** - In a D Latch when the Q output follows the D input the device is said to be transparent. [sec.5.8]
- **Trigger** - Input signal to a Flip-Flop which causes its output to change states depending on the conditions of the control signals. [sec.5.5]

Problems

SECTIONS 5.1-5.3 *NAND Gate Latch/NOR Gate Latch/ Troubleshooting Case Study*

5.1 Complete the following statements.

(a) The output of a ________ gate latch is invalid, when the SET and CLEAR inputs are HIGH.

(b) The Q output of a NAND gate latch is LOW, when the SET input is ______ and the CLEAR input is _______.

(c) The output of a ________ gate latch will not change if both inputs are HIGH.

5.2 Refer to the TTL logic circuit and associated input waveforms of Figure 5.1(a) and (b) respectively. Assume that initially Q=K=1. Apply the waveforms X, Y and Z to the circuit, and draw the timing diagram for the outputs Q and K.

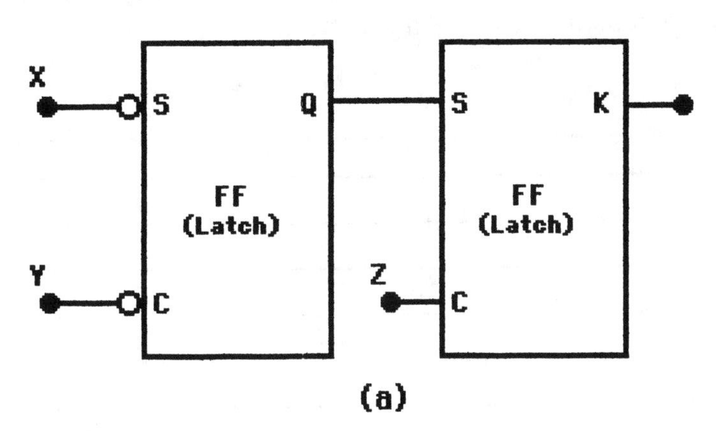

Figure 5.1: NAND/NOR gate flip-flop circuit.

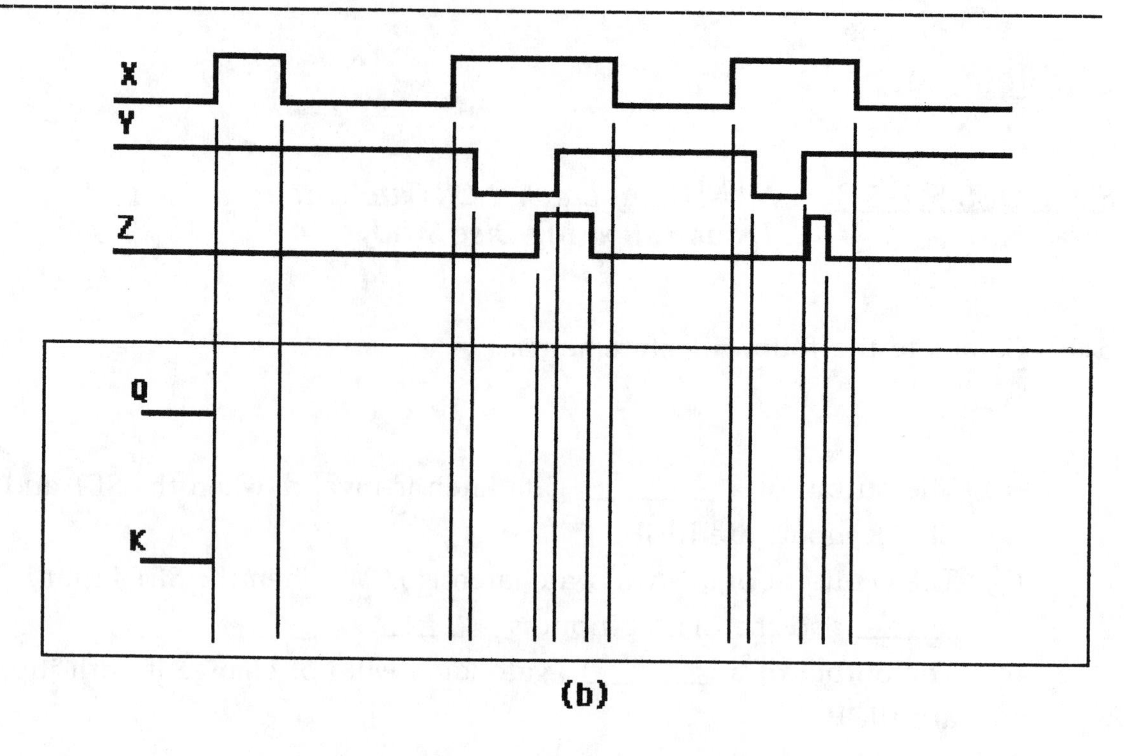

Figure 5.1: Input waveforms

5.3 An oscilloscope is used to test the circuit of Figure 5.1(a). The technician notices that the K output is always HIGH. Determine some possible reasons for the malfunction.

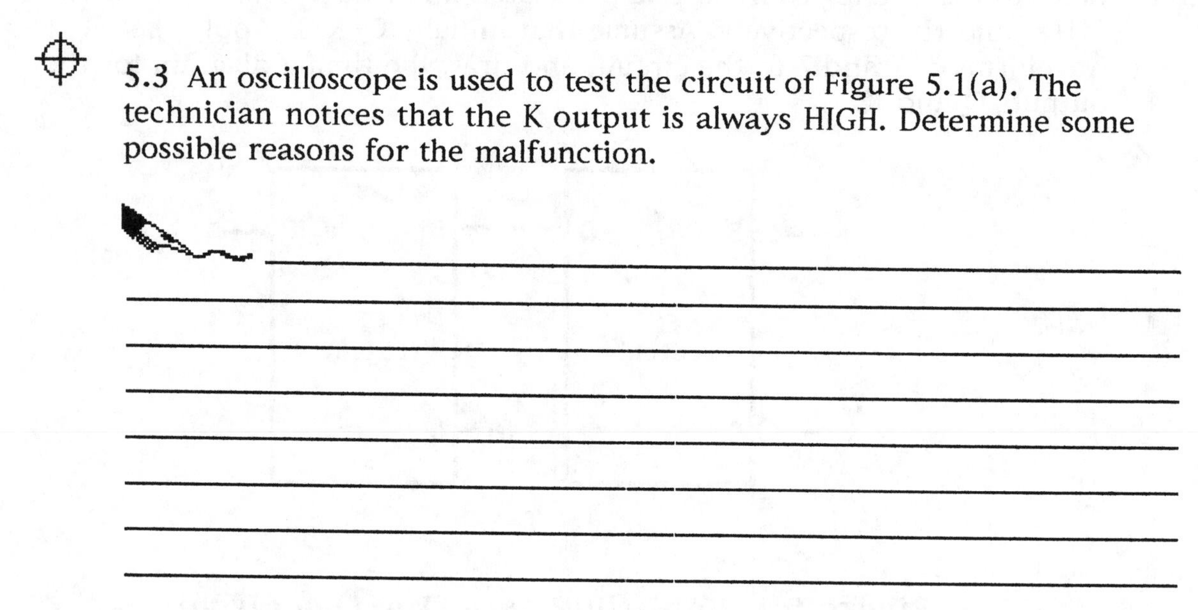

SECTIONS 5.4-5.5 *Clock Signals and Clocked Flip-Flops/ Clocked S-C Flip-Flop*

5.4 Apply waveforms of Figure 5.2(b) to the S (SET), C (CLEAR) and CLK (CLOCK) inputs of the negative edge triggered S-C Flip-Flop of Figure 5.2(a). Assume that Q=0 initially. Draw the output waveform Q.

(The output of a negative edge triggered S-C Flip-Flop changes only when the clock signal goes from HIGH to LOW.)

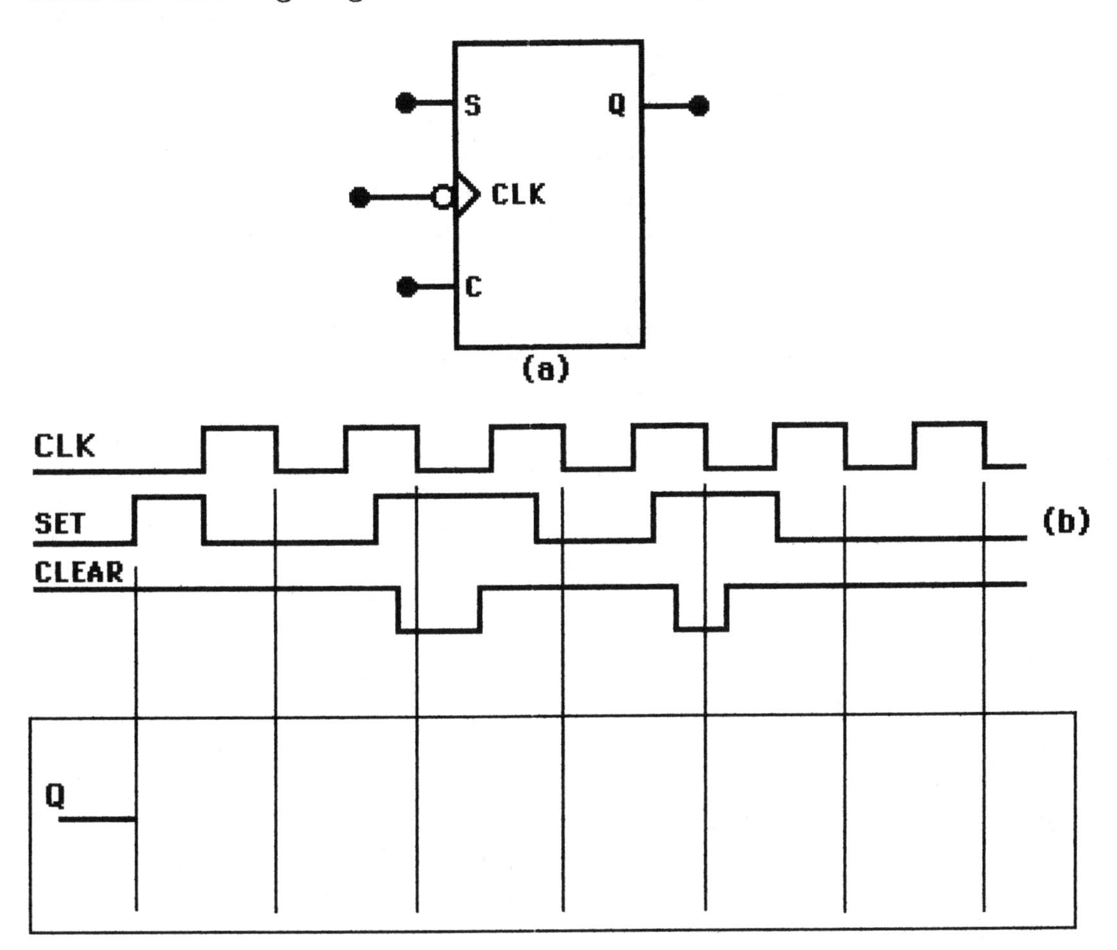

**Figure 5.2: (a) Negative-edge triggered S-C flip-flop.
(b) Set and Clear input waveforms.**

5.5 Change ONE word in each of the following two statements in order to make each one a true statement.

(a) In asynchronous digital systems, the time at which the output of a circuit can change, is determined by a signal called CLOCK.

(b) The Set-Up time requirement of a Flip-Flop, is that time interval immediately following the active transition of the CLK signal during which the synchronous inputs have to be maintained at the proper level.

SECTIONS 5.6-5.8; 5.25 Clocked J-K Flip-Flop / Clocked D Flip-Flop / D Latch (Transparent Latch / Troubleshooting Flip-Flop Circuits.

5.6 A technician uses waveforms of Figure 5.3(b) to test FF Q of Figure 5.3(a). With an oscilloscope, she measures a 5KHz 50% duty-cycle waveform at output Q. Consider the following list of possible causes for the problem. For each one indicate 'yes' or 'no'. Explain your reasoning for each response. (Determine waveform Q to justify each of your answers. Assume Q=0 initially)

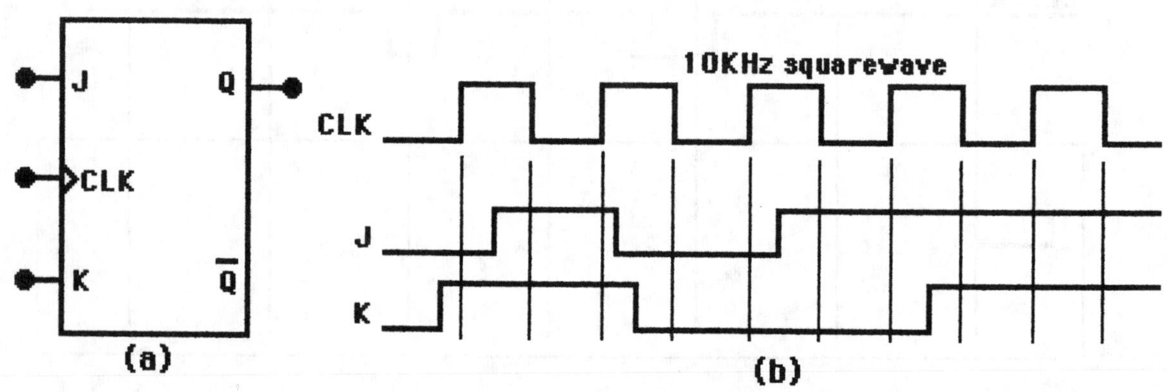

Figure 5.3: (a) Positive edge triggered J-K FF; (b) Waveforms

(a) The technician unintentionally used a negative edge-triggered J-K Flip-Flop.

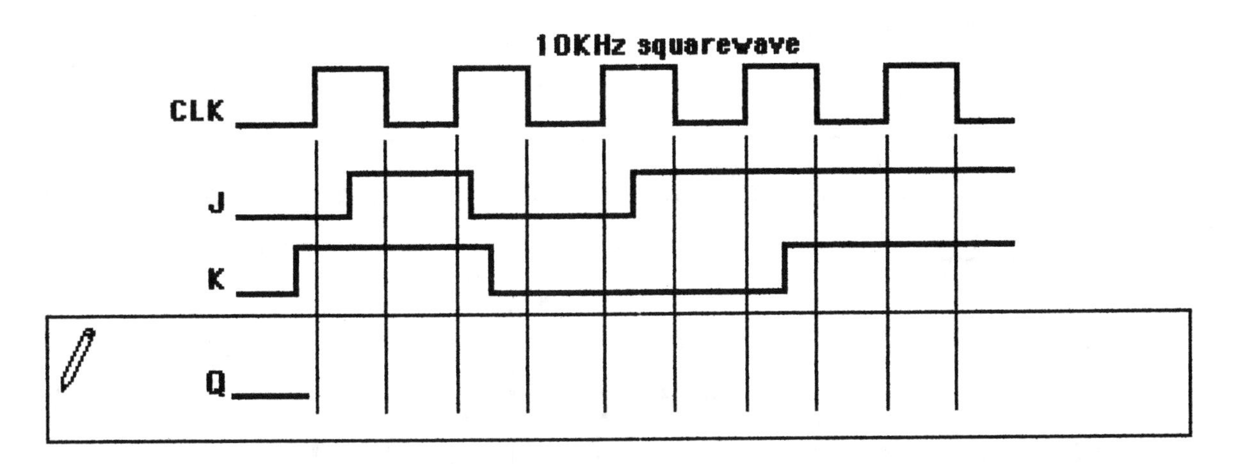

(b) Input J is shorted to ground.

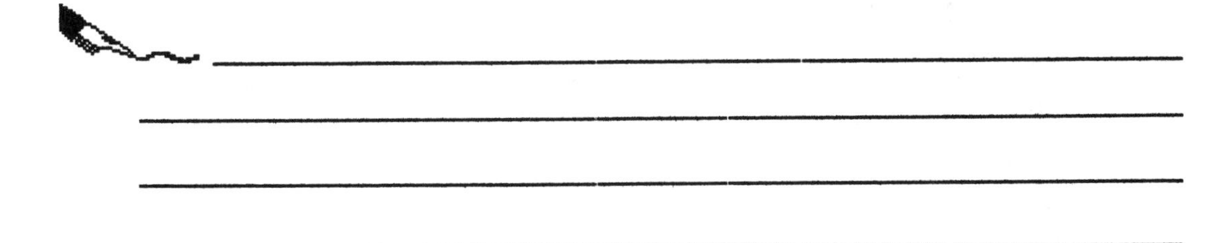

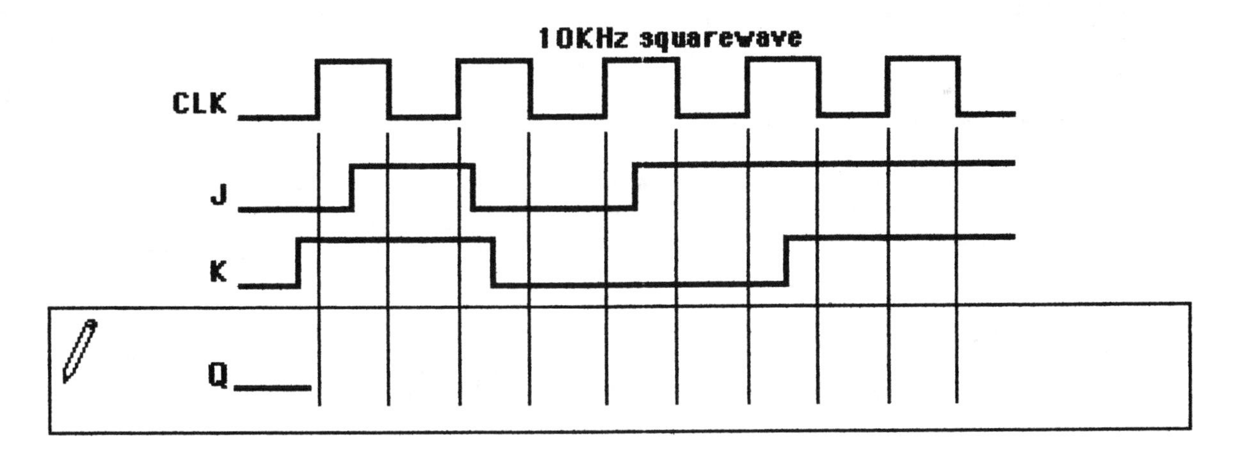

(c) Input K is internally shorted to Vcc.

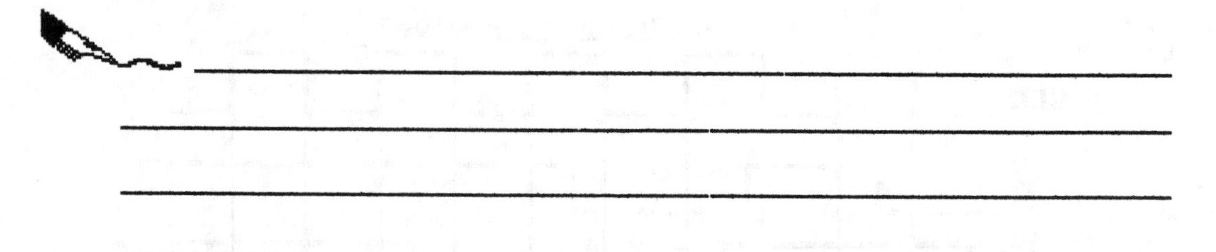

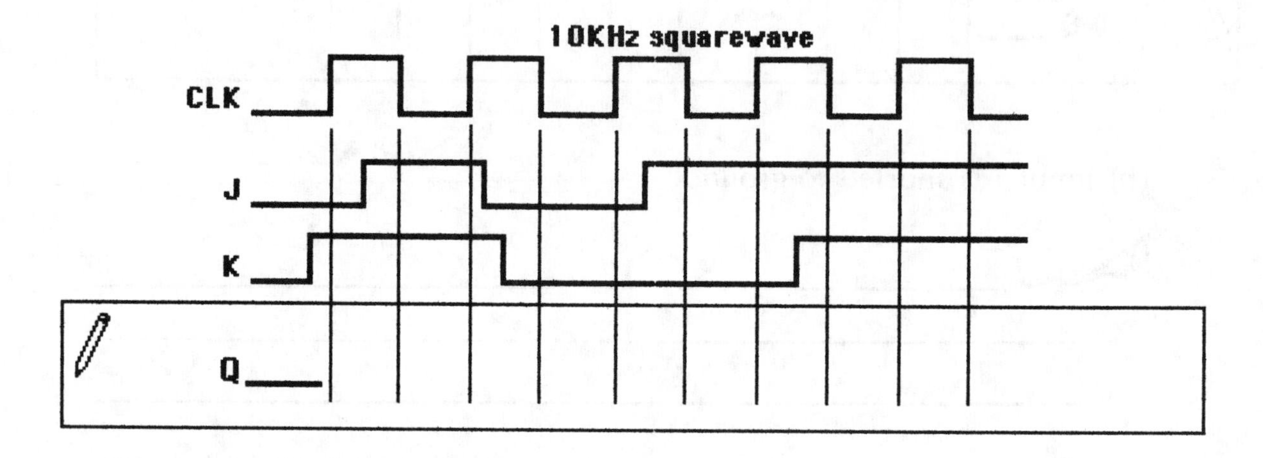

(d) Output Q is shorted to $\overline{Q}$.

5.7 Refer to Flip-Flops of Figure 5.4(a), (b) and waveforms (c). Sketch the output waveforms X and L. Assume that both Flip-Flops are initially cleared.

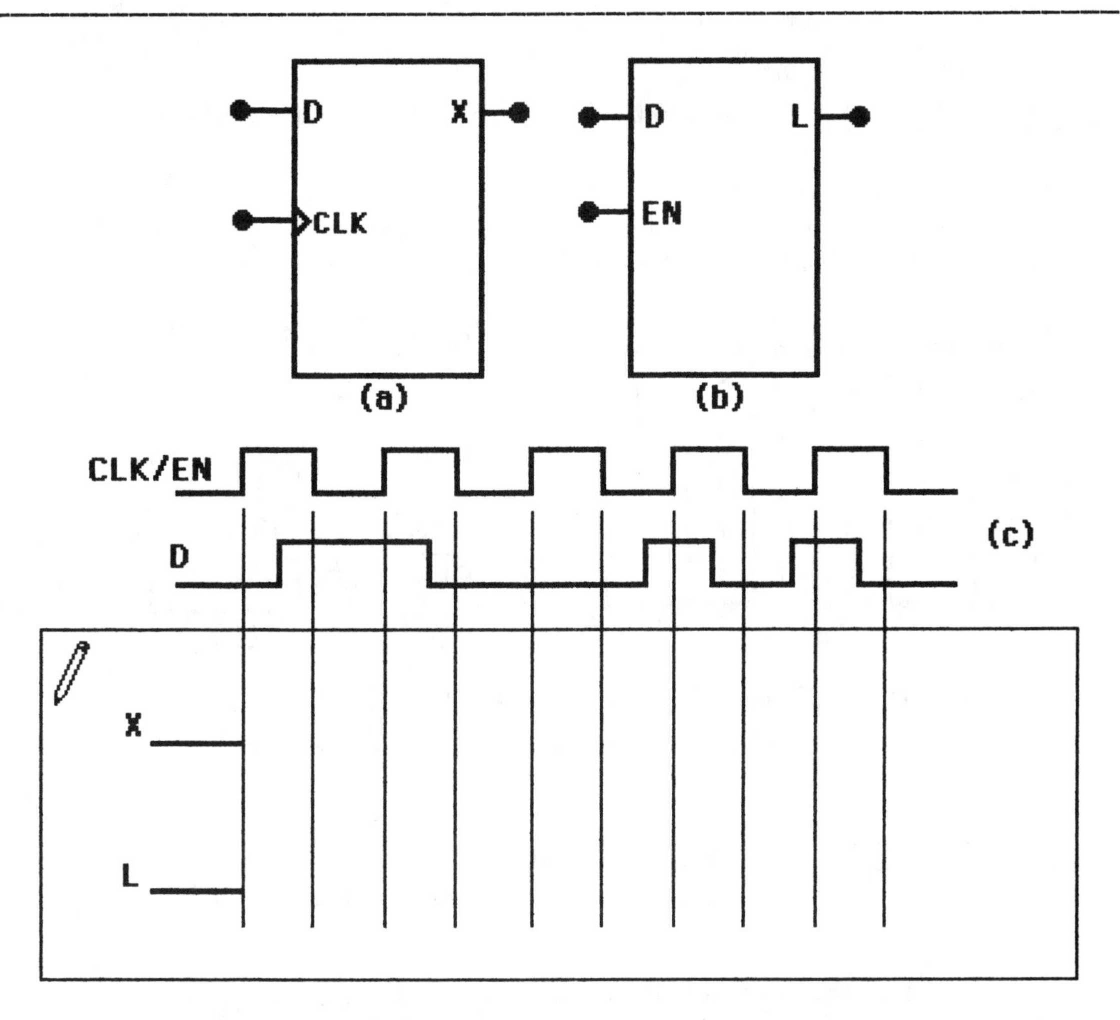

Figure 5.4: (a) Positive edge triggered D FF; (b) D Latch; (c) Waveforms

5.8 In problem 5.7, the technician notices that output L always follows input D. Determine a malfunction that could have caused the D Latch to be permanently in the 'transparent' mode.

SECTIONS 5.9-5.10 *Asynchronous Inputs/IEEE-ANSI Symbols*

5.9 Refer to the circuit represented by the IEEE/ANSI symbol of Figure 5.5(b). Complete the state table of Figure 5.5(a) for each NGT of the clock. Assume that initially Q=0 for each condition.

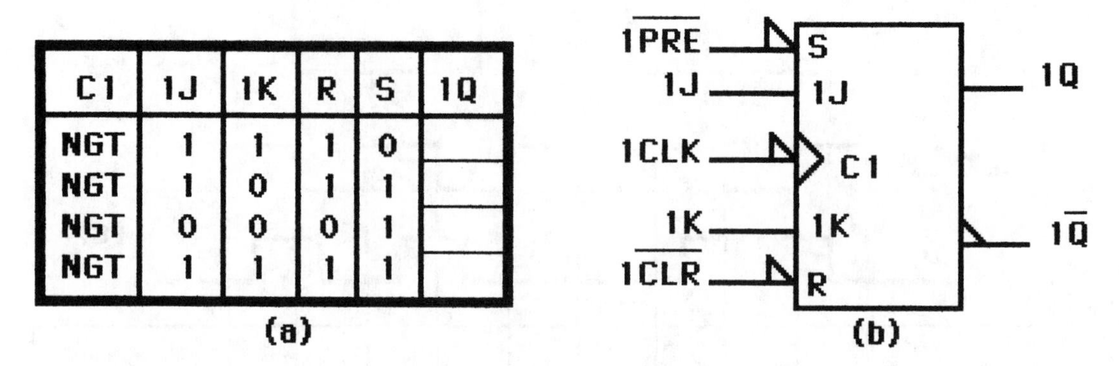

C1	1J	1K	R	S	1Q
NGT	1	1	1	0	
NGT	1	0	1	1	
NGT	0	0	0	1	
NGT	1	1	1	1	

Figure 5.5: (a) J-K FF State Table;
(b) IEEE/ANSI symbol of a J-K FF

SECTIONS 5.11-5.12 *Flip-Flop Timing Considerations/ Potential Timing Problem in FF Circuits*

5.10 Refer to the table of Flip-Flop values of Figure 5.6.

Flip-Flop timing values (ns)	4013B	7474
t_S	60	20
t_H	0	5
t_{pHL}- from CLK to Q	200	40
t_{pLH} - from CLK to Q	200	25
t_{pHL} - from $\overline{CLR}$ to Q	225	40
t_{pLH} - from $\overline{PRE}$ to Q	225	25
$t_{W(L)}$ - CLK LOW time	100	37
$t_{W(H)}$ - CLK HIGH time	100	30
$t_{W(L)}$ - at $\overline{PRE}$ or $\overline{CLR}$	60	30
f_{MAX} - in MHz	5	15

Figure 5.6

A CMOS 4013B Flip-Flop is wired as a toggle Flip-Flop. The clock signal used is a 1MHz, 7.5% Duty Cycle. On the oscilloscope, output Q of the Flip-Flop is not a 500KHz squarewave, but is instead a non-repetitive ambiguous waveform. Why do you think the reason is for the apparent malfunction?

___

___

___

5.11 The Flip-Flops of Figure 5.7(a) and (b) have the parameters listed in the table of Figure 5.6. The waveforms of Figure 5.7(c) are applied to both Flip-Flops. Assume that Z=M=0 initially. Sketch outputs Z and M.

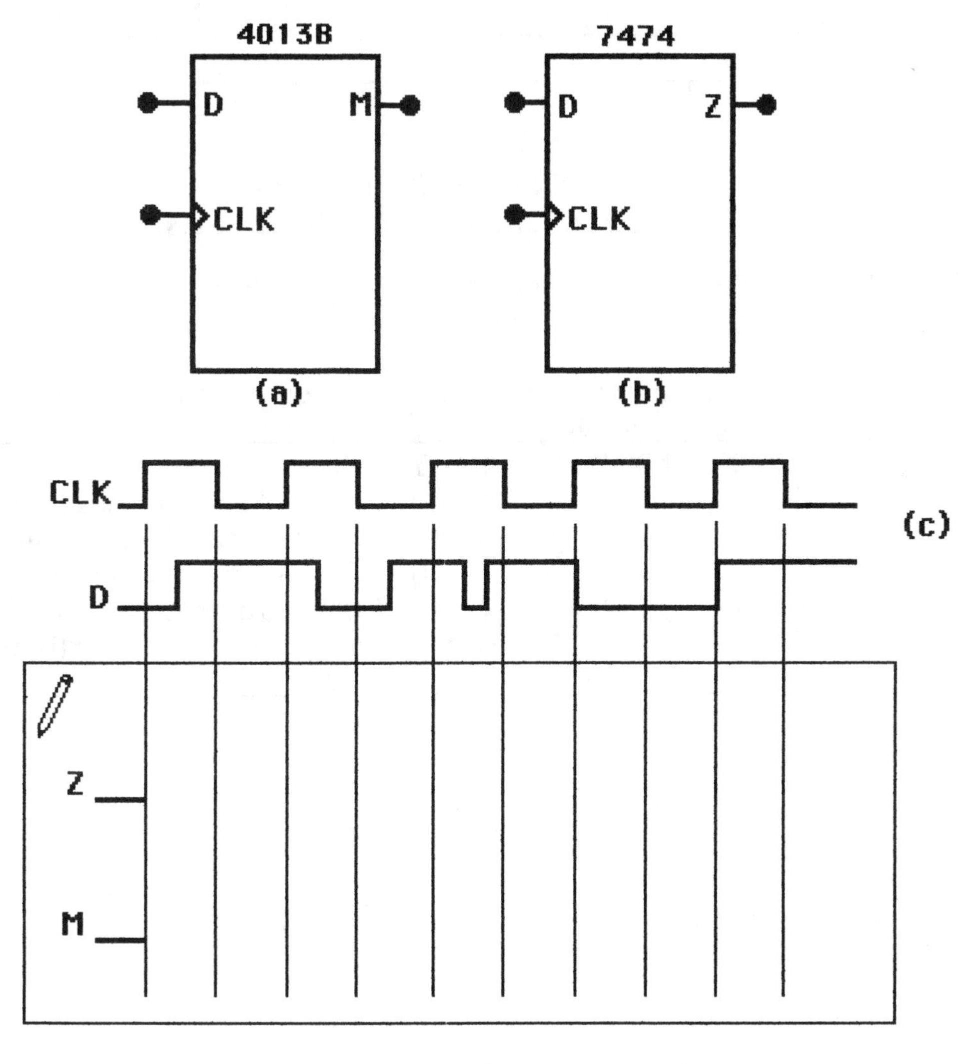

Figure 5.7: (a) 4013B FF; (b) 7474 FF; (c) Waveforms

SECTION 5.13 *Master-Slave Flip-Flops*

5.12 State the difference between a M/S Flip-Flop and a M/S Flip-Flop with Data Lock-Out.

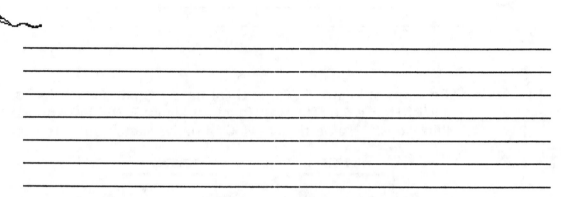

SECTIONS 5.17-5.18 *Data Storage and Transfer/ Serial Data Transfer: Shift Registers*

5.13 The table below was observed as shift pulses were applied to the shift register of Figure 5.8.

(a) Determine the most probable cause for the malfunction.

X3	X2	X1	X0	Y3	Y2	Y1	Y0	
1	0	1	1	0	0	0	0	<-- Before pulses applied.
0	1	0	1	1	0	0	0	<-- After the first pulse.
0	0	1	0	1	1	0	0	<-- After the second pulse.
0	0	0	1	0	1	1	0	<-- After the third pulse.
0	0	0	0	1	0	1	0	<-- After the fourth pulse.

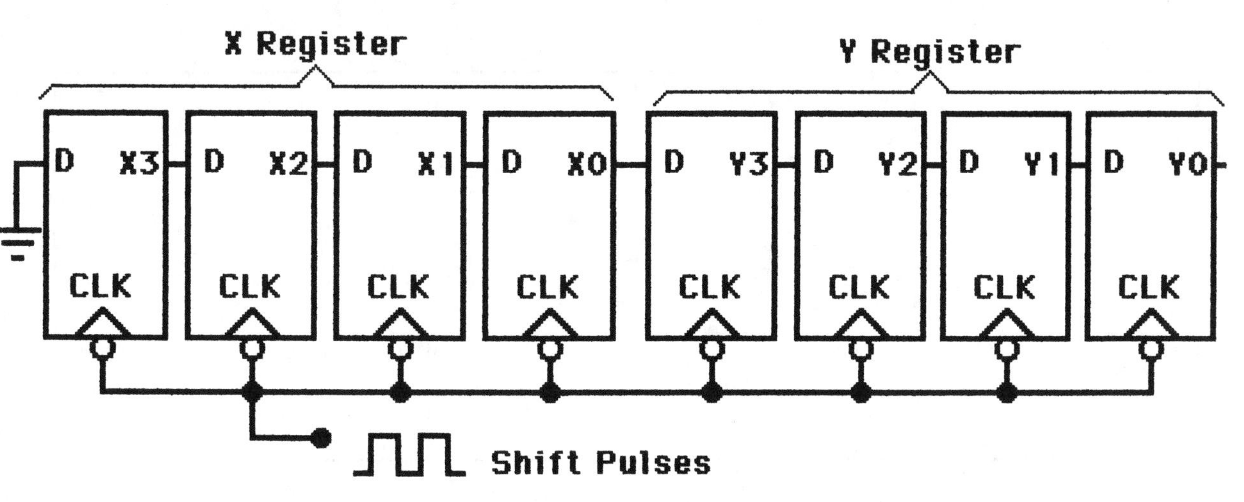

Figure 5.8: Serial data transfer.

(b) Disconnect the D input of Flip-Flop X_3 (MSB of the X-Register) from Ground, and connect it to the output Y_1. Complete the table below.

X3	X2	X1	X0	Y3	Y2	Y1	Y0	
1	0	0	1	0	0	0	0	<-- Before pulses applied.
_	_	_	_	_	_	_	_	<-- After the first pulse.
_	_	_	_	_	_	_	_	<-- After the second pulse.
_	_	_	_	_	_	_	_	<-- After the third pulse.
_	_	_	_	_	_	_	_	<-- After the fourth pulse.

SECTION 5.19 *Frequency Division and Counting*

5.14 Draw the state transition diagram for a three bit counter that has the following counting sequence: 0,1,2,4,6,7,3,5,0,1...repeats.

5.15 What is the MOD number of the counter described in problem 5.14?

5.16 If the counter of problem 5.14 is initially at 101_2, what count will it hold after 3673 pulses?

5.17 The input clock to the counter of Figure 5.9 is a 100KHz, 10% Duty-Cycle waveform.

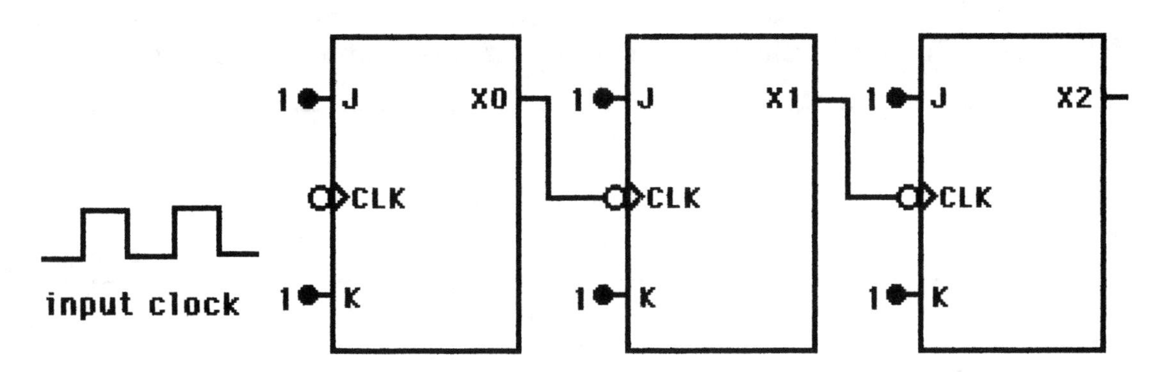

Figure 5.9: J-K FFs wired as a 3-bit binary counter

(a) What is the frequency and Duty-Cycle of the waveforms at the outputs X0, X1 and X2?

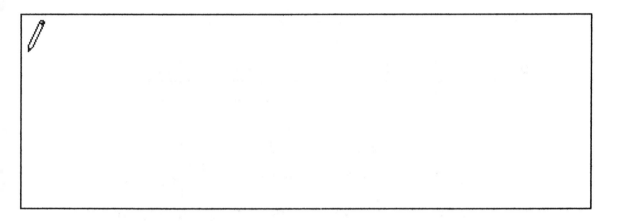

(b) Assume that the counter has eight J-K Flip-Flops (X0-X7). What is the total number of different states that the counter can go through?

(c) What is the maximum binary count that this counter can reach?

SECTION 5.20 *Microcomputer Applications*

5.18 Add the necessary "Decoding Logic" circuitry to the microcomputer of Figure 5.10 so that the four decoded outputs X, Y, Z, and W correspond to the MPU addresses $\overline{0F}xx_{16}$, $0Exx_{16}$, $0Dxx_{16}$, and $0Cxx_{16}$ respectively.

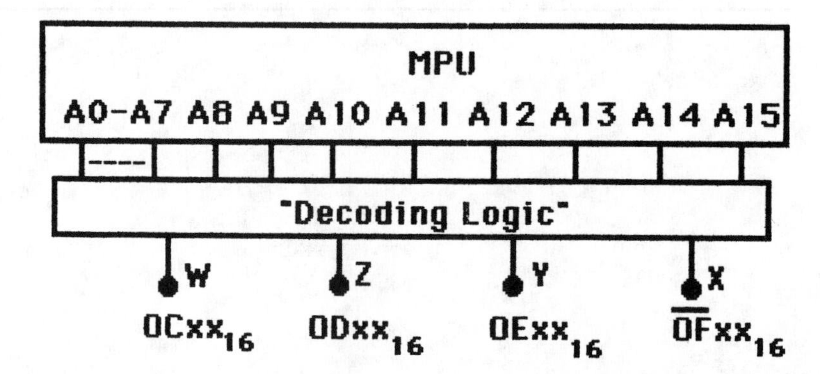

Figure 5.10: Microcomputer decoding application

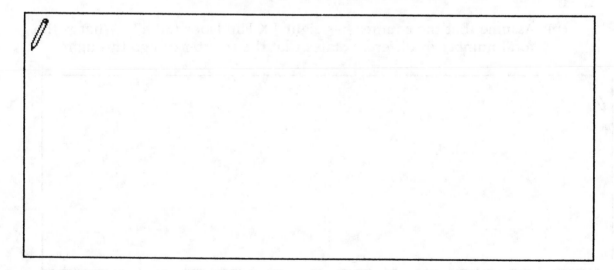

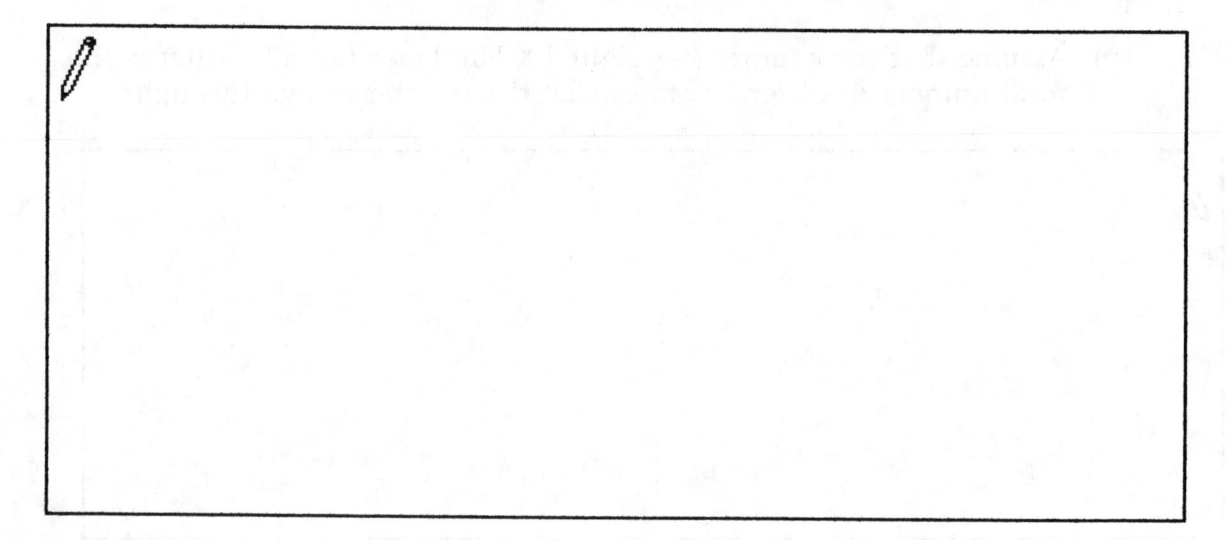

5.19 Describe the operational affects of an external open on address bus line 9 of Figure 5.10.

(Assume that the external logic circuitry to the MPU is TTL.)

____

____

____

____

____

____

SECTION 5.22 *One-Shot (Monostable Multivibrator)*

5.20 State the major operational difference between a retriggerable and a non-retriggerable One-Shot.

____

____

____

____

____

____

____

____

SECTION 5.25 *Troubleshooting Flip-Flop Circuits*

(Refer to Figure 5.11 on for the next four problems)

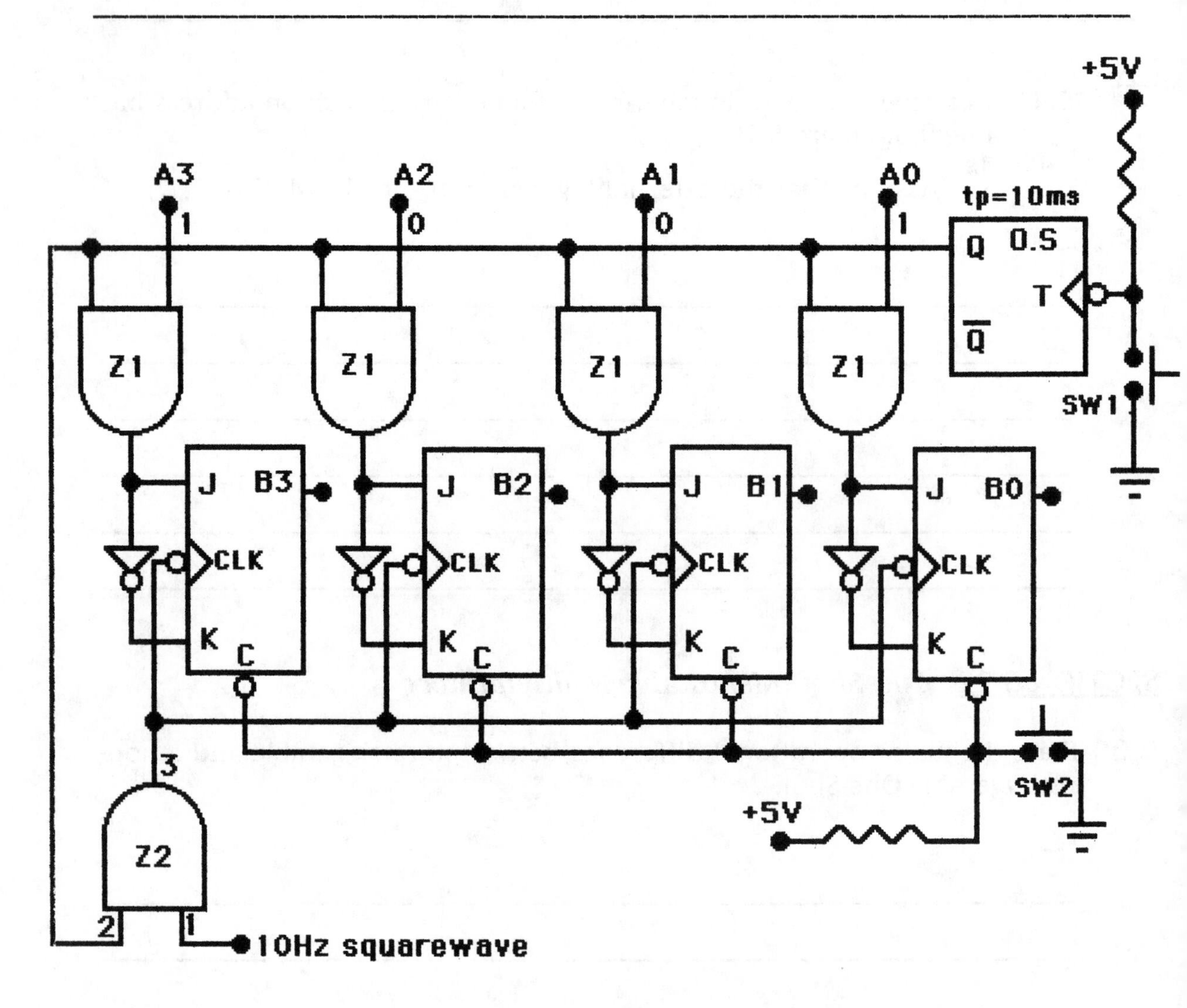

Figure 5.11: Parallel shift register

5.21 The TTL shift register circuit of Figure 5.11 was designed so that the parallel transfer of input data A_0-A_3 to outputs B_0-B_3 would occur upon actuation of switch SW1. When switch SW2 is momentarily depressed, the parallel register is cleared.

(a) What binary data will be present at outputs B_0-B_3 after SW1 is depressed?

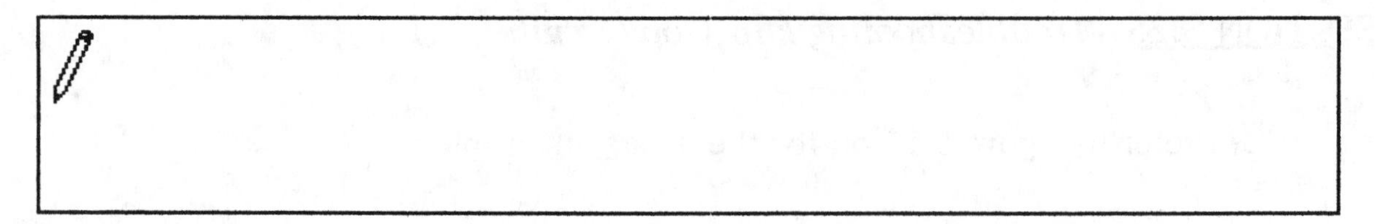

(b) Why are the Inverters needed between the J and the K inputs of each Flip-Flop?

5.22 During the testing of the circuit, the technician records the following results:

A_3	A_2	A_1	A_0	B_3	B_2	B_1	B_0	
1	0	0	1	0	0	0	0	after SW2 is depressed.
1	0	0	1	1	0	0	1	after SW1 is depressed.
1	1	0	0	1	0	0	1	after # A_0-A_3 is changed.
1	1	0	0	0	1	0	0	after SW1 is depressed.
1	1	0	0	1	1	0	0	after SW1 is depressed again.
1	1	1	1	1	1	0	0	after # A_0-A_3 changed.
1	1	1	1	0	1	1	1	after SW1 is depressed.
1	1	1	1	1	1	1	1	after SW1 is depressed again.
1	1	1	1	0	0	0	0	after SW2 is depressed.

Determine the cause for the malfunction observed in the recorded data.

5.23 After the technician repaired the problem that caused the malfunction observed in the previous question, he discovered that sometimes the binary number A_0-A_3 doesn't get transferred to outputs B_0-B_3.

(a) Analyze the circuit design and offer a possible explanation as to the reason why data don't always get transferred.

(Describe the necessary steps required for a data transfer to occur.)

(b) List two different ways of fixing the problem.

1.

2.

5.24 Once in a while, it appears that pressing SW1 has the same affect as pressing SW2, that is, the shift register gets cleared. At other times, what gets transferred to the output does not reflect the data at the inputs. Determine what is the most probable cause for the problem and how it can be fixed.

TEST 5

1. The two possible output states of a Flip-Flop, regardless of its type, is:

   (a) The Q output is LOW and the $\overline{Q}$ output is LOW.
   (b) The Q output is LOW and the $\overline{Q}$ output is HIGH.
   (c) The Q output is HIGH and the $\overline{Q}$ output is HIGH.
   (d) The Q output is HIGH and the $\overline{Q}$ output is LOW.
   (e) Both b and d.

2. If a D-type Flip-Flop is clocked with a 10KHz signal with a duty-cycle of 30%, and its $\overline{Q}$ output is tied to the D input, then the signal at the output Q is:

   (a) A 10KHz signal with a 30% duty-cycle.
   (b) A 10KHz squarewave.
   (c) A 5KHz signal with a 30% duty-cycle.
   (d) A 5KHz squarewave.
   (e) None of the above.

3. The major difference between a J-K Flip-Flop and any other type of Flip-Flop is that the J-K Flip-Flop:

   (a) has asynchronous inputs. (b) has a toggle mode. (c) has a t_H=0.
   (d) has both synchronous and asynchronous inputs.

4. Which of the following require an external resistor and capacitor for proper operation?

   (a) Master slave J-K Flip-Flop. (b) Schmitt-trigger. (c) One-shot.
   (d) NOR latch.

5. Which of the following Flip-Flop timing parameters indicate the time requirements of the synchronous inputs prior and after the clocking of the Flip-Flop?

(a) t_S (b) t_{PHL} (c) $t_{W(L)}$ (d) t_H (e) a and d (f) b and c

6. Which of the following are true about a MOD 32 counter?

(a) It has a maximum count of 32.
(b) It has 31 possible counts.
(c) It will divide an input frequency by a factor of 31.
(d) None of the above.

7. Determine the output frequency for a 4-bit serial binary counter with an input clock of 160KHz.

(a) 40KHz (b) 10.66KHz (c) 20KHz (d) 10KHz (e) None of the above.

8. Which of the following is true about a retriggerable one-shot.

(a) Its normal state is the CLEARED state.
(b) Its quasi-stable state is the SET state.
(c) It can be retriggered while it is in its quasi-stable state.
(d) If it is triggered by a waveform that has a period (T) less than the one-shot's pulse duration (t_P), then the one-shot will always be in its quasi stable state.
(e) All of the above.

9. How many shift pulses will be needed to serially shift the contents of a 4-bit shift register to another 4-bit shift register?

(a) 1 (b) 2 (c) 4 (d) 8 (e) None of the above.

10. The D latch can only change states when either a positive or a negative going transition occurs at its Enable input.

(a) [TRUE] (b) [FALSE]

6 DIGITAL ARITHMETIC: OPERATIONS AND CIRCUITS

Objectives

Upon completion of this chapter, you will be able to:

- Perform binary addition, subtraction, multiplication, and division on two binary numbers.
- Add and subtract hexadecimal numbers.
- Know the difference between binary addition and OR addition.
- Compare the advantages and disadvantages among three different systems of representing signed binary numbers.
- Manipulate signed binary numbers using the 2's-complement system.
- Understand the BCD adder circuit and the BCD addition process.
- Describe the basic operation of an arithmetic-logic unit.
- Employ full adders in the design of parallel binary adders.
- Cite the advantages of parallel adders with the look-ahead carry feature.
- Analyze the operation of a serial binary multiplier circuit.
- Describe some of the operations performed by an arithmetic processing unit.
- Read and understand the IEEE/ANSI symbol for a parallel adder.
- Analyze several troubleshooting case studies of adder/subtractor circuits.

Glossary of key terms covered in this chapter:

- **Accumulator Register** - The principal register of an Arithmetic Logic Unit. [sec.6.9]
- **Addend** - A number to be added to another. [sec.6.3]
- **Arithmetic Logic Unit (ALU)** - A digital circuit used in computers to perform various arithmetic and logic operations. [sec.6.9/6.18]
- **Arithmetic Processing Unit (APU)** - A Large Scale Integration IC capable of performing many complex arithmetic operations on 16-bit or 32-bit binary numbers. [sec.6.18]
- **Augend** - The number to which an addend is added. [sec.6.3]
- **BCD Adder** - A special adder containing two 4-bit parallel adders and a correction detector circuit. Whenever the addition of two BCD code groups is greater than 1001_2 (9_{10}), the correction detector circuit senses it, adds to the result the correction factor 0110_2 (6_{10}) and generates a carry to the next decimal position. [sec.6.16]
- **Binary Multiplier** - A special digital circuit capable of performing the arithmetic operation of multiplication on two binary numbers. Binary multipliers can be either serial or parallel. [sec.6.17]
- **Carry** - A digit or a bit that is generated when two numbers are added and the result is greater than that of the base for that number system. [sec.6.1]
- **Carry Propagation** - It is the intrinsic circuit delay of some parallel adders that prevents the Carry bit (C_{OUT}) and the result of the addition from appearing at the output simultaneously. [sec.6.13]
- **Carry Ripple** - See Carry Propagation. [sec.6.13]
- **Dividend** - The number to be divided. [sec.6.6]
- **Divisor** - The number by which a dividend is divided. [sec.6.6]

- **Full Adder** - Logic circuit with three inputs and two outputs. The inputs are a carry bit (C_{IN}) from a previous stage, a bit from the augend, and a bit from the addend respectively. The outputs are the sum bit produced by the addition of the bit from the addend with the bit from the augend and the resulted carry (C_{OUT}) bit which will be added to the next stage. [sec.6.10]

- **Function Generator** - See Arithmetic Logic Unit (ALU). [sec.6.18]

- **Half Adder** - Logic circuit with two inputs and two outputs. The inputs are a bit from the augend, and a bit from the addend respectively. The outputs are the sum bit produced by the addition of the bit from the addend with the bit from the augend and the resulted carry (C_{OUT}) bit which will be added to the next stage. [sec.6.11]

- **Look-Ahead Carry** - The ability of some parallel adders to predict, without having to wait for the carry to propagate through the Full Adders, whether or not a carry bit (C_{OUT}) will be generated as result of the addition, thus reducing the overall propagation delays. [sec.6.13]

- **Minuend** - The number from which the subtrahend is to be subtracted. [sec.6.4]

- **Negation** - The operation of converting a positive number to its negative equivalent or a negative number to its positive equivalent. [sec.6.2]

- **1's-Complement Form** - The result obtained when each bit of a binary number is complemented. [sec.6.2]

- **Overflow** - When in the process of adding two binary numbers a 1 is generated from the MSB position of the number into the sign bit position. [sec.6.4/6.12]

- **Parallel Adder** - A digital circuit made from full adders and used to add all the bits from the addend and the augend together and simultaneously. [sec.6.10]

- **Serial Multiplier** - See Binary Multiplier.

- **Sigma** (Σ) - A Greek letter which represents addition and is often used to label the Sum Output bits of a parallel adder. [sec.6.14/6.19]

- **Sign Bit** - A binary bit that is added to the leftmost position of a binary number to indicate whether that number represents a positive or a negative quantity. [sec.6.2]

- **Sign-magnitude system** - Number system consisting of a sign-bit and the necessary magnitude bits. In this system the magnitude bits are the true binary equivalent of the decimal value being represented. [sec.6.2]

- **Software Multiplication** - Multiplication accomplished by using the program (software) that does the multiplication through the repeated execution of addition and shifting instructions. [sec.6.17]

- **Subtrahend** - The number that is to be subtracted from a minuend. [sec.6.4]

- **2's-Complement Form** - The result obtained when a 1 is added to the least significant bit position of a binary number in the 1's-complement form. [sec.6.2]

- **2's-Complement System** - See 2's-complement form.

Problems

SECTION 6.1 *Binary Addition*

6.1 Add the following groups of binary numbers using binary addition:

(a) $110110_2+10010_2$ (b) 10101.001_2+1011_2 (c) $11101_2+10000_2$

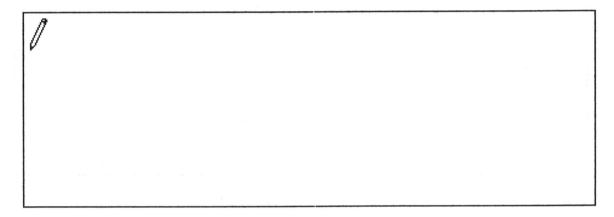

SECTION 6.2 *Representing Signed Numbers*

6.2 Using the 2's-complement system, represent each of the following signed decimal numbers:

(a) +14 (b) -32 (c) -8 (d) +55 (e) -355 (f) +123

SECTIONS 6.3-6.4 *Addition in the 2's-Complement System/ Subtraction in the 2's-Complement System*

6.3 The following binary numbers are expressed in the 2's-complement system. If the MSB is the sign bit, determine the equivalent decimal number.

(a) 010110_2 (b) 110111_2 (c) 01110011_2 (d) 11000_2 (e) 111001_2

6.4 Perform the following operations using the 2's Complement system:

(a) [8+5] (b) [-12+15] (c) [+32-(-8)] (d) [9-13]

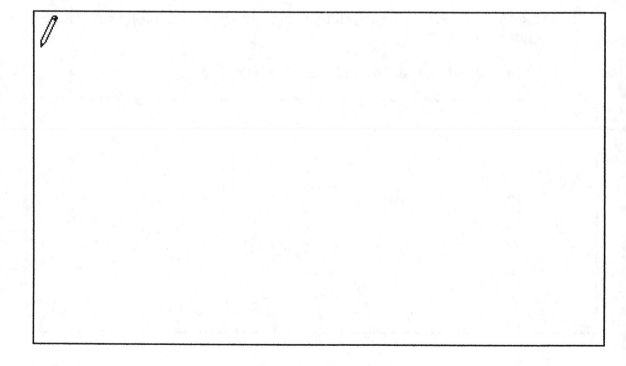

SECTIONS 6.5-6.6 *Multiplication of Binary Numbers/Binary Division*

6.5 Perform the following arithmetic operations using binary numbers:

(a) 23 x 5 (b) 12.5 x 2.25 (c) 35/5 (d) 135/15

SECTIONS 6.7-6.8 *BCD Addition/Hexadecimal Arithmetic*

6.6 Perform the following additions after converting each decimal number to its BCD code:

(a) 25+26 (b) 50+32 (c) 2357+1250 (d) 9+9

6.7 Add or subtract the following hexadecimal numbers:

(a) $75F_{16}+32D_{16}$ (b) $12A_{16}-FF_{16}$ (c) $7834_{16}+ABCD_{16}$

SECTIONS 6.9-6.15 ***Arithmetic Circuits/Parallel Binary Adder/ Complete Parallel Adder with Registers/ Integrated-Circuit Parallel Adder/ 2's Complement System***

6.8 When the contents of register [A] (A_0-A_3) and register [B] (B_0-B_3) of the TTL circuit of Figure 6.1 are added, the following results are obtained. Determine the cause for the malfunction.

A_3	A_2	A_1	A_0	B_3	B_2	B_1	B_0	C_4	S_3	S_2	S_1	S_0
1	1	0	1	0	1	1	0	1	0	0	1	1
1	0	0	0	0	1	1	1	1	0	1	1	1
0	1	0	0	0	1	0	1	0	1	0	0	1
0	0	1	1	0	1	0	0	0	1	1	1	1

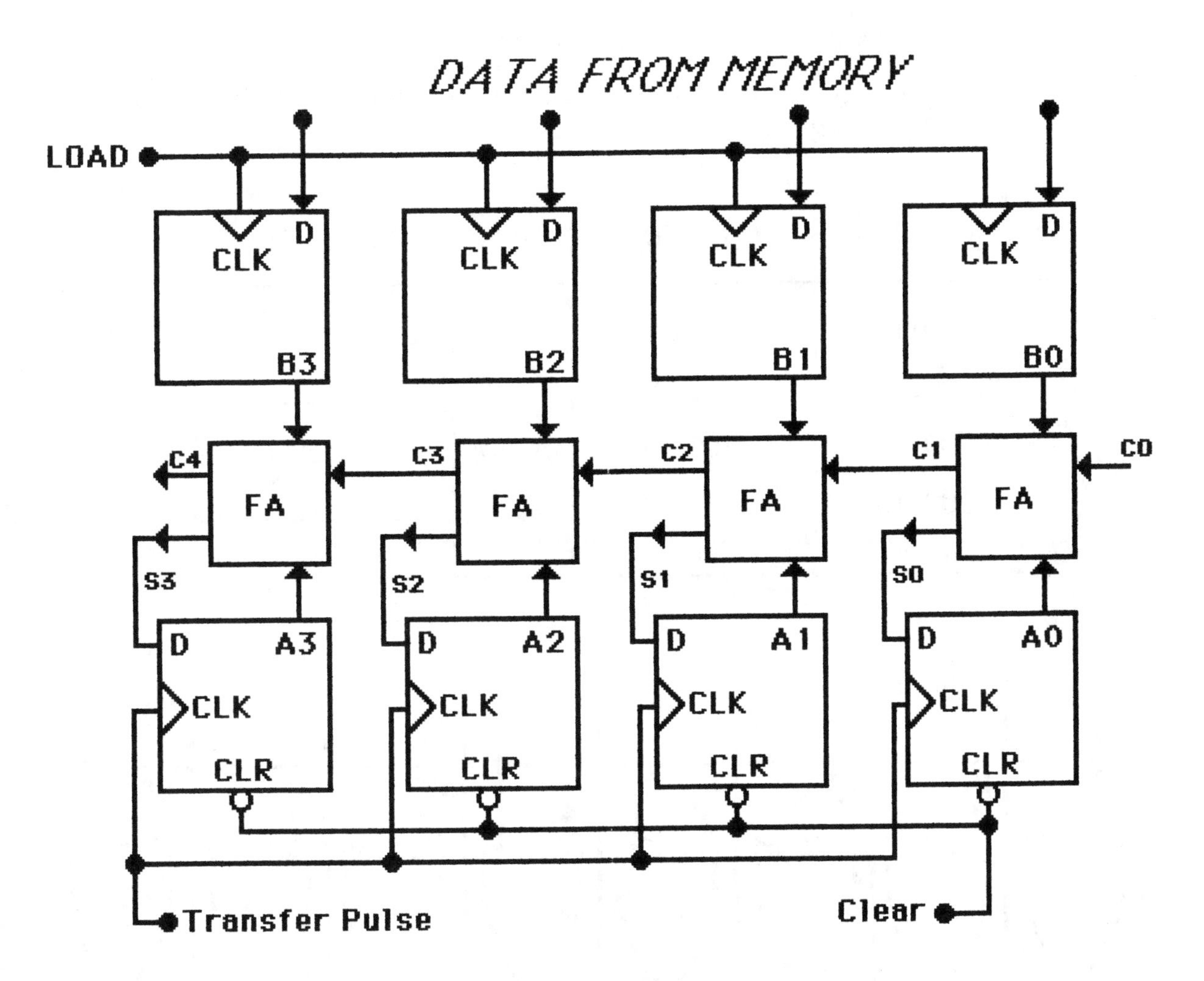

Figure 6.1:Complete 4-bit TTL parallel adder with registers

6.9 Refer to the circuit of Figure 6.2 and complete the table below for each of the following set of conditions:

A7-A0	B7-B0	SW1	OUTPUTS
01100111	10001011	GND	______
11000001	10100010	+5V	______
10100101	11000010	GND	______
11100110	11000110	+5V	______

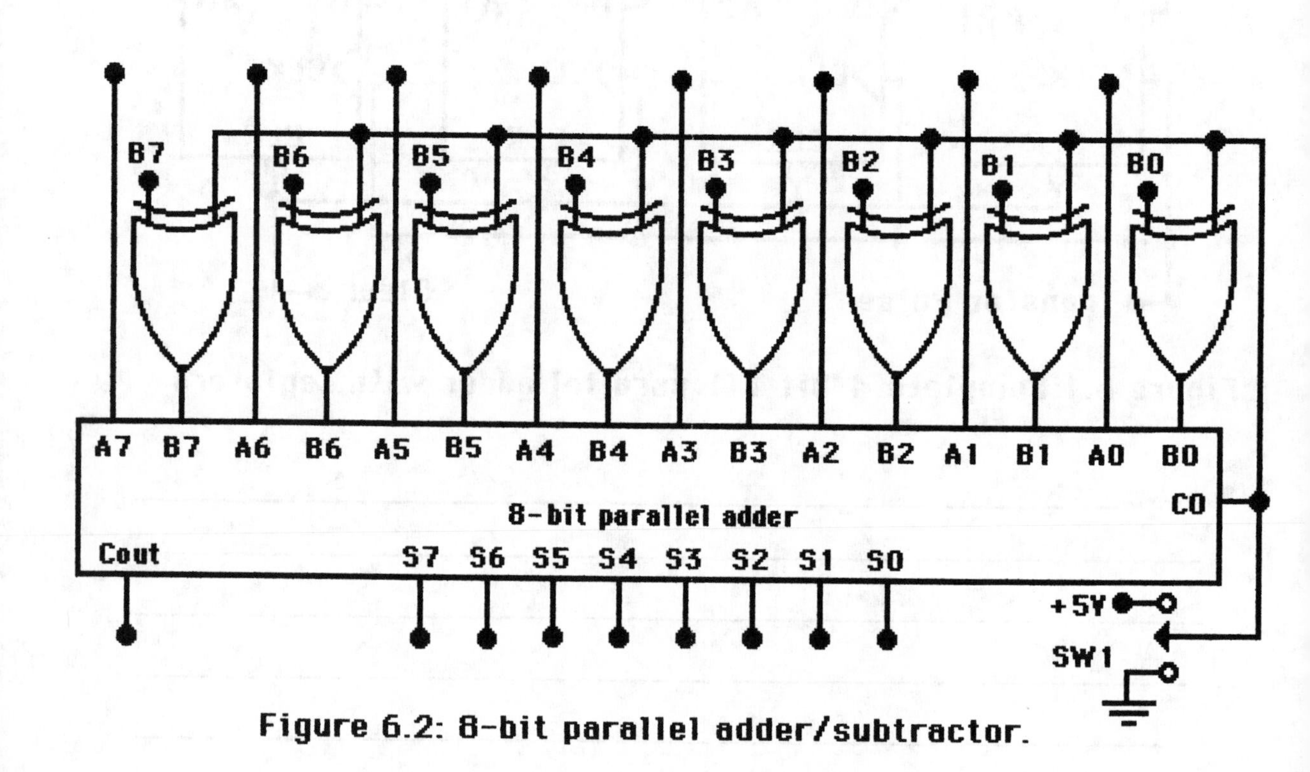

Figure 6.2: 8-bit parallel adder/subtractor.

SECTION 6.16 *BCD Adder*

Refer to Figure 6.3 for all the problems in this section.

6.10 Complete the table below for each of the following conditions of the BCD circuit of Figure 6.3:

$A_3A_2A_1A_0$	$B_3B_2B_1B_0$	$S_3S_2S_1S_0$	X	$\Sigma_3\Sigma_2\Sigma_1\Sigma_0$
1001	1001	______	______	______
1000	0011	______	______	______
0111	0101	______	______	______
0110	0011	______	______	______

6.11 How would the results of the BCD sum be affected, if input C_0 of the 'Correction adder' became open?

(A floating TTL input acts as a logic HIGH.)

____

____

____

____

____

____

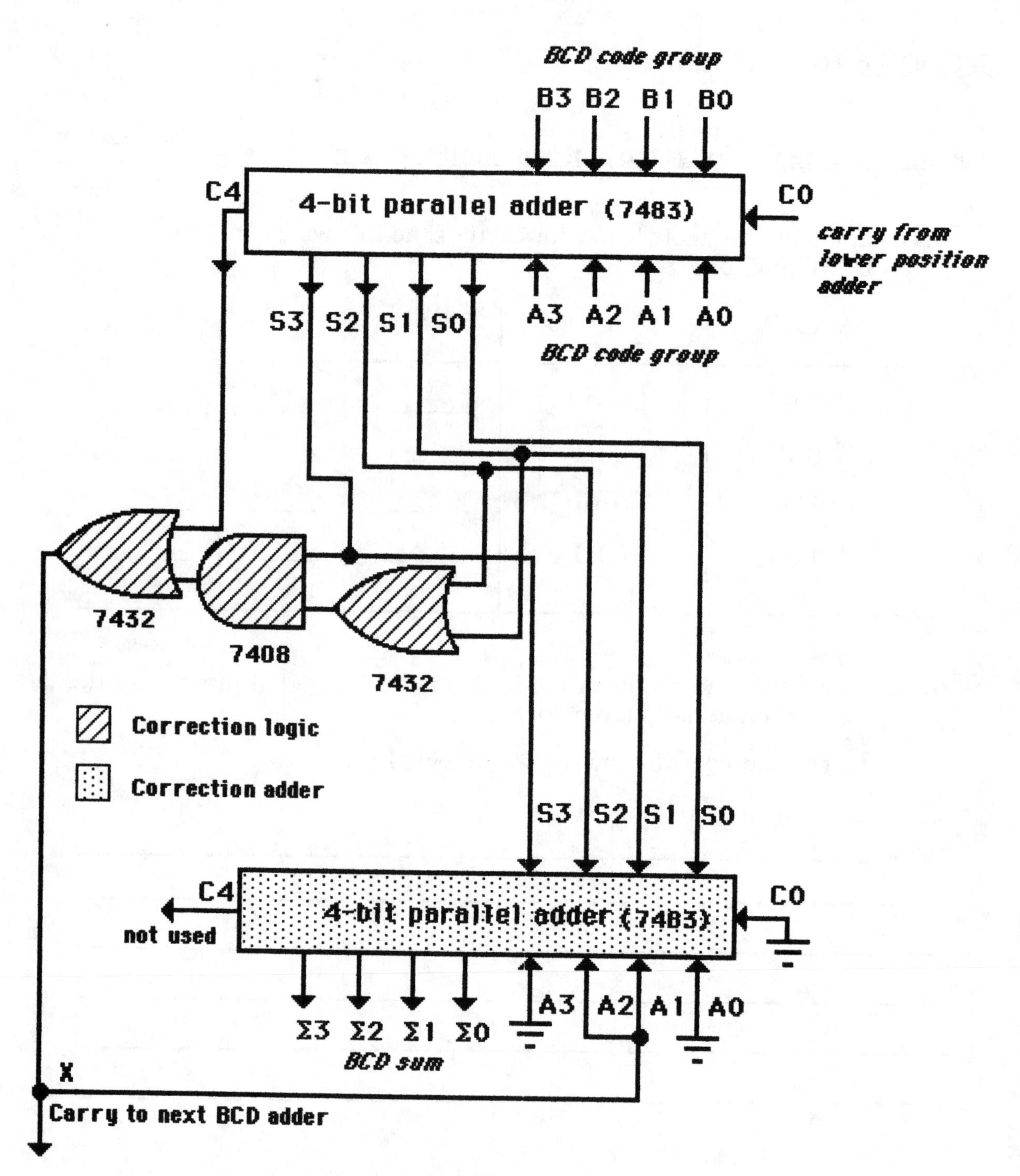

Figure 6.3: BCD adder

6.12 Redraw the 'Correction adder' circuit using only one Full-Adder and two Half-Adders.

6.13 The following table of results were obtained by a technician while testing the BCD adder circuit of Figure 6.3. Determine the cause for the malfunction.

B_3 B_2 B_1 B_0	A_3 A_2 A_1 A_0	X	BCD sum
0 1 0 1	0 1 0 1	1	0000
0 0 1 0	0 1 1 1	0	1101
0 1 1 1	0 1 0 1	1	0010
0 0 1 0	0 0 1 1	0	1001
0 0 0 0	0 0 0 0	0	0100

SECTIONS 6.17,6.19,6.20 *Binary Multipliers/IEEE-ANSI Symbols/ Troubleshooting Case Study*

6.14 In the laboratory, a technician builds the 4-bit serial multiplier circuit of Figure 6.4 and the following data are taken. Describe what is the most probable fault with the circuit.

Multiplicand	Multiplier	Result
11	13	143
10	13	129
3	3	10
6	2	12
2	2	4
5	2	9

6.15 The circuit of Figure 6.4 is properly constructed and tested. The circuit does not work once the frequency of the clock pulses exceeds a certain value. As the frequency is reduced, the circuit resumes proper operation. What is the most likely reason for the malfunction?

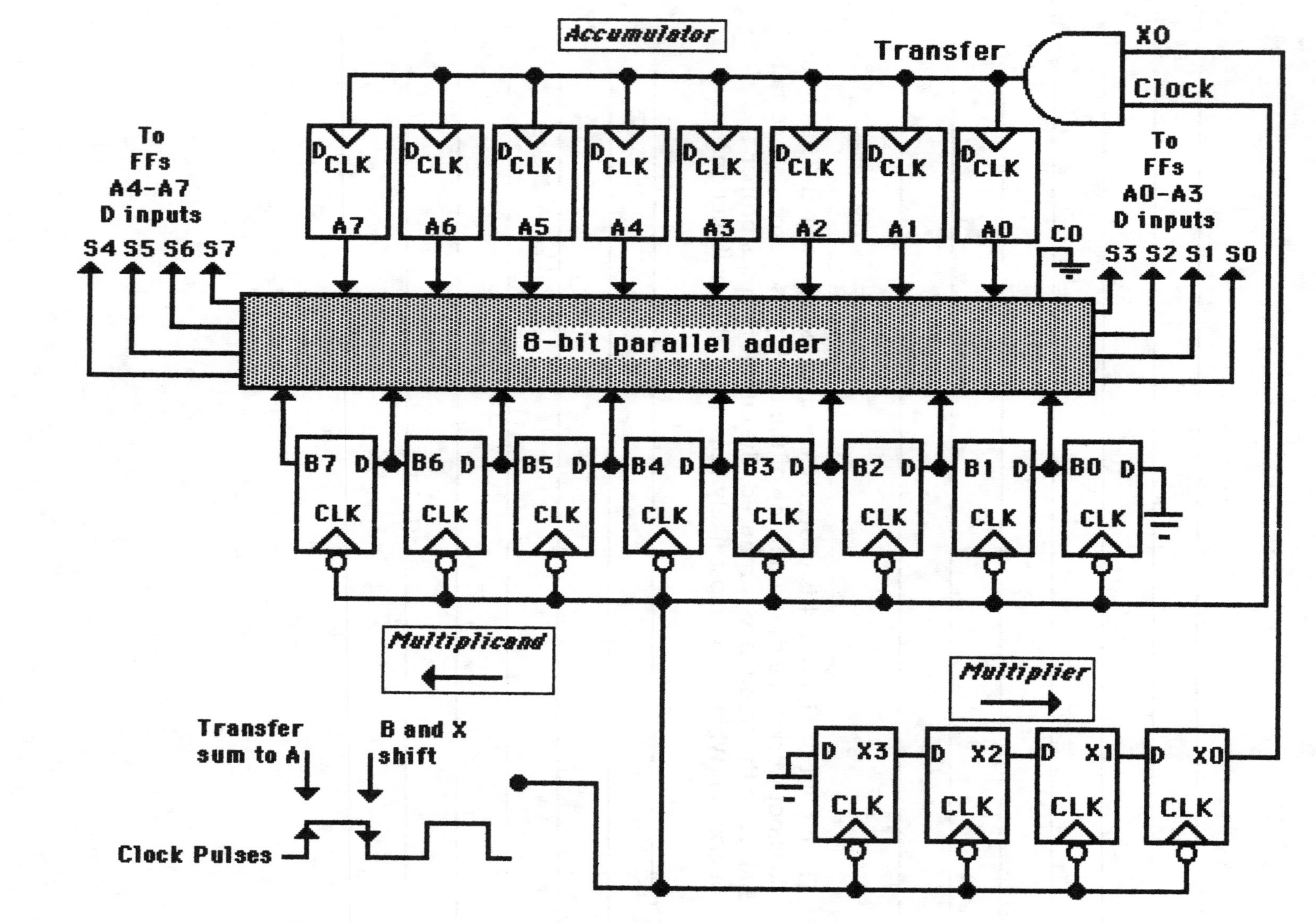

Figure 6.4: Circuit for a 4-bit serial multiplier that uses shift registers for the multiplier and multiplicand.

6.16 For the circuit of Figure 6.4, describe what would happen if the Ground connection to the D input of Flip-Flop X3 became open.

6.17 Describe the function performed by the IEEE/ANSI symbol depicted in Figure 6.5.

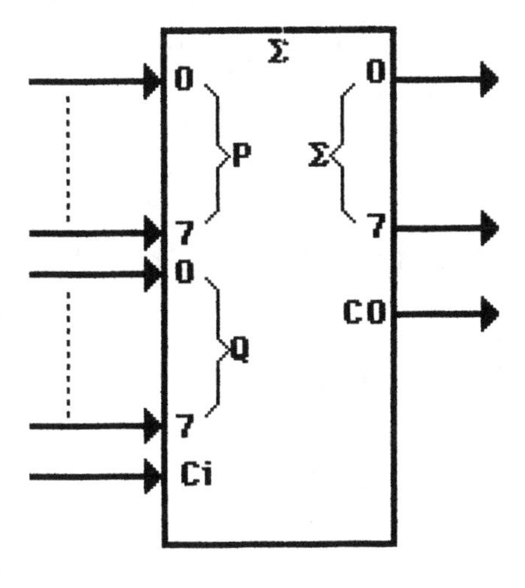

Figure 6.5: IEEE/ANSI Symbol

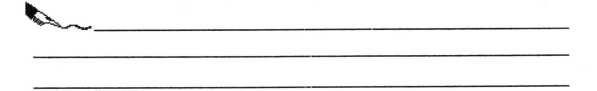

TEST 6

1. What are the minimum number of bits required to represent -150_{10} as a signed binary number is the 2's-complement form?

   (a) 10 (b) 9 (c) 8 (d) 7

2. What is the decimal value of the 2's-complement system signed binary number 11111111_2?

   (a) -127 (b) -255 (c) 255 (d) -1

3. When performing subtraction by addition in the 2's-complement system:

   (a) Both the minuend and the subtrahend are changed to the 2's-complement form.
   (b) The minuend is changed to the 2's-complement form and the subtrahend is left unchanged.
   (c) Both the minuend and the subtrahend are left unchanged.
   (d) The subtrahend is changed to the 2's-complement form and the minuend is left unchanged.

4. Which of the following represents the sum of $4CA5_{16}$+$FEA5_{16}$?

   (a) $15B4A_{16}$ (b) $2FB4A_{16}$ (c) $14B4A_{16}$ (d) $14B40_{16}$

5. Convert the numbers 255_{10} and 989_{10} to BCD and then add them together. Select the BCD code groups that reflect the final answer.

   (a) 0001 0010 0100 0100 (b) 0010 0100 0100
   (c) 1011 1101 1110 (d) 1111 0010 0100 0100

6. Multiplication of the unsigned binary numbers 1011_2 and 1101_2 will yield a product of:

   (a) 00111111_2 (b) 10001111_2 (c) 10110000_2 (d) 00110111_2

7. Which of the following represents the four possible conditions for the addition of two binary numbers?

(a) 0+0=0, 1+0=1, 1+1=0+carry, 1+1+1=0+carry
(b) 0+0=0, 1+0=1, 1+1=1+carry, 1+1+1=1+carry
(c) 0+0=0, 1+0=1, 1+1=0+carry, 1+1+1=1+carry
(d) 0+0=0, 1+0=1, 1+1=1+carry, 1+1+1=1+carry

8. After two signed binary numbers are added, the final result in the accumulator is 11001011_2. What is the decimal equivalent of the result?

(a) 203 (b) -53 (c) -203 (d) 52

9. Negation is the operation of converting a positive number to its negative equivalent.

(a) [TRUE] (b) [FALSE].

10. APU is a digital circuit used in 8-bit μcomputers to perform various arithmetic and logic operations.

(a) [TRUE] (b) [FALSE].

7 COUNTERS AND REGISTERS

Objectives

Upon completion of this chapter, you will be able to:

- Understand the operation and characteristics of synchronous and asynchronous counters.
- Construct counters with MOD numbers less than 2^N.
- Identify IEEE/ANSI symbols used in IC counters and registers.
- Construct both up and down counters.
- Connect up multi-stage counters.
- Analyze and evaluate various types of presettable counters.
- Design arbitrary-sequence synchronous counters.
- Understand several types of schemes used to decode different types of counters.
- Eliminate decoder spikes by employing a technique called strobing.
- Compare the major differences between the ring and Johnson counters.
- Analyze the theory of operation of a frequency counter and of a digital clock.
- Recognize and understand the operation of various types of registers.
- Apply existing troubleshooting techniques used for combinational logic systems to troubleshoot sequential logic systems.

Glossary of key terms covered in this chapter:

- **Active-HIGH decoder** - A decoder which produces a logic HIGH at the output when detection occurs. [sec.7.11]
- **&** - When used inside an IEEE/ANSI symbol, it indicates an AND gate or AND function. [sec.7.3]
- **Asynchronous Counter** - Type of serial counter where each Flip-Flop output serves as the Clock input signal for the next Flip-Flop in the chain. [sec.7.1]
- **BCD Counter** - A binary counter that counts from 0000_2 to 1001_2 before it recycles. [sec.7.2]
- **Buffer Register** - A register that holds digital data temporarily. [sec.7.15]
- C - When used inside an IEEE/ANSI symbol, the letter C in the label for an input indicates that, the input controls the entry of data into a storage element. [sec.7.10]
- C1 - When used inside an IEEE/ANSI symbol, the designation C1 in the label indicates that, this input controls the entry of data into any storage element that has the prefix 1 in its label. [sec.7.22]
- **Circulating Shift Register** - A shift register where one of the outputs of the last Flip-Flop in the shift register is connected to the input of the first Flip-Flop in the shift register. [sec.7.14]
- CT=0 - When used inside an IEEE/ANSI symbol, CT=0 in the label for an input indicates that the counter will clear when that input goes active. [sec.7.3]
- CTR - When used inside an IEEE/ANSI symbol, it indicates that the IC is a counter. [sec.7.3]
- D - When used inside an IEEE/ANSI symbol, the letter D in the label indicates Data. [sec.7.10]
- **Decade Counter** - Any counter which is capable of going through ten different logic states. [sec.7.2]
- **DIVn** - When used inside an IEEE/ANSI symbol, it indicates that it is a MODn counter. [sec.7.3/7.10]

- **Down Counter** - A counter that counts downward from a maximum count to zero. [sec.7.4]

- **Frequency Counter** - A circuit that can measure and display the frequency of a signal. [sec.7.15]

- G - When used inside an IEEE/ANSI symbol, the letter G in the label for an input indicates AND dependency. [sec.7.10]

- **Glitch** - Momentary narrow spurious and sharply defined change in voltage. [sec.7.2]

- **Johnson Counter** - A shift register where the inverted output of the last Flip-Flop in the shift register is connected to the input of the first Flip-Flop in the shift register. [sec.7.14]

- LED - Light Emitting Diode. [sec.7.11]

- - When used inside an IEEE/ANSI symbol and on a clock input, it indicates that the counter will be decremented by 1 when clocked. [sec.7.3/7.10]

- **Parallel Counter** - See Synchronous counter. [sec.7.6]

- **Parallel In/Parallel Out Register** - A type of register that can be loaded with parallel data and has parallel outputs available. [sec.7.18]

- **Parallel In/Serial Out** - A type of register that can be loaded with parallel data and has only one serial output. [sec.7.20]

- + - When used inside an IEEE/ANSI symbol and on a clock input, it indicates that the counter will be incremented by 1 when clocked. [sec.7.3/7.10]

- **Presettable Counter** - A counter which can be preset to any starting count either synchronously or asynchronously. [sec.7.8]

- R - When used inside an IEEE/ANSI symbol, the letter R in the label for an input indicates a Reset function. [sec.7.22]

- **Ring Counter** - A shift register where the output of the last Flip-Flop in the shift register is connected to the input of the first Flip-Flop in the shift register. [sec.7.14]

- **Ripple Counter** - See Asynchronous Counter. [sec.7.1]

- **Sampling Interval** - The time window during which a frequency counter samples and thereby determines the unknown frequency of a signal. [sec.7.15]

- **Sequential Logic Systems** - A logic system in which its logic outputs' states and sequence of operations depend on both the present and past input conditions. [sec.7.23]

- **Serial In/Parallel Out** - A type of register that can be loaded with data serially and has parallel outputs available. [sec.7.21]

- **Serial In/Serial Out** - A type of register that can be loaded with data serially and has only one serial output. [sec.7.19]

- / - When used inside an IEEE/ANSI symbol, the slash (/) in the label for an input indicates the separation of two functions. [sec.7.22]

- **Spike** - See Glitch. [sec.7.2]

- SRG 8 - When used inside an IEEE/ANSI symbol, SRG 8 in the common-control block indicates that this IC is an 8-bit shift register. [sec.7.22]

- **Strobing** - A technique often used to eliminate decoders' spikes. [sec.7.12]

- **Synchronous Counter** - A counter where all of its Flip-Flops are clocked simultaneously. [sec.7.6]

- **Twisted Ring Counter** - See Johnson Counter. [sec.7.14]

- **Up Counter** - A counter that counts upward from zero to a maximum count. [sec.7.4]

- ⟶ - When used inside an IEEE/ANSI symbol, ⟶ in the label for an input indicates that when an active transition exists at that input a Shift-Right operation will result. [sec.7.22]

Problems

SECTION 7.1 *Asynchronous (Ripple) Counters*

7.1 Figure 7.1 (a) is a general four-bit asynchronous counter circuit. Complete table (b) as it pertains to similar counters by filling in the blanked spaces.

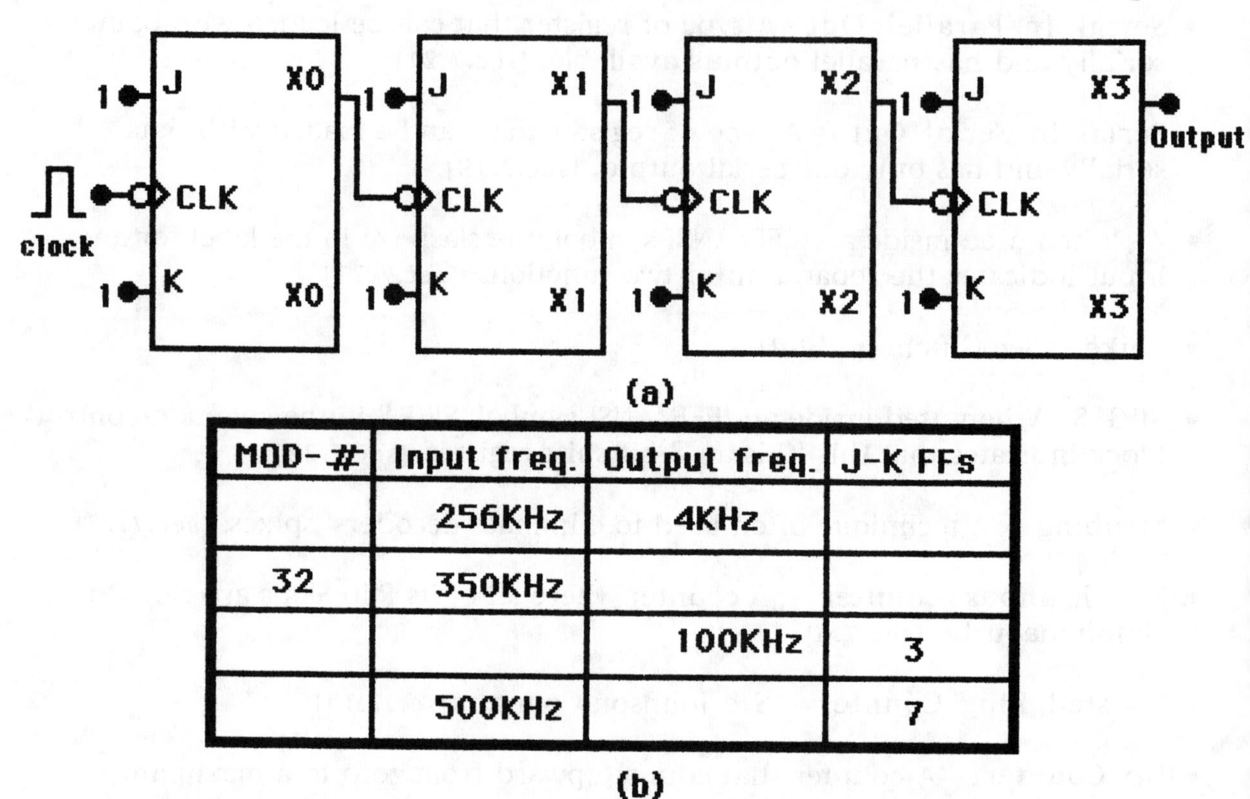

MOD-#	Input freq.	Output freq.	J-K FFs
	256KHz	4KHz	
32	350KHz		
		100KHz	3
	500KHz		7

Figure 7.1: (a) Four-bit asynchronous (ripple) counter; (b) Table.

SECTIONS 7.2-7.3,7.24 *Counters With MOD Numbers <2^N/ IC Asynchronous Counters/ Troubleshooting.*

7.2 For each of the following modifications in the circuit of Figure 7.2, determine the new MOD number of the ripple counter:

(a) The inputs to gate Z1 are A,B,C. [MOD number is _____]
(b) The inputs to gate Z1 are A,C,D. [MOD number is _____]
(c) The inputs to gate Z1 are A,B,+Vcc. [MOD number is _____]

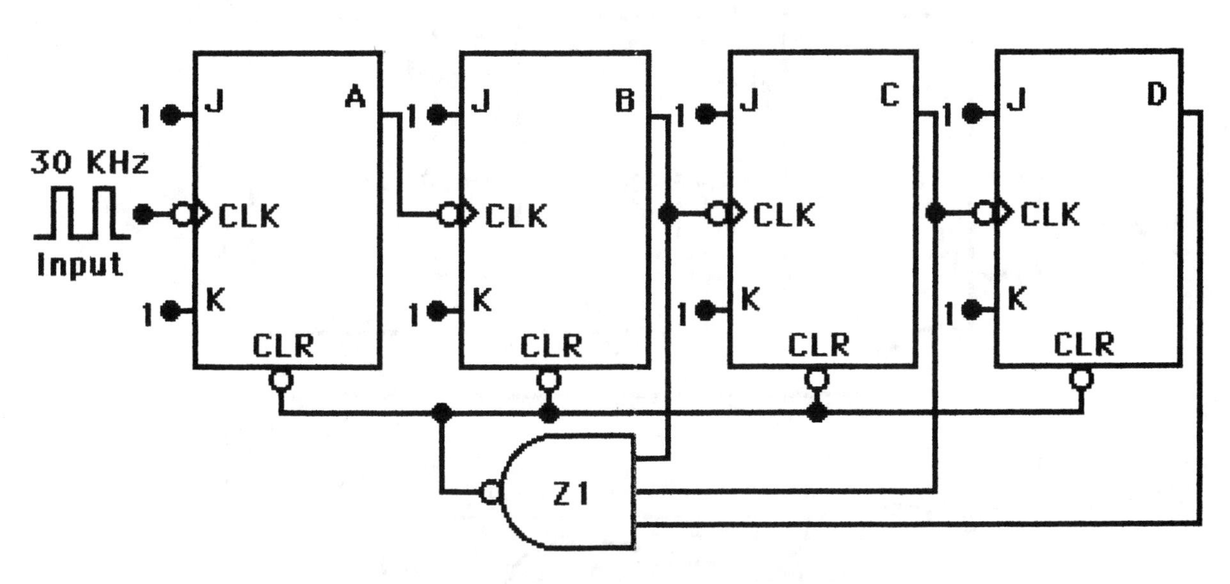

Figure 7.2: MOD-14 ripple counter.

7.3 The asynchronous counter of Figure 7.3 is tested and the technician determines that it is working as a MOD-8 counter (000_2-111_2). Find <u>two</u> different possible faults for the malfunction.

1. ____

____

____

____

____

2. ____

____

____

____

____

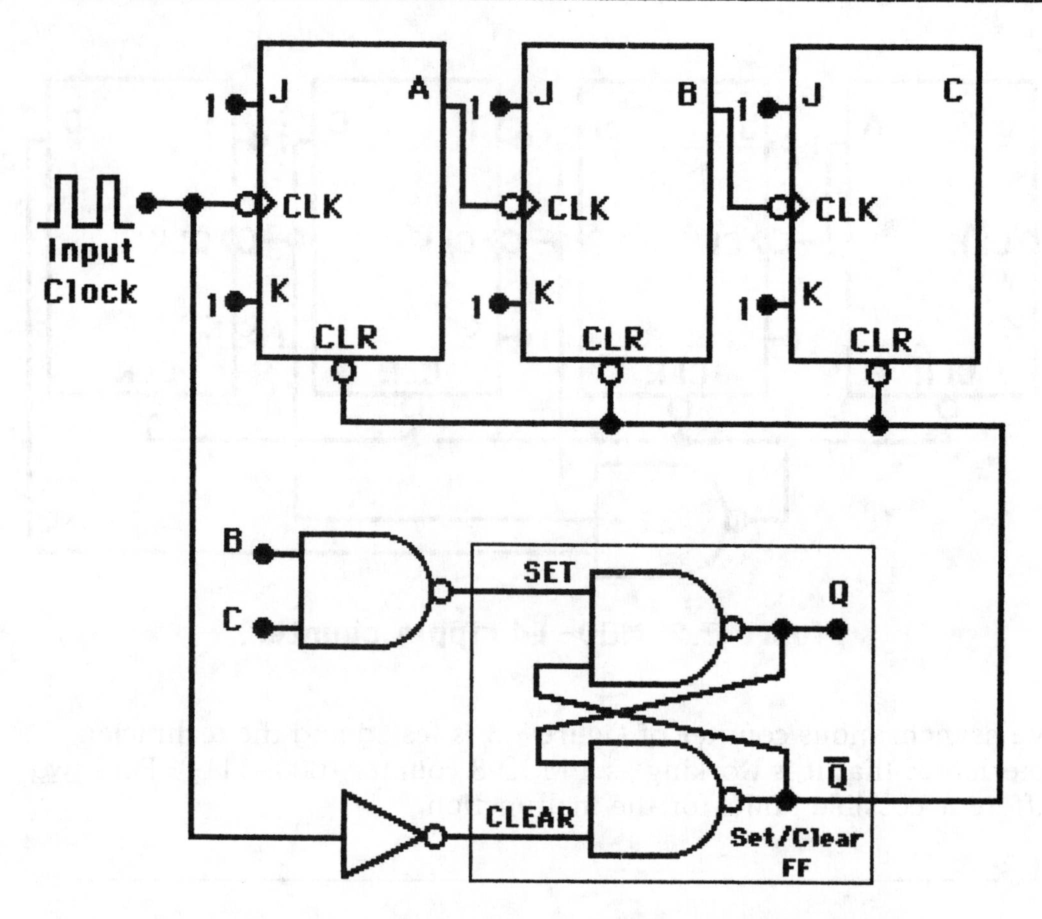

Figure 7.3: Asynchronous Counter.

7.4 A technician wires and tests the circuit of Figure 7.4. When its output X is displayed on the oscilloscope, the technician finds that it is a 54Hz squarewave. Determine the most probable cause for the malfunction.

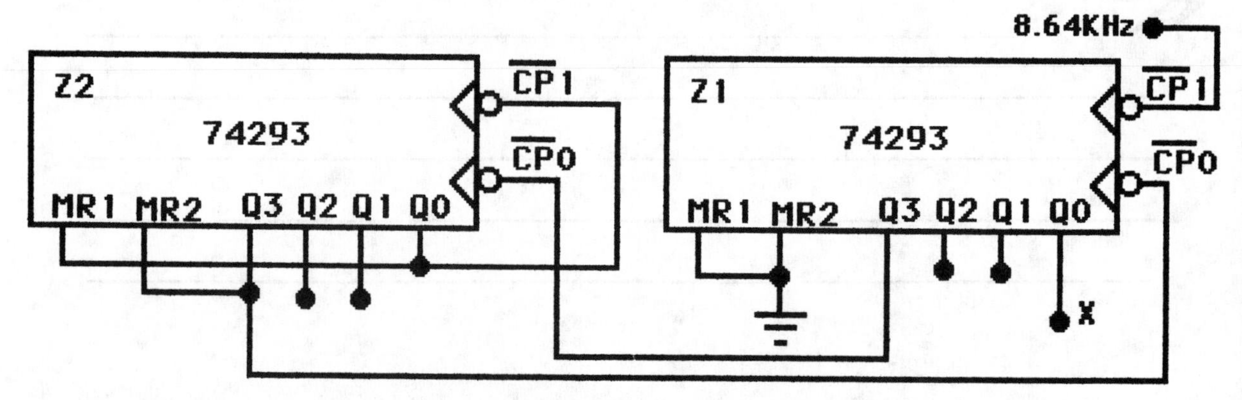

Figure 7.4: 8-bit Binary Counter using two 74293s.

SECTIONS 7.5,7.24 *Propagation Delay in Ripple Counters/ Troubleshooting*

7.5 If the counter of Figure 7.5 holds count 7 (111_2) and the Clock input goes through a negative-going transition, how long will it take for FF C to change states?

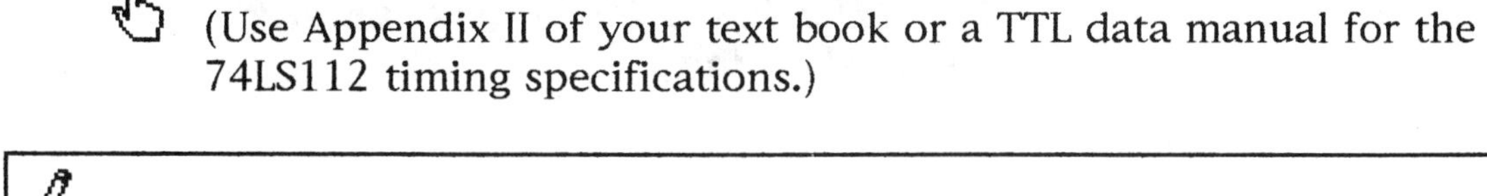

(Use Appendix II of your text book or a TTL data manual for the 74LS112 timing specifications.)

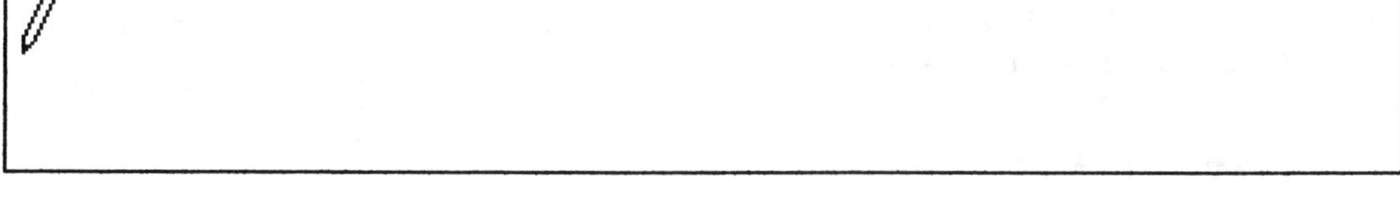

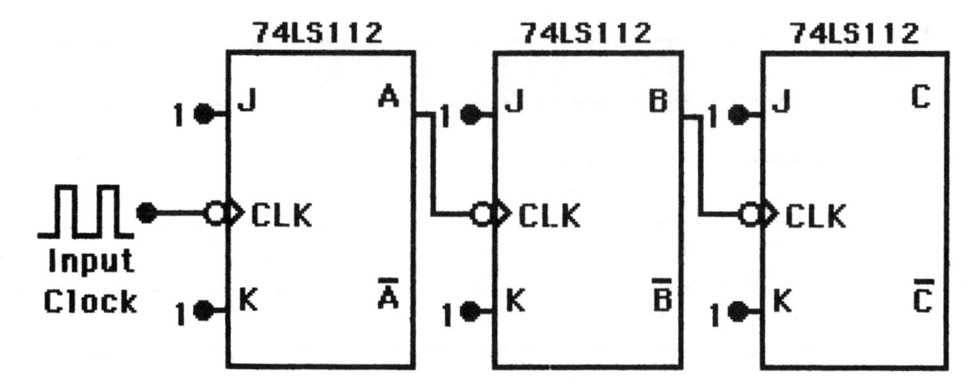

Figure 7.5: Ripple Counter.

7.6 The circuit of Figure 7.5 is built and tested. As the technician applies clock pulses, he notices that the counter goes through the following binary counting sequence: 000_2, 001_2, 110_2, 111_2, 100_2, 101_2, 010_2, 011_2, 000_2 ... (repeats). Why do you think the counter behaves this way?

____

____

____

____

____

____

SECTIONS 7.6-7.9,7.24 — *Synchronous (Parallel) Counters/ Synchronous Down and Up/Down Counters Presettable Counters/ The 74193 (LS193/HC193) Counter/ Troubleshooting.*

7.7 State the major advantages and disadvantages between synchronous and asynchronous counters.

Asynchronous Counter:

Advantage- ______________________________

____

____

____

Disadvantage- ______________________________

____

____

____

Synchronous Counter:

Advantage- ______________________________

____

____

____

Disadvantage- ______________________________

____

____

____

7.8 Complete the timing diagram for a SN74193 counter that goes through the sequence of operation listed below.

(Use Appendix II of your text book or a TTL data manual for the 74193 timing specifications.)

1. Clear the counter.
2. Load the counter with binary fourteen.
3. Count up by four counts.
4. Count down by three counts.

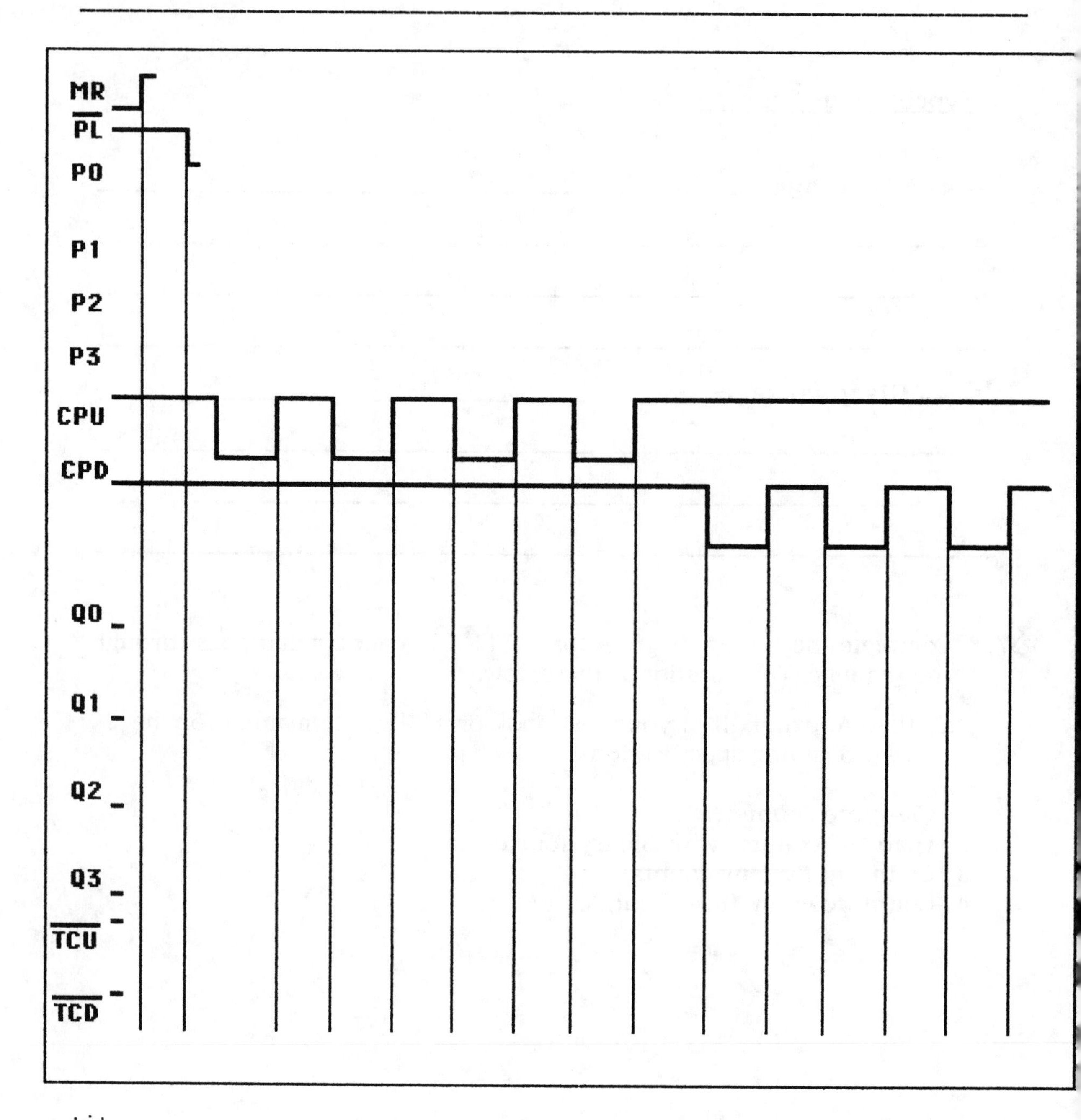

7.9 Design an UP/DOWN counter that operates according to the following set of specifications:

1. SN74193s are to be used in its design.
2. The counter is a MOD-256 when counting UP.
3. The counter is a MOD-34 when counting DOWN.
4. A switch (SW1) controls whether the counter counts UP or DOWN.
5. A maximum of three external gates can be used in the design of this counter.

7.10 While testing the counter built for problem 7.9, the technician applies a 10KHz signal to the input of the counter and notices that the frequency at the MSB of the counter is around 39Hz regardless of the position of SW1. Determine three possible causes for the malfunction.

1. ______________________________

____

2. ______________________________

____

3. ______________________________

____

SECTION 7.10 *More on the IEEE-ANSI Dependency Notation*

7.11 Refer to the IEEE/ANSI symbol of Figure 7.6 and answer the following questions:

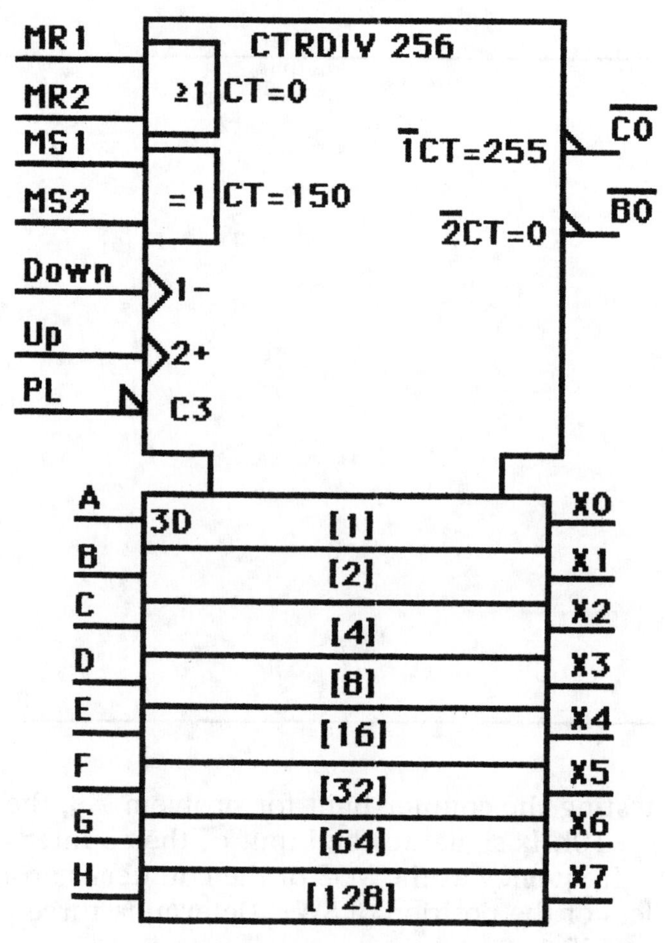

Figure 7.6: IEEE/ANSI Circuit Symbol.

(a) What is the maximum count that this counter can reach?

____

(b) What happens when MS1=MS2=1?

____

(c) What happens when MS1=MS2=0?

____

(d) What does '3D' mean at input A?

____

(e) What happens at count 255?

____

(f) What must be done to clear this counter?

___

(g) What does it mean when output $\overline{BO}$=0?

___

(h) What would have to be changed in this IEEE/ANSI symbol, if MR1=MR2=1 was needed to clear the counter?

___

(i) What does the '+' symbol at the UP input mean?

___

SECTIONS 7.11-7.13, 7.24 *Decoding a Counter/Decoding Glitches/ Cascading BCD Counters/ Troubleshooting*

⊕7.12 The inputs to the circuit of Figure 7.7 come from the ripple counter of Figure 7.5. The clock input frequency for this counter is a 1MHz squarewave. Output Q of flip-flop Z2 is to be Set whenever the binary count of zero (000_2) is decoded and Cleared whenever the binary count of three (011_2) is decoded.

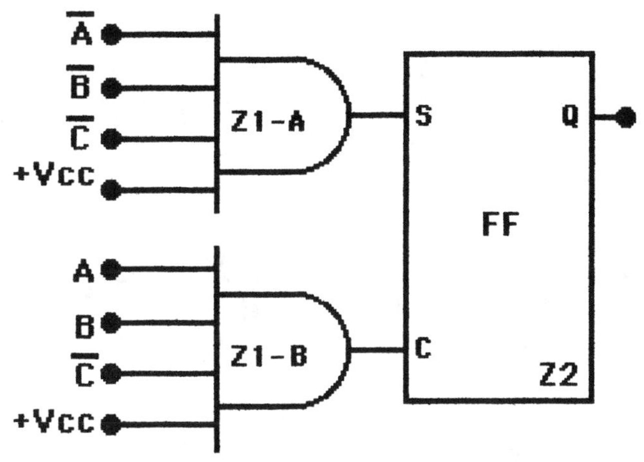

Figure 7.7: Logic Circuit used to detect 000_2 and 011_2.

(a) Draw the expected Q waveform, keeping in mind the effects of propagation delays of the flip-flops of figure 7.5. (Assume that initially: Q=1 and the counter of figure 7.5 is clear).

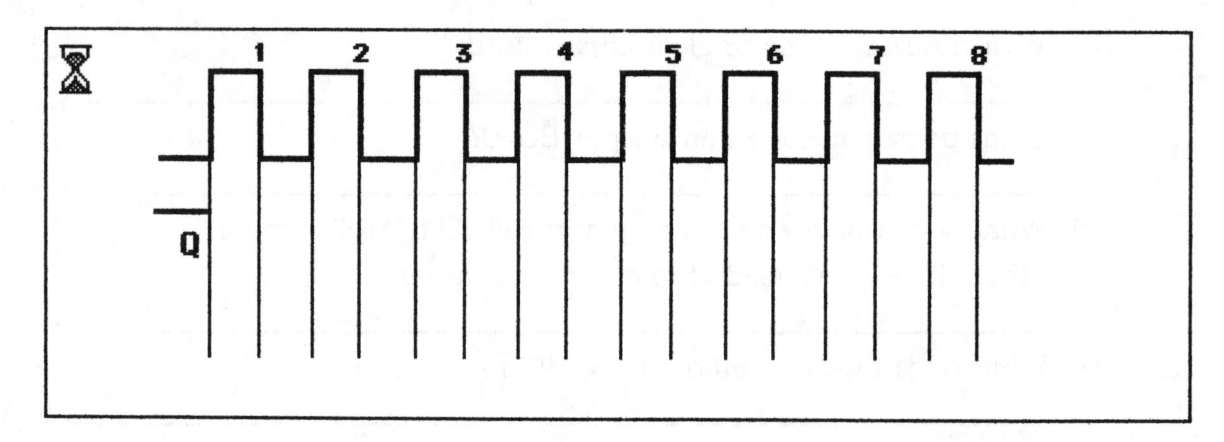

(b) Modified the circuit of Figure 7.7 so that it operates as intended?

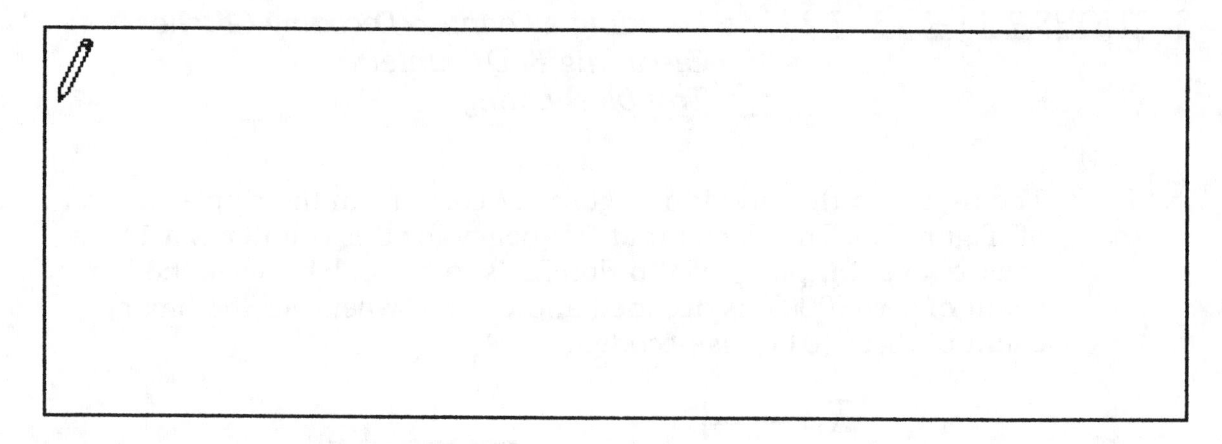

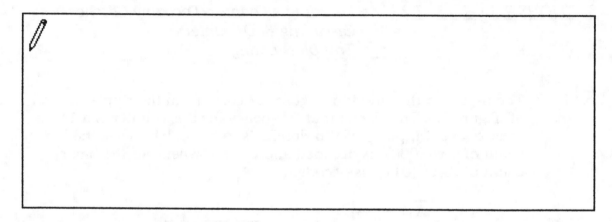

7.13 A technician builds the circuit of Figure 7.8 (a). She tests the circuit and records the results shown in table (b). Find the cause for the circuit's malfunction.

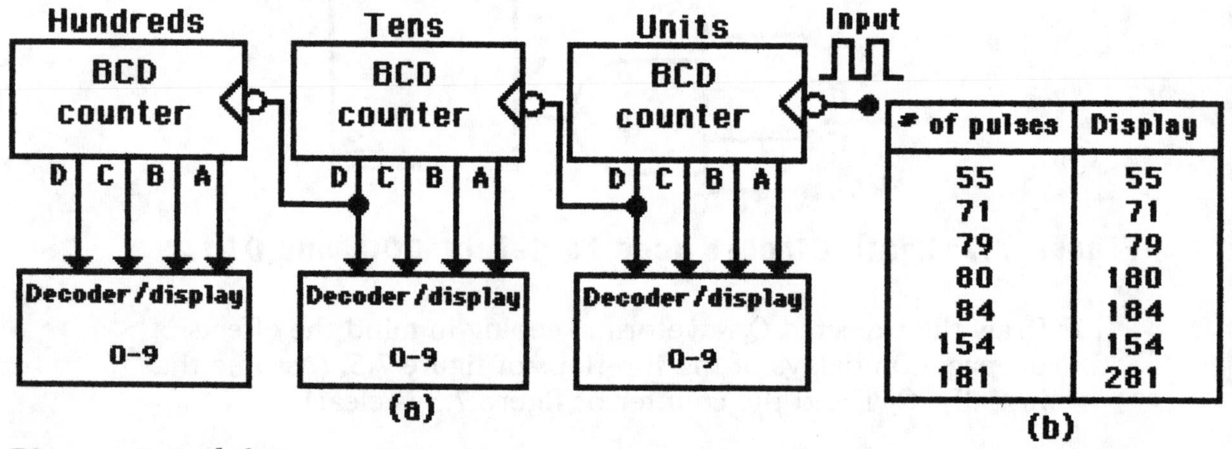

# of pulses	Display
55	55
71	71
79	79
80	180
84	184
154	154
181	281

Figure 7.8: (a) Cascading BCD counters; (b) Troubleshooting table.

SECTION 7.14 *Synchronous Counter Design*

7.14 (a) Design a synchronous counter that has the following counting sequence: 001_2, 010_2, 110_2, 111_2 and then repeats. The undesired (unused) states are: 000_2, 011_2, 100_2, 101_2. The counter must always go to 001_2 on the NEXT clock pulse after an undesired state. To design the synchronous counter do each of the following steps:

Step 1: Determine the desired number of bits and the desired counting sequence.

Step 2: Draw the state transition diagram showing all possible states, including those that are not part of the desired counting sequence.

Step 1	Step 2

Step 3: Use the state transition diagram to set up a table that lists all present states and their next states.

Step 4: Add a column to this table for each J and K input. For each present state, indicate the levels required at each J and K input in order to produce the transition to the next state.

Step 5: Design the logic circuits to generate the levels required at each J and K input.

Step 6: Implement the final expression.

(b) Using the procedure described in problem 7.14 (a), redesign the counter without any requirement on the unused states; that is, their NEXT states can be "don't care" states.

<u>SECTIONS 7.15-7.17, 7.24</u> *Shift-Register Counters/ Counter Applications: Frequency Counter and Digital Clock/Troubleshooting*

7.15 The waveforms of Figure 7.9 represent the input clock and the output Q_0 of a ring counter. How many flip-flops are being used by the counter?

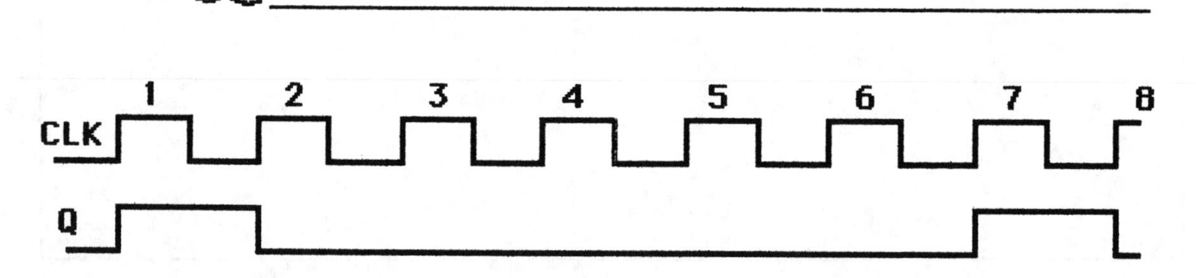

Figure 7.9: Clock waveform and the response at the Q0 output of a Ring Counter.

7.16 Complete each of the following statements:

(a) To build a MOD-____ Ring Counter, it requires 23 Flip-Flops.

(b) To decode any state of a MOD-14 Johnson Counter, it requires a _____input AND gate.

(c) If the frequency of the input clock of a 6-bit Johnson Counter is _____ KHz, then the output signal at any of the Flip-Flops in the counter is equal to 5KHz,___ Duty Cycle.

7.17 When the circuit of Figure 7.10 was tested, it showed that, regardless of the SAMPLE pulses width (t_p), the Decoder/Display always displayed a number which was twice as large as the actual input frequency. What could have caused the malfunction?

____

____

____

____

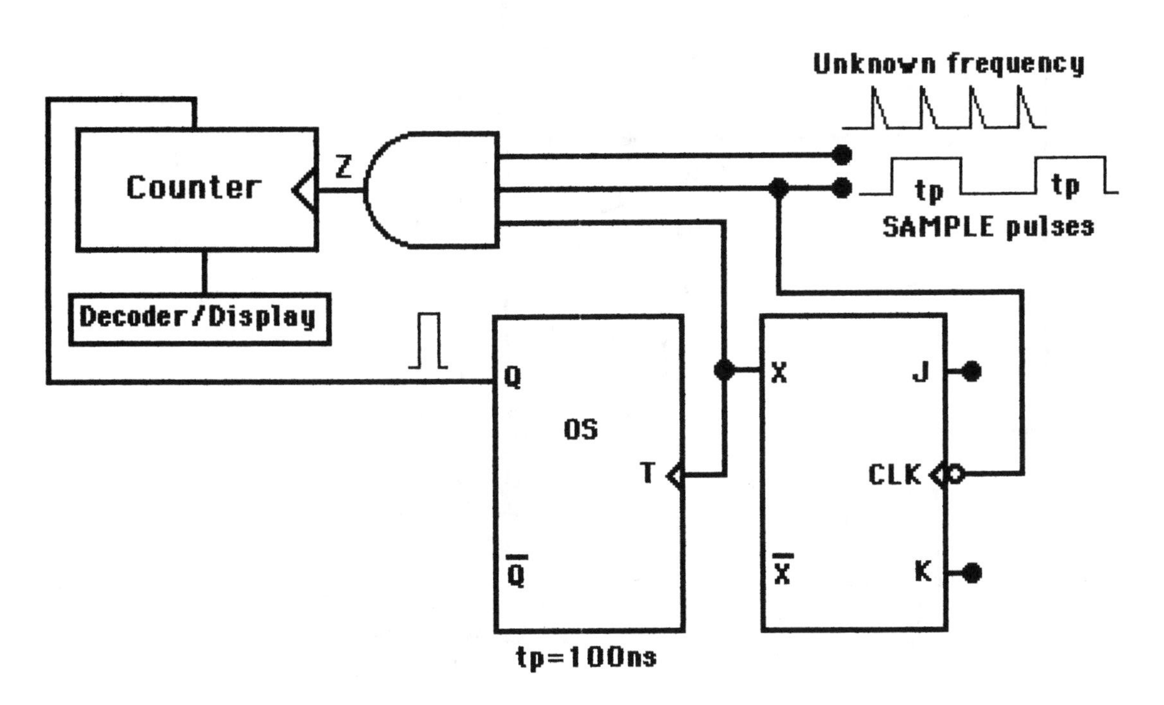

Figure 7.10: Frequency counter.

7.18 The digital clock of figure 7.11 is built and tested. After observing the clock's operation for a couple of hours, it is noticed that the minutes section works properly but the hours section of the clock changes back and forth between 11:00 and 12:00 with each "Pulse/Hour"pulse. Determine the cause for the malfunction.

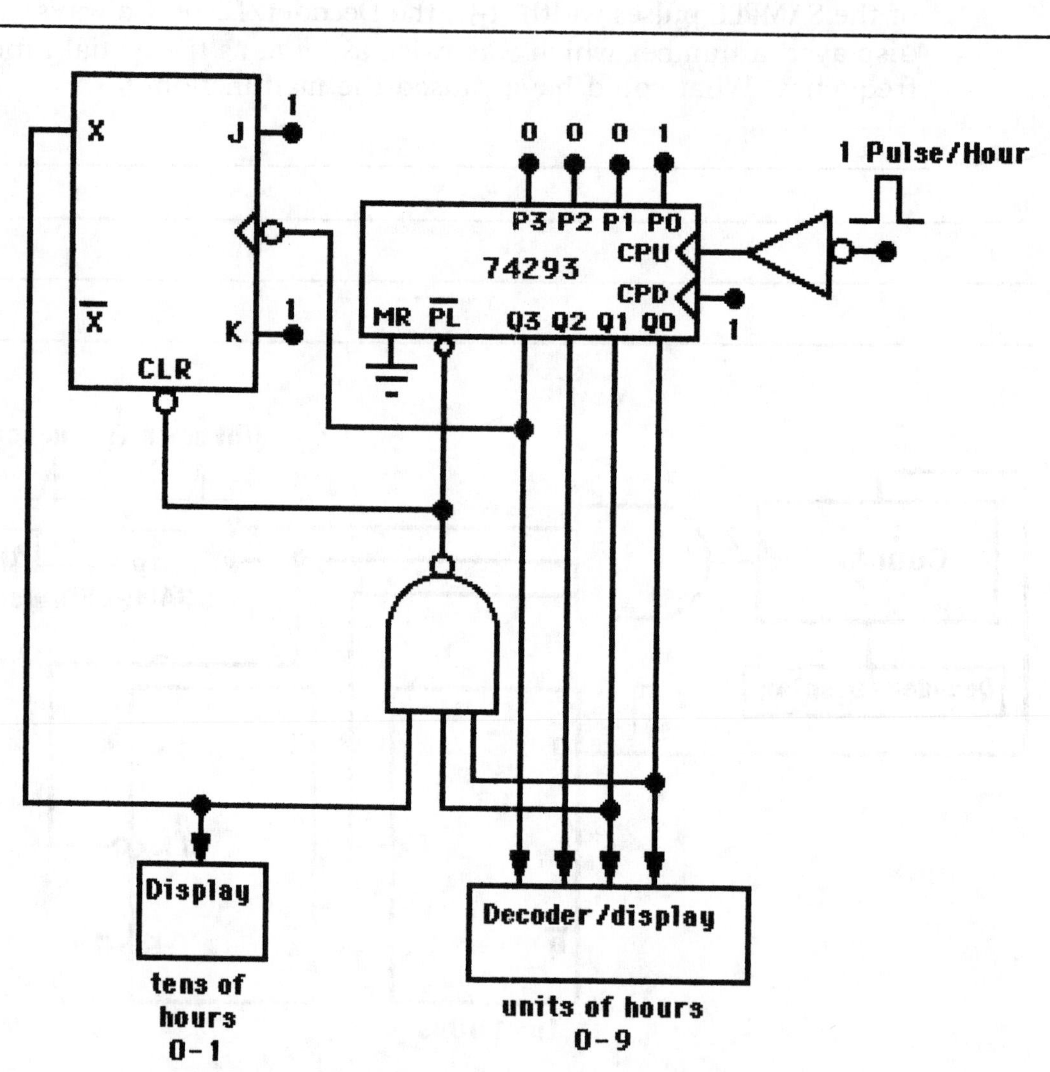

Figure 7.11: Detailed cicuitry for the hours section of a digital clock.

SECTION 7.21, 7.24 *Parallel In/Serial Out-(The 74165/74LS165/74HC165/) Troubleshooting.*

7.19 Refer to the parallel in/serial out 8-bit register of Figure 7.12 and the associated waveforms. Complete the timing diagram by drawing the output waveform Q7. (Assume that Q7=0 initially.)

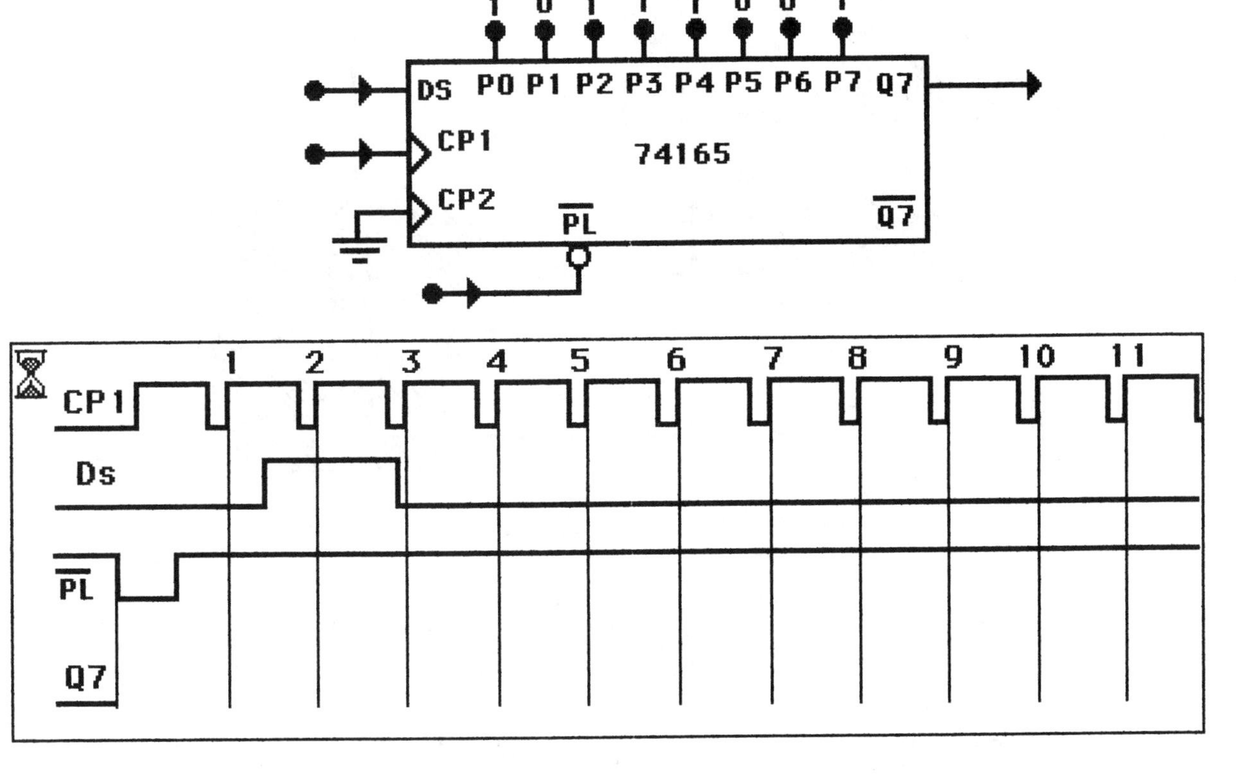

Figure 7.12: Parallel In/Serial Out 8-bit register.

SECTION 7.23 *IEEE-ANSI Register Symbols.*

7.20 Describe in detail the function of each of the 5 blocks shaded of the IEEE/ANSI symbol of Figure 7.13.

1. ____________________
2. ____________________
3. ____________________
4. ____________________
5. ____________________

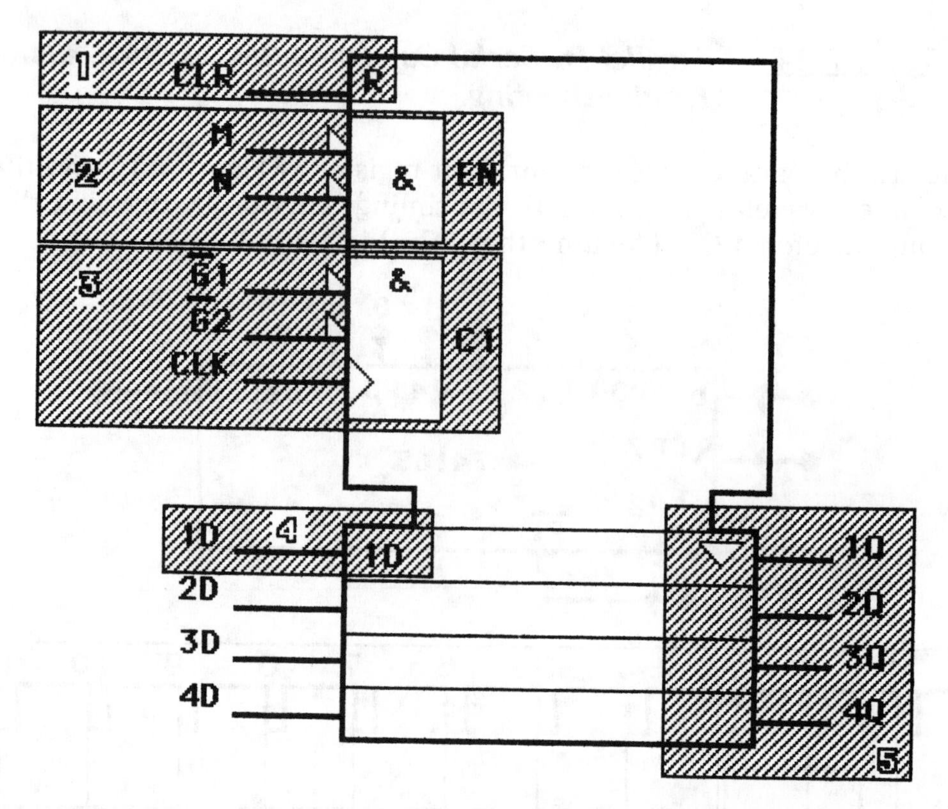

Figure 7.13: IEEE/ANSI Symbol.

TEST 7

1. How many Flip-Flops are needed to make a MOD-64 ripple counter?

   (a) 4 (b) 5 (c) 6 (d) 7

2. An asynchronous counter has an input clock frequency of 256KHz and an output frequency of 64KHz. What is the MOD number of this counter?

   (a) 4 (b) 25 (c) 3 (d) 2

3. What would the dependency notation CTRDIV 256 mean on a IEEE/ANSI counter symbol?

   (a) It indicates a counter that divides the input frequency by 256.
   (b) It indicates a MOD-256 counter.
   (c) It indicates a counter with a F_{max} of 256KHz.
   (d) (a) and (b).

4. Synchronous counters eliminate the delay problems found with asynchronous counters because the:

   (a) The individual Flip-Flops used in this type of counter are much faster than those used by asynchronous counters.
   (b) Input pulses are applied simultaneously to each Flip-Flop.
   (c) Input pulses are applied simultaneously to both the least and the most significant Flip-Flops.
   (d) Input pulses are applied only to the least significant Flip-Flop of the counter.

5. How many BCD counters are needed to provide a decimal count of 9,999?

   (a) 4 (b) 14 (c) 2^4 (d) 5

6. A ten bit ring counter is clocked with a 200KHz. The output frequency is:

   (a) 195.5Hz (b) 195.3Hz (c) 20KHz (d) 10KHz

7. A ten bit Johnson counter is clocked with a 200KHz. The output frequency is:

   (a) 195.5Hz (b) 195.3Hz (c) 20KHz (d) 10KHz

8. A MOD-16 asynchronous counter uses Flip-Flops that have "worst case" propagation delay of t_{PHL}=15ns. The maximum input frequency for this counter is:

   (a) 4.1MHz (b) 16.6MHz (c) 66.6MHz (d) 22.2MHz

9. All the Flip-Flops change states at the same time in a synchronous counter.

   (a) [TRUE] (b) [FALSE]

10. The + sign on a counter symbol by the $\overline{CP}$ indicates a count-down operation.

   (a) [TRUE] (b) [FALSE]

8 INTEGRATED-CIRCUIT LOGIC FAMILIES

Objectives

Upon completion of this chapter, you will be able to:

- Read and understand digital IC technology as specified in manufacturers' data sheets.
- Compare the characteristics of standard TTL and the various TTL series.
- Determine the fan-out for a particular logic device.
- Use logic devices with open-collector outputs.
- Analyze circuits containing tristate devices.
- Compare the characteristics of the various CMOS series.
- Analyze circuits which use a CMOS bilateral switch.
- Describe the major characteristics and differences among TTL, ECL, MOS, and CMOS logic families.
- Cite and implement the various considerations that are required when interfacing digital circuits from different logic families.
- Use a logic pulser and current tracer as digital circuit troubleshooting tools.

Glossary of key terms covered in this chapter:

- **Advanced Schottky TTL (AS-TTL)** - This is the fastest TTL series and its speed-power product is significantly lower than the Schottky series. [sec.8.4]
- **Advanced Low-Power Schottky TTL (ALS-TTL)** - This is an improvement over the Low-Power Schottky series. It is faster and it dissipates less power. The ALS series has the lowest speed-power product of all the TTL series. [sec.8.4]
- **Bilateral Switch** - A CMOS circuit which acts as a single-pole, single-throw switch (SPST) controlled by an input logic level. [sec.8.17]
- **Buffer/Driver** - A circuit designed to have a greater output current and/or voltage capability than an ordinary logic circuit. [sec.8.7]
- **Bus Contention** - Situation in which the outputs of two or more active devices are placed on the same bus line at the same time. [sec.8.8]
- **CML** - Current-Mode Logic. Also referred to as Emitter-Coupled Logic. [sec.8.9]
- **CMOS** - Complementary Metal-Oxide-Semiconductor. [sec.8.14]
- **Current-Sinking Action** - Term attributed to a logic family in which the output of a logic circuit sinks current from the input of a different logic circuit. [sec.8.1/8.2]
- **Current-Sinking Transistor** - Name given to the output transistor (Q_4) of a TTL circuit. This transistor is turned on when the output logic level is Low. [sec.8.2]
- **Current-Sourcing Action** - Term attributed to a logic family in which the output of a logic circuit sources or supplies current to the input of a different logic circuit. [sec.8.1/8.2]
- **Current-Sourcing Transistor** - Name given to the output transistor (Q_3) of most TTL circuits. This transistor is conducting when the output logic level is HIGH. [sec.8.2]
- **Current Tracer** - A testing tool that detects a changing current in a wire or PC-board trace. [sec.8.22]

- **Current Transients** - Current spikes (30 to 50mA) generated by the Totem-Pole output structure of a TTL circuit, and caused when both transistors are simultaneously turned on. [sec.8.6]

- **ESD (Electrostatic Discharge)** - When a small static charge flows into a high input impedance present at a MOSFET it results in the build-up of a large and harmful voltage. Even though most modern ICs have on-chip resistor-diode networks to protect inputs and outputs from the effects of ESD, special precautions must be used to minimize the damages caused by the build-up of these charges. [sec.8.13]

- ECL- Emitter-Coupled Logic. Also referred to as Current-Mode Logic. [sec.8.9]

- **Fan-Out** - The maximum number of standard logic inputs that the output of a digital circuit can drive reliably. [sec.8.1]

- **Fast TTL** - The newest TTL series. New IC fabrication techniques are employed resulting in a decrease of the inter-device capacitances. Therefore, reduced propagation delays are achieved. [sec.8.4]

- **Floating inputs** - Any digital circuit input that is left disconnected. [sec.8.6]

- **High-Speed TTL** - An old TTL subfamily which uses the basic TTL standard circuit except that all the resistor values are decreased and the emitter-follower transistor (Q_3) is replaced by a Darlington pair. High-Speed TTL is not used in new circuit designs since its performance has been surpassed by newer TTL series. [sec.8.4]

- **Interfacing** - When the output of a system is connected to the input of a different system with different electrical characteristics. [sec.8.18]

- **Invalid Voltage Levels** - Logic voltage levels that are higher than $V_{IL(max)}$ or lower than $V_{IH(min)}$. [sec.8.1]

- **Latch-up** - Condition that occurs when parasitic PNP and NPN transistors embedded in the substrate of CMOS ICs are triggered into conduction and stay permanently on (latch-up). [sec.8.15]

- **Loading Factor** - See Fan-Out. [sec.8.1]

- **Logic Pulser** - A testing tool that generates a short-duration pulse when manually actuated. [sec.8.22]

- **Low-Power Schottky TTL (LS-TTL)** - A TTL subfamily which uses the identical Schottky TTL circuit except that uses larger resistor values. [sec.8.4]

- **Low-Power TTL** - An old TTL subfamily which uses the basic TTL standard circuit except that all the resistor values are increased. Low-Power TTL is not used in new circuit designs since its performance has been surpassed by newer TTL series. [sec.8.4]

- **Low-Voltage Technology (LVT)** - A new technology that uses a nominal 3.3V as its operating power supply voltage in order to increase the overall chip density, increase the circuit's operating speed, and decrease the power dissipation. [sec.8.21]

- **MOS** - Metal-Oxide-Semiconductor. [sec.8.10]

- **Multiple-Emitter Transistor** - The type of input transistor used in the design of TTL AND and NAND gates. [sec.8.2]

- **Noise Immunity** - A circuit's ability to tolerate noise without causing spurious changes in the output voltage. [sec.8.1]

- **Noise Margin** - A quantitative measure of Noise Immunity. [sec.8.1]

- **Open-Collector Output** - A type of output structure of some TTL circuits in which only one transistor with a floating collector is used. [sec.8.7]

- **Power-Supply Decoupling** - When a small RF capacitor is connected between Ground and Vcc near each TTL IC on a circuit board. [sec.8.6]

- **Propagation Delays** - Delay encounter by a signal as it goes through a circuit. There are two types of propagation delays as far as logic circuits are concerned: t_{PLH} (delay time in going from logical LOW to a logical HIGH state) and t_{PHL} (delay time in going from logical HIGH to a logical LOW state). [sec.8.1]

- **Pull-Down Transistor** - See Current-Sinking Transistor. [sec.8.2]

- **Pull-Up Transistor** - See Current-Sourcing Transistor. [sec.8.2]

- **SBD** - Schottky Barrier Diode. [sec.8.4]

- **Schottky TTL** - A TTL subfamily which employs the same basic TTL standard logic circuit with the exception that it uses a Schottky barrier diode (SBD) connected between the base and collector of each transistor. [sec.8.4]
- **Static Sensitivity** - See Electrostatic discharge (ESD). [sec.8.13]
- **Speed-Power product** - A numerical value (in Joules) often used to compare different logic families. It is obtained by multiplying the propagation delay by the power dissipation of a logic circuit. [sec.8.1]
- **Totem-Pole Output** - A term used to describe the way in which two bipolar transistors are arranged at the output of most TTL circuits. [sec.8.2]
- **Transmission Gate** - Same as Bilateral Switch. [sec.8.17]
- **Tristate TTL** - A type of TTL output structure which allows three types of output states - HIGH, LOW and High-Impedance (Hi-Z). [sec.8.8]
- **Unit Load (UL)** - Way which some manufacturers specify a device's input and output currents. In a standard TTL circuit, 1UL in the HIGH state is equal to 40μA, and in the LOW state is equal to 1.6mA. [sec.8.5]
- **Voltage Level-Translator** - Circuit that takes a low-voltage input and translates it to a high-voltage output. [sec.8.19/8.20]
- **Wired-AND Connection** - Term used to describe the logic function created when Open-Collector outputs are tied together. [sec.8.7]

Problems

SECTIONS 8.1-8.3 *Digital IC Terminology/The TTL Logic Family/ Standard TTL Series Characteristics.*

8.1 Complete each of the following definitions:

(a) ______, is the voltage level at the output of a TTL logic circuit in the LOW state.

(b) I_{IL}, is the current that flows _______ a TTL input when a specified ________ -level voltage is applied to that input.

(c) ______, is the delay time from a logic 0 to a logic 1.

(d) _________ , is the maximum number of standard logic inputs that an output can reliably drive.

8.2 Refer to the circuit of Figure 8.1. Use the appropriate maximum ratings from data tables in Appendix II, and determine the following:

(a) the circuit's average current and power dissipation.
(b) the average propagation delay of a single 7404 Inverter.
(c) the speed-power product of a 7404 Inverter.

SECTIONS 8.4-8.6 *Improved TTL series/TTL Loading and Fan-Out/ Other TTL Characteristics.*

Refer to Figure 8.1 for the next two problems.

8.3

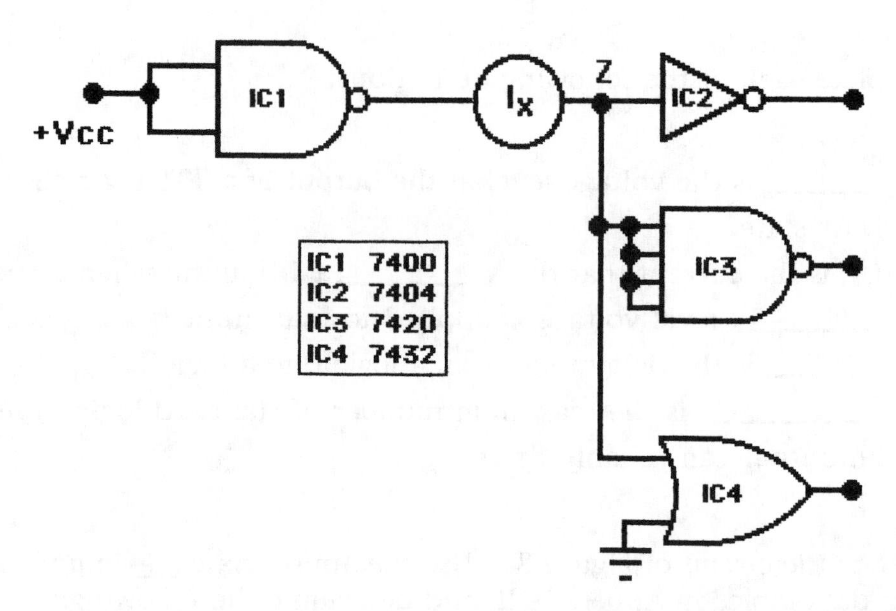

Figure 8.1: TTL gate driving various TTL loads.

(a) Calculate the value and direction of current Ix.

(b) Connect IC1's input to ground and repeat (a).

(c) Change IC1 to a 54LS04 and repeat (a). Would this situation cause any problems with the circuit's operation?

8.4 Determine the number of Unit Loads that IC1 is driving.

8.5 The 74F04 IC has the following voltage parameters: $V_{OH}(min)$=2.5V, $V_{OL}(max)$=0.5V, $V_{IH}(min)$=2.0V, $V_{IL}(max)$=0.8V. Calculate the dc noise margins for the 74F04 IC. How does it compare with the 74LS04 dc noise margins?

<u>SECTIONS 8.7-8.8</u> *Connecting TTL Outputs Together/ Tristate (3-State) TTL.*

8.6 Calculate the minimum values for resistors Ra and Rb of Figure 8.2.

(Consult the IC data sheets of Appendix II in your text book or your TTL data manual.)

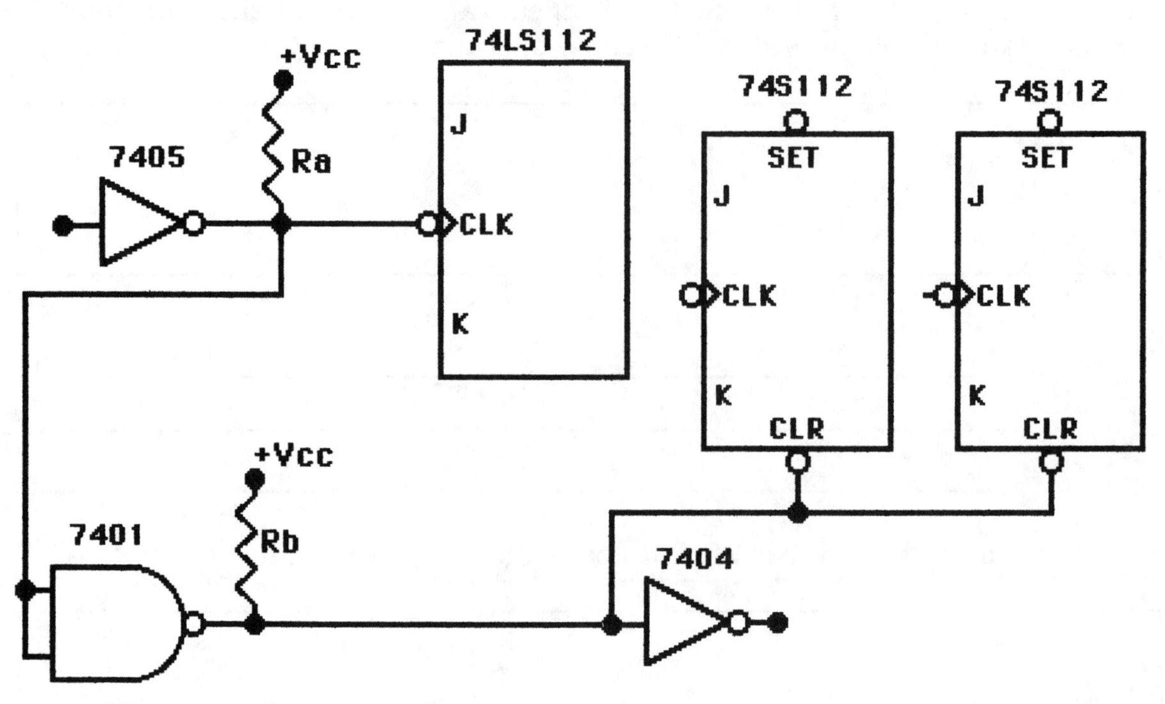

Figure 8.2: Open-collector outputs driving other TTL loads.

8.7 A technician builds the TTL circuit of Figure 8.2 and uses a 220Ω resistor for Ra. While testing the circuit, the technician notices that the 74LS112 Flip-Flop toggles randomly even without having a signal present at its clock input. In fact, the occurrence of this unusual circuit behavior gets worse when a nearby air conditioning unit starts. Determine why this problem exists with the circuit.

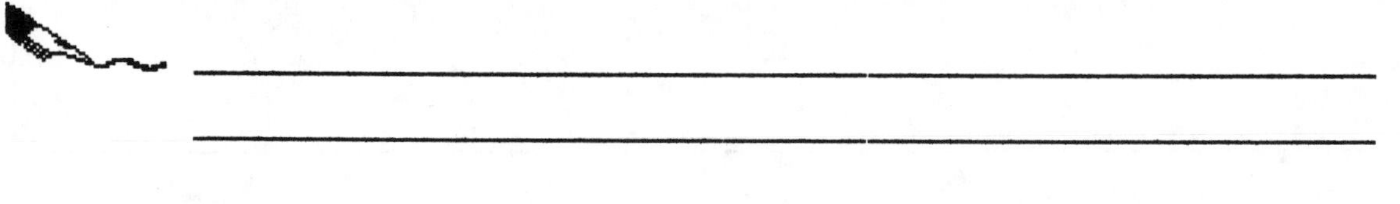

8.8 The TTL circuit of Figure 8.3 shows three MOD-8 counters (Z1-Z3) which are connected to a common 3-bit data bus (A,B,C) through tristate buffers. Ring counter Z4 selects which MOD-8 counter places data on the data bus via the tristate buffers (Z5-Z7). A technician uses a logic probe to obtain the table of Figure 8.4. Determine a cause for the circuit's malfunction.

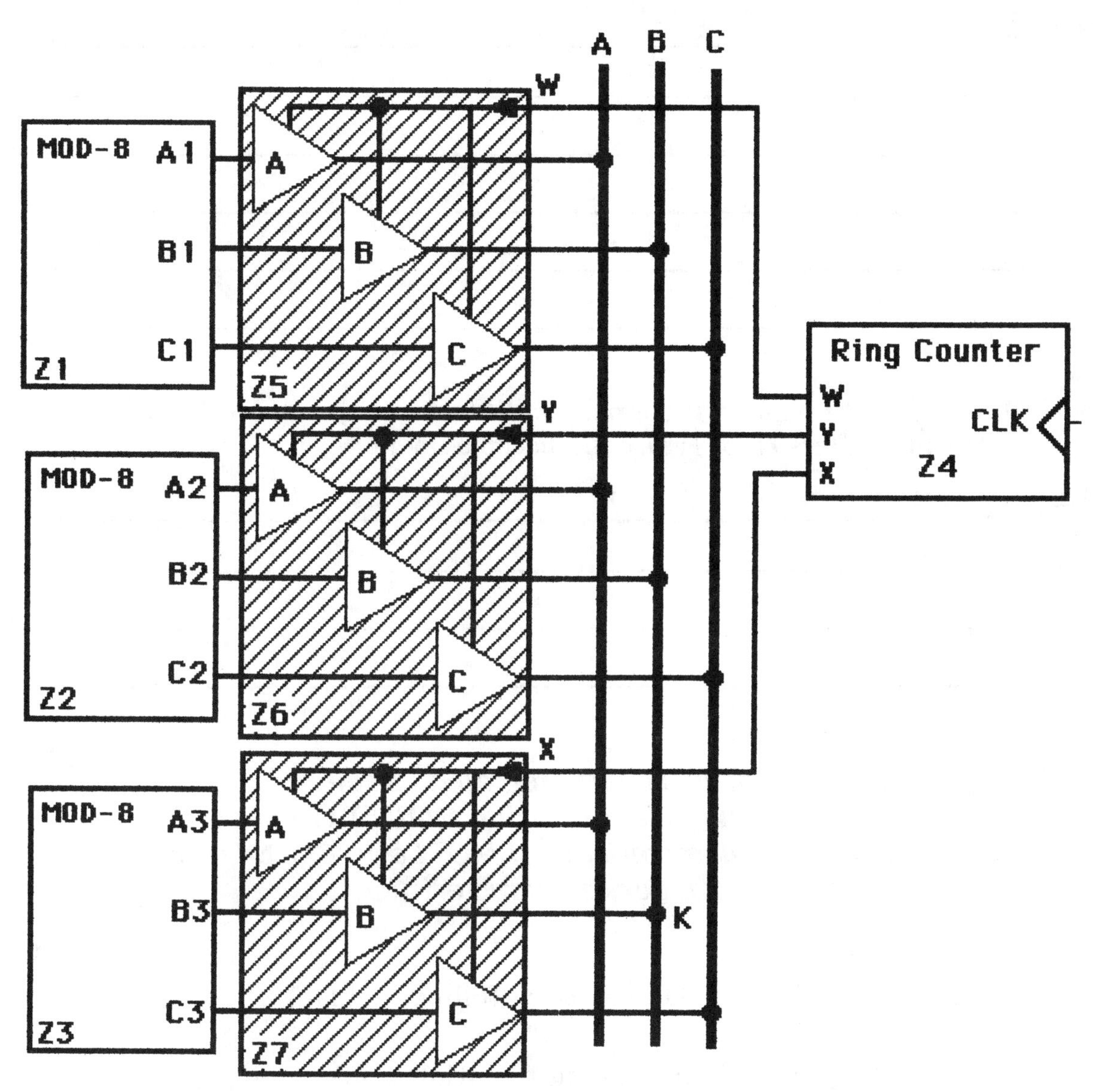

Figure 8.3: Tristate buffers used to connect several signals to a common bus.

Ring Counter W Y X	Z1 A_1 B_1 C_1	Z2 A_2 B_2 C_2	Z3 A_3 B_3 C_3	Data Bus A B C
0 0 1	1 1 0	0 1 0	0 1 1	0 1 1
0 1 0	1 1 1	0 0 1	1 0 0	0 0 1
1 0 0	1 1 0	1 1 0	0 1 0	1 1 0
0 0 1	1 1 0	0 0 0	1 0 1	1 1 1
0 0 1	0 1 1	1 1 1	0 0 0	0 1 0

Figure 8.4: Operating table for the circuit of Figure 8.3.

8.9 What would happen if connection W from Ring counter Z4 to the tristate buffers Z5 of Figure 8.3 became open?

SECTIONS 8.10-8.15 *MOS Digital Integrated Circuits/ The MOSFET/Digital MOSFET Circuits/ Characteristics of MOS Logic/CMOS Logic/ CMOS Series Characteristics.*

8.10 Complete the following statements by filling the blanks:

(a) ______ logic uses bipolar transistors, while CMOS uses ______ ______ transistors.

(b) In a circuit which utilizes ______ logic, the power dissipation is directly proportional to the frequency at which it operates.

(c) CMOS should be chosen over ______, if the main consideration is packing density.

(d) Among all the MOSFET circuits, ______ - ______ has the lowest packing density.

(e) The usage of negative voltages is a disadvantage of ___ - ______ MOSFETs.

(f) Current spikes are drawn from biasing power supply each time a CMOS output switches from __________ to __________.

(g) In both the HIGH and LOW states, the ____ _______ ________ of _____ logic can be determined by multiplying V_{DD} by 30%.

(h) Damage can occur to a ______circuit, if its inputs are left floating.

(i) _____ - ______ _____ is an improved version of the 74C series which has a tenfold increase in switching speed.

(j) _______ logic combines the best of features of bipolar and CMOS logic.

(k) Because of parasitic PNP and NPN transistors embedded in the substrate of CMOS ICs, _______ - ____ can occur under certain circumstances.

<u>SECTIONS 8.16-8.21</u> *CMOS Open Drain and Tristate Outputs/ CMOS Transmission Gate (Bilateral Switch)/ IC Interfacing/TTL Driving CMOS/ CMOS Driving TTL/Low-Voltage Technology - LVT*

8.11 For the circuit of Figure 8.5, the logic conditions were recorded in the truth table of Figure 8.6. Point out the fault with the circuit, and explain the most probable reason for the malfunction.

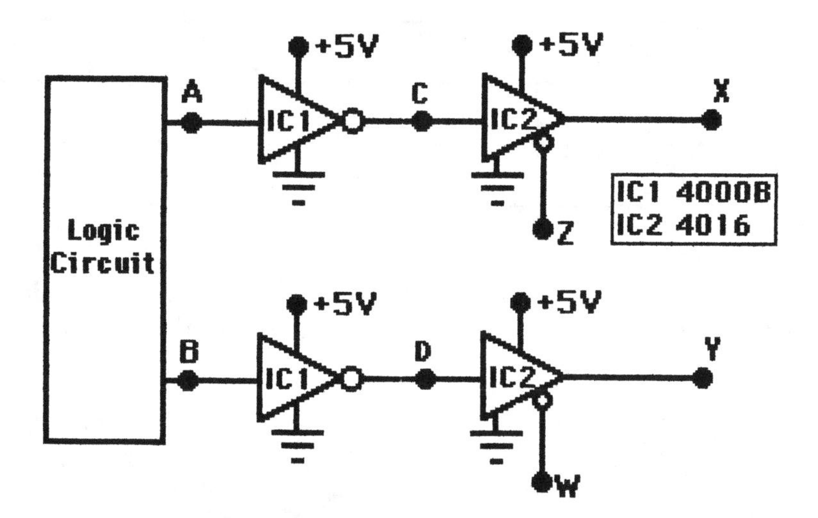

Figure 8.5: CMOS inverters driving CMOS transmission gates.

A	B	Z	W	C	D	X	Y
0	0	0	0	1	1	1	1
0	0	0	1	1	1	1	HI-Z
0	0	1	0	1	1	HI-Z	1
0	0	1	1	1	1	HI-Z	HI-Z
0	1	0	0	?	?	?	?
0	1	0	1	?	?	?	HI-Z
0	1	1	0	?	?	HI-Z	?
0	1	1	1	?	?	HI-Z	HI-Z
1	0	0	0	?	?	?	?
1	0	0	1	?	?	?	HI-Z
1	0	1	0	?	?	HI-Z	?
1	0	1	1	?	?	HI-Z	HI-Z
1	1	0	0	0	0	0	0
1	1	0	1	0	0	0	HI-Z
1	1	1	0	0	0	HI-Z	0
1	1	1	1	0	0	HI-Z	HI-Z

Shaded areas represent indeterminate logic levels.

Figure 8.6: Truth table for the circuit of Figure 8.5.

8.12 Explain why the circuit of Figure 8.7 would not function properly? Add the necessary circuitry in order to fix the designer's mistakes.

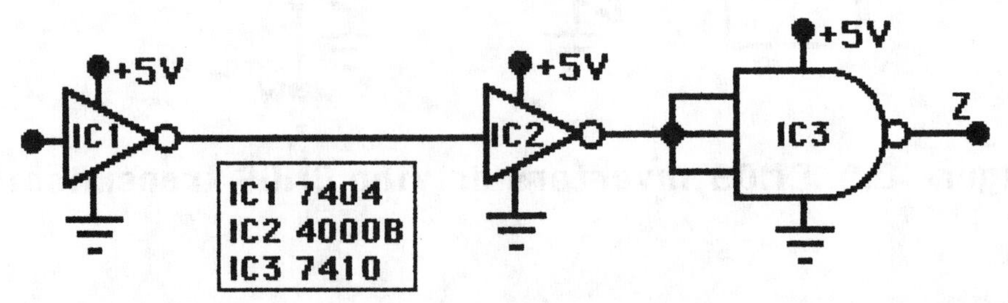

Figure 8.7: TTL driving CMOS, and CMOS driving TTL.

8.13 What must be done to the circuit of Figure 8.8 for it to work properly?

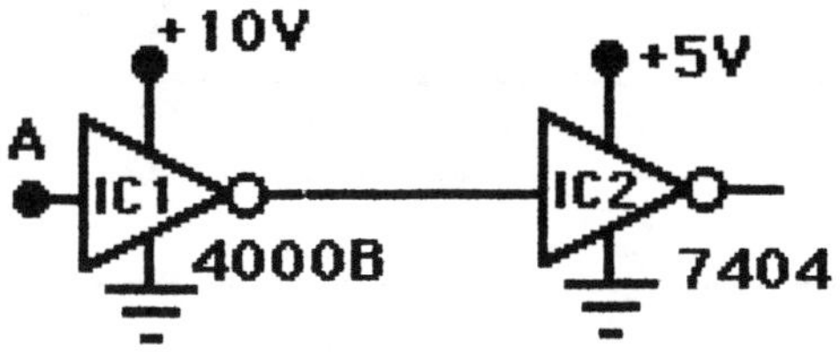

Figure 8.8: High voltage CMOS driving TTL.

8.14 How many standard TTL unit loads can a typical 74LVT device drive in the LOW state? In the HIGH state?

SECTION 8.22 *Troubleshooting.*

The Logic Probe, Logic Pulser and Current Tracer will be used as troubleshooting tools for the problems in this Section:

8.15 For problem 8.11, describe a procedure for using a logic pulser and a logic probe to verify the suspected fault.

8.16 In the circuit of Figure 8.3 bus line B is shorted to Ground at node K. Develop a procedure for using the current tracer and the logic pulser to isolate the short.

8.17 After the circuit of Figure 8.7 is redesigned properly (problem 8.12), the technician finds that output Z is always HIGH. List at least three possible causes for the circuit's malfunction.
Using the troubleshooting tools mentioned at the beginning of this section, how would you find the problem with the circuit?

1.
2.
3.

TEST 8

1. The typical dc noise margin for the standard TTL series is:

   (a) > 3.5V (b) 0.4V (c) 200mV (d) 1.5V

2. Which of the following statements is not true about the standard TTL logic family?

   (a) Totem-pole outputs cannot be tied together.
   (b) When the current-sinking transistor is ON it may sink as much as I_{OL} without exceeding $V_{OL(max)}$.
   (c) Whenever a totem-pole TTL output goes from HIGH to LOW, a high amplitude current spike is drawn from the Vcc supply.
   (d) When NOR gate inputs are tied together, they are always treated as a single load on the signal source.

3. A certain IC is rated at $I_{OH(max)}$=640μA and I_{OL}=46.4mA. Express the IC's fan-out in terms of standard TTL unit loads.

   (a) HIGH state = 64 UL, LOW state = 4.6 UL.
   (b) HIGH state = 400 UL, LOW state = 1.125 UL.
   (c) HIGH state = 16 UL, LOW state = 29 UL.
   (d) HIGH state = 8 UL, LOW state = 14.5 UL.

4. The unused input of a 3-input CMOS AND gate should:

   (a) be left floating since it will assume a logic HIGH.
   (b) be tied to Ground to prevent it from picking up noise.
   (c) any unused CMOS input should be tied to ground.
   (d) be tied to V_{DD}.

5. A certain TTL output is LOW. The IC parameter that describes the current that flows from a TTL input to that output is:

   (a) I_{OL} (b) I_{IL} (c) I_{OH} (d) I_{IH}

6. Tristate buffers are commonly used to:

   (a) Simultaneously connect multiple signals to a common bus line.
   (b) Sequentially connect multiple signals to a common bus line.
   (c) Randomly connect multiple signals to a common bus line.
   (d) A new replacement for open-collector devices.

7. A current tracer:

   (a) Cannot be used to manually inject pulses into a circuit in order to test the circuit's operation.
   (b) Can be used in conjunction with a logic pulser to trace the precise location of shorts to ground or Vcc.
   (c) Can detect a changing current in a wire or printed circuit-board trace without breaking the circuit.
   (d) All of the above.

8. The totem-pole outputs of various AND gates can be tied together to form a "wired AND" connection.

   (a) [TRUE] (b) [FALSE]

9. Bus contention is a term used to describe a condition where two or more active devices are placed on the same bus line at the same time.

   (a) [TRUE] (b) [FALSE]

10. LVT series is fabricated using BiCMOS technology.

   (a) [TRUE] (b) [FALSE]

9 MSI LOGIC CIRCUITS

Objectives

Upon completion of this chapter, you will be able to:

- Analyze and use decoders and encoders in various types of circuit applications.
- Compare the advantages and disadvantages of LEDs and LCDs.
- Utilize the observation/analysis technique for troubleshooting digital circuits.
- Understand the operation of multiplexers and demultiplexers by analyzing several circuit applications.
- Compare two binary numbers by using the magnitude comparator circuit.
- Understand the function and operation of code converters.
- Cite the precautions that must be considered when connecting digital circuits using the data-bus concept.
- Interpret the notation used on the IEEE/ANSI symbols for various MSI devices.

Glossary of key terms covered in this chapter:

- **Back-lit LCDs** - LCDs that use a small light source which is part of the display unit. [sec.9.3]
- **Backplane** - An electrical connection common to all segments of a LCD. [sec.9.3]
- **Bidirectional Data Line** - Term used when a data line functions either as an Input or an Output line depending on the states of the enable inputs. [sec.9.16]
- **BIN/OCT** - When used inside an IEEE/ANSI symbol, it indicates a Binary-to Octal decoding function. [sec.9.5]
- **Binary-to-octal Decoder** - Decoder that takes a 3-bit binary input code and activates one of the eight outputs corresponding to that code. [sec.9.1]
- **BCD/DEC** - When used inside an IEEE/ANSI symbol, it indicates a BCD-to-Decimal decoding function. [sec.9.5]
- **BCD-to-Decimal Decoder** - Decoder that converts a BCD quantity into a single decimal output equivalence. [sec.9.1]
- **BCD-to-7-segment Decoder/Driver** - Digital circuit that takes a 4-bit BCD input and activates the required outputs to display the equivalent decimal digit on a LED. [sec.9.2]
- **Bus Contention** - When two or more signals tied to the same bus line are active and are essentially fighting each other. [sec.9.14]
- **Bus Drivers** - When a large number of devices have to be connected to a common bus, the devices' outputs have to be buffered by circuits called bus drivers. [sec.9.16]
- **Code Converter** - This is a logic circuit that changes data presented in one type of binary code to another type of binary code. [sec.9.13]
- **Common-anode LED displays** - 7-segment LEDs where the anodes of all segments are tied together and connected to Vcc. [sec.9.2]
- **Common-cathode LED displays** - 7-segment LEDs where the cathodes of all segments are tied together and connected to ground. [sec.9.2]

- **Data Bus** - A collection of conducting paths over which digital data is transmitted from one device to another. [sec.9.14]

- **Data Distributors** - See Demultiplexers. [sec.9.9]

- **Data Selectors** - See Multiplexers. [sec.9.7]

- **Decoder** - A digital circuit that converts an input binary code into a corresponding single numeric output. [sec.9.1]

- **Decoder/Demultiplexer** - Type of decoder where an ENABLE input has been added to the decoder gates. This type of decoder can be used either as a decoder or as a demultiplexer. Hence, manufacturers often refer to this type of device as a decoder/demultiplexer. [sec.9.9]

- **Demultiplexer (DEMUX)** - A logic circuit that depending on the status of its Select inputs will channel its data input to one of several data outputs. [sec.9.9]

- DEMUX - See Demultiplexer. [sec.9.9]

- **Driver** - Technical term sometimes added to IC descriptions to indicate that the IC's outputs can operate at higher current and/or voltage limits than a normal standard IC. [sec.9.1]

- **8-line-to-3-line Encoder** - Digital circuit which generates a different 3-bit code depending on which one of the eight inputs is activated. [sec.9.4]

- **Encoder** - A digital circuit that produces an output code depending on which of its inputs is activated. [sec.9.4]

- **4-to-10 Decoder** - See BCD-to-Decimal Decoder. [sec.9.1]

- HPRI/BCD - When used inside an IEEE/ANSI symbol, it indicates that the function of this IC is to convert the active input with the highest priority to its corresponding BCD code. [sec.9.5]

- LCD - Liquid Crystal Display. This type of display controls the reflection of available light. [sec.9.3]

- **Logic Function Generation** - When a logic function is implemented directly from a truth table by using a digital IC such as a Multiplexer. [sec.9.8]

- **Magnitude Comparator** - A digital circuit that compares two input binary quantities and generates outputs to indicate whether the inputs are equal or which of the two has the greater magnitude. [sec.9.12]

- **Multiplexers (MUX)** - A logic circuit that depending on the status of its Select inputs will channel one of several data inputs to its output. [sec.9.7]

- **Multiplexing** - The process of selecting one of several input data sources and transmitting the selected data to a single output channel. [sec.9.7]

- **Observation/Analysis** - Process used to troubleshoot circuits or systems in order to predict the possible faults before ever picking up a troubleshooting instrument. When this process is used the troubleshooter must understand the circuit operation, observe the symptoms of the failure and then reason through the operation. [sec.9.6]

- **Octal-to-Binary Encoder** - See 8-Line-to-3 Line Encoder. [sec.9.4]

- **1-of-8 Decoder** - See Binary-to-octal decoder - [sec.9.1]

- **1-of-10 Decoder** - See BCD-to-Decimal Decoder. [sec.9.1]

- **Parallel-to-Serial Conversion** - Process by which data are presented simultaneously to a circuit's input and then transmitted one bit at a time to its output. [sec.9.8]

- **Priority Encoders** - A special type of encoder that senses when two or more inputs are activated simultaneously and then generates a code corresponding to the highest-numbered input. [sec.9.4]

- **Reflective LCDs** - LCDs that use ambient light such as sunlight or normal room lighting. [sec.9.3]

- **3-line-to-8-line decoder** - See Binary-to-octal decoder - [sec.9.1]

- **Time-Division-Multiplexing** - When several different sets of data appear on the same output line at different times. [sec.9.9]

- **Time-Share** - Term used when access to a circuit or system from different sources is possible but simultaneous processing of the data is not required. [sec.9.8]

- **Word** - This is a unit of data whose size depends on the number of lines of the computer's data bus. [sec.9.16]

Problems

SECTIONS 9.1-9.2 *Decoders/BCD-to-7-Segment Decoder-Drivers.*

9.1 Complete each of the following statements:

(a) A 1-of-16 decoder is a circuit which accepts _____ binary inputs and has ______ outputs.

(b) A BCD-to-Decimal decoder can also be called a __________decoder.

9.2 What must be the levels of inputs A_0-A_4 in order for output $\overline{O}_{13}$ of Figure 9.1 to be active?

9.3 How would the operation of the circuit in Figure 9.1 (next page) be affected, if all the connections to Address lines A_3 and A_4 were reversed during the circuit construction?

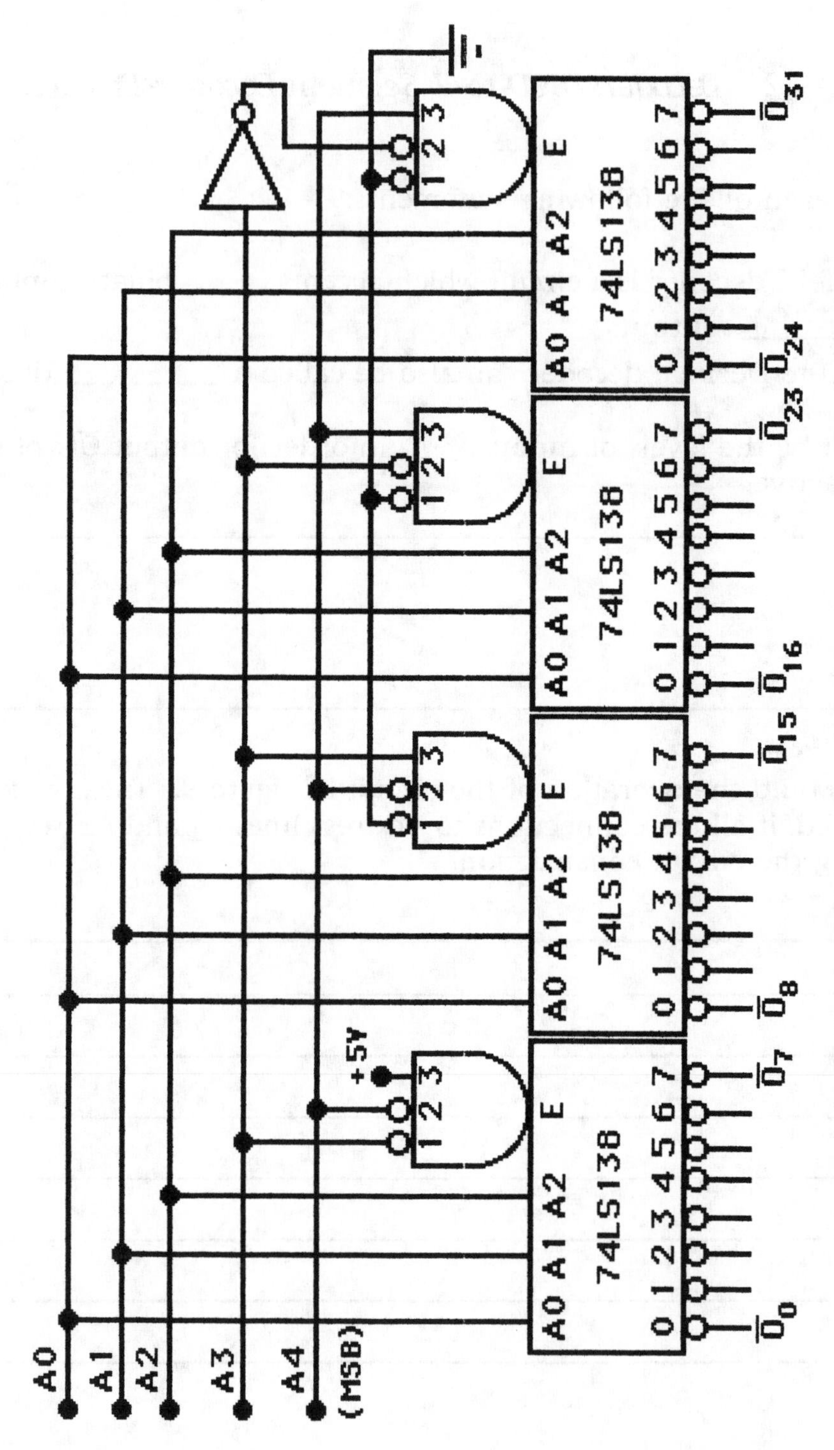

Figure 9.1: Four 74LS138s forming a 1-of-32 decoder.

SECTIONS 9.3-9.4 *Liquid Crystal Displays/Encoders.*

9.4 (a) Why can't LCDs and LEDs displays be interchangeable?

____

____

____

____

(b) Complete the following statement:

Reflective LCDs use ________ light, while *back-lit LCDs* use a ________.

9.5 State two possible reasons why segment 'a' of the LCD in the circuit of Figure 9.2 (next page) would be lit intermittently during the circuits' operation.

1. ______________________________
2. ______________________________

9.6 Refer to the circuit of Figure 9.3.

(a) What are the logic levels at the outputs of the 74147 encoder if SW8 is depressed?

(b) What would these output levels be if switches SW3 and SW8 were simultaneously depressed?

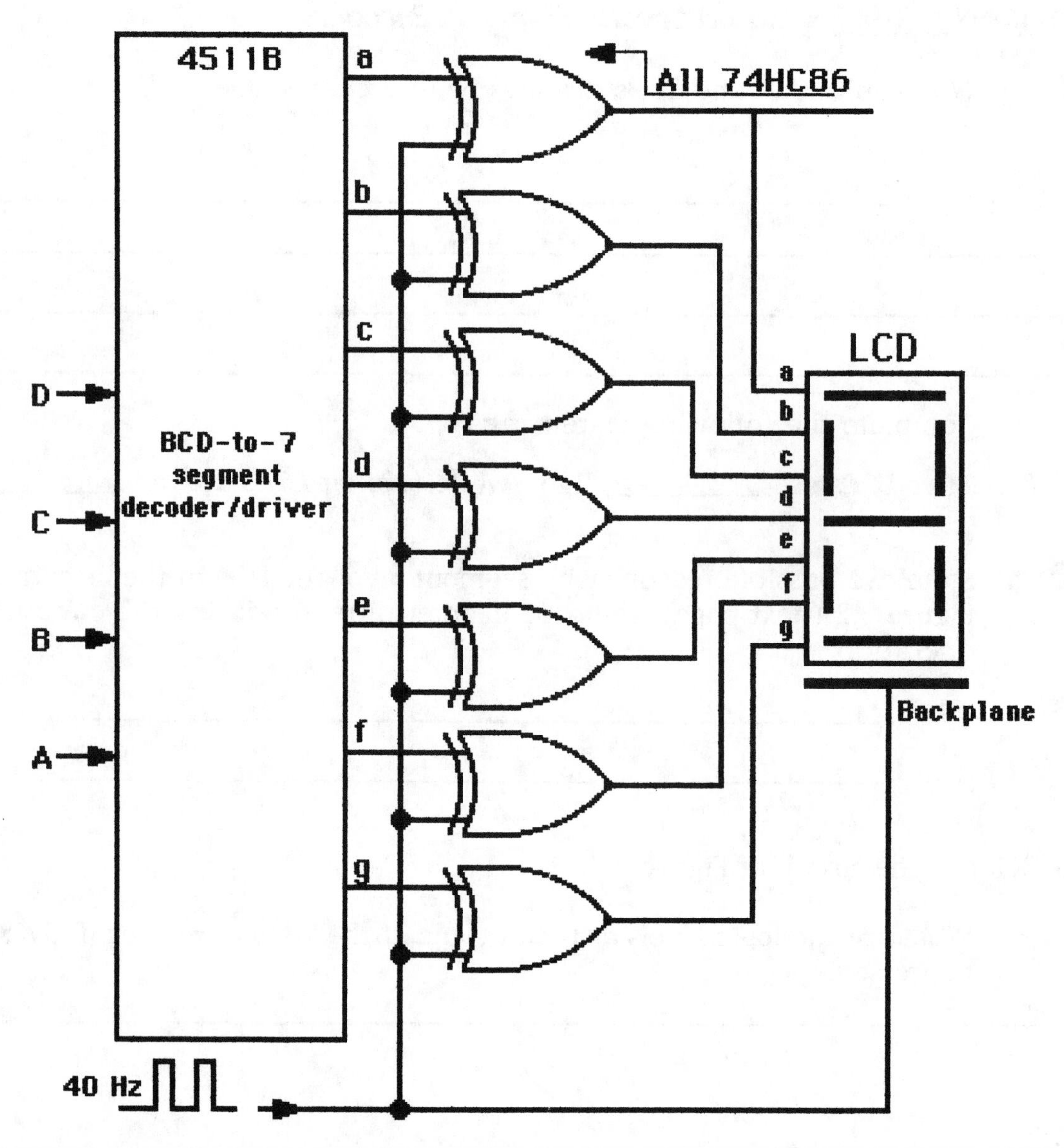

Figure 9.2: Method for driving 7-segment LCD.

9.7 Explain the major difference between priority and non-priority encoders.

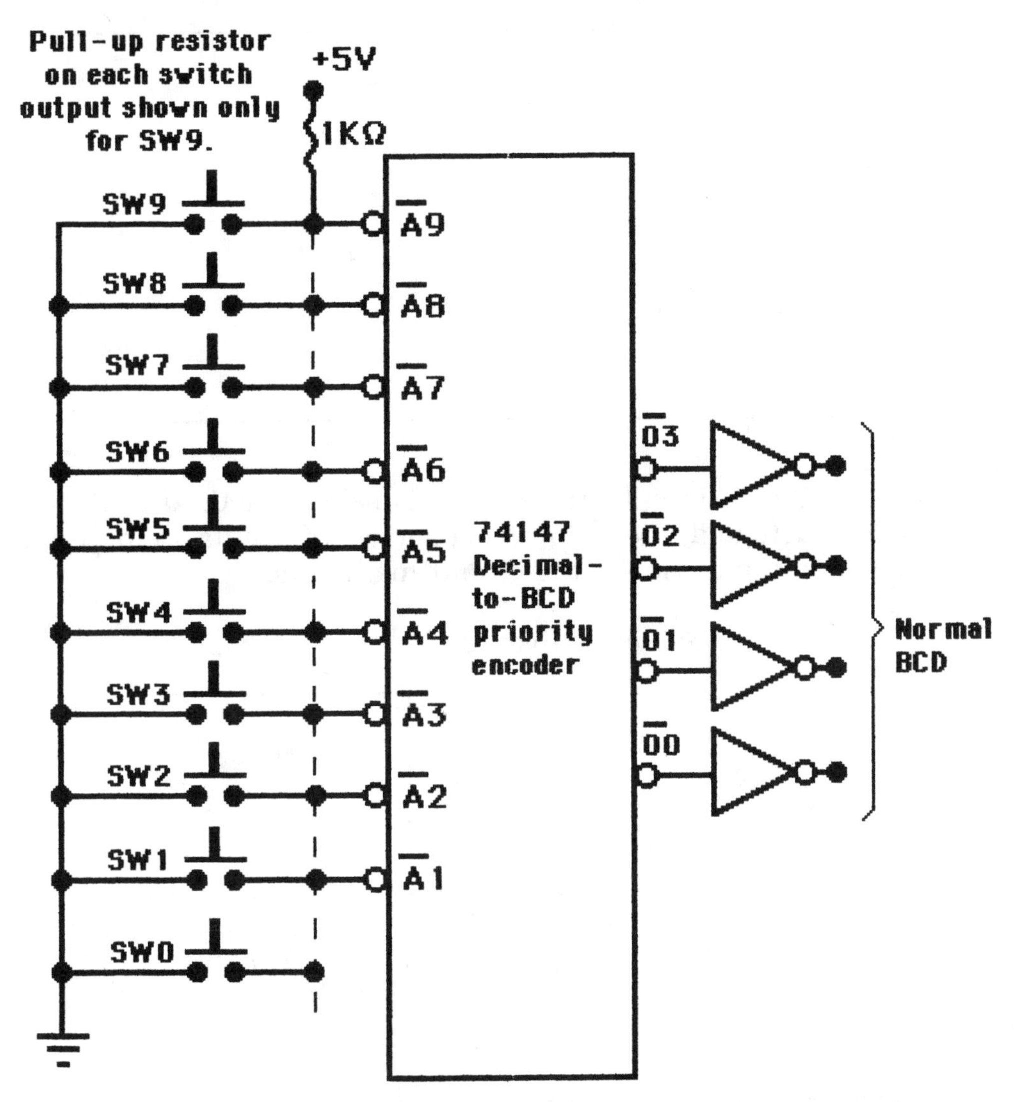

Figure 9.3:Decimal-to-BCD switch encoder.

<u>SECTIONS 9.5-9.6</u> *IEEE-ANSI Symbols/Troubleshooting.*

9.8 Draw the IEEE/ANSI symbol that will reflect the following set of functions:

(i) Binary-to-Octal decoder.
(ii) The decoder should be enabled when either $\overline{E1}$, $\overline{E2}$, or E3 is active.
(iii) The outputs are active HIGH.

9.9 The circuit of Figure 9.4 (next page) is being tested. After the CLEAR switch is actuated, the following sequence of events are observed. Determine the cause for the circuit malfunction.

Switch activated	BCD Display
SW8	800
SW6	806
SW4	846
SW5	546
SW2	542
SW7	572
SW3	372

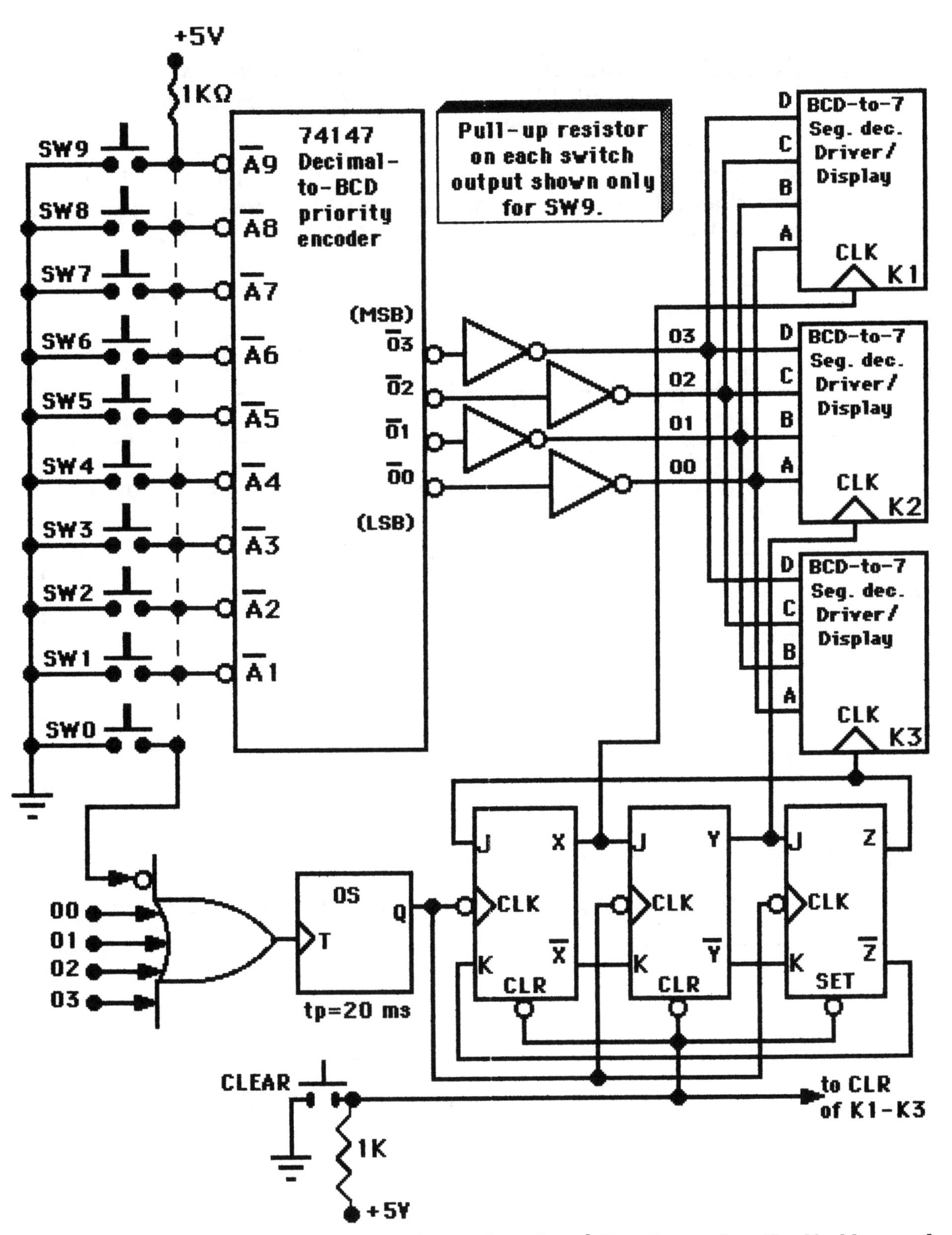

Figure 9.4: Circuit for keyboard entry/display of a 3-digit number.

<u>SECTIONS 9.7-9.11</u> *Multiplexers/Multiplexers Applications/ Demultiplexers/More IEEE-ANSI Symbology/ More Troubleshooting.*

9.10 The circuit of Figure 9.5 (a) (next page) is built using CMOS integrated circuits and the waveform Z of Figure 9.5 (b) is obtained. (*Assume that counter ABC is clear before the first NGT of the CLK is applied.*)

(a) Determine the most probable cause for the circuit malfunction.

(b) By analyzing waveform Z, determine what the contents of X_0-X_7 are most likely to be.

9.11 Implement the function $Z=\overline{C}\overline{B}A+C\overline{B}\overline{A}+CBA$ using a 74HC151 Multiplexer.

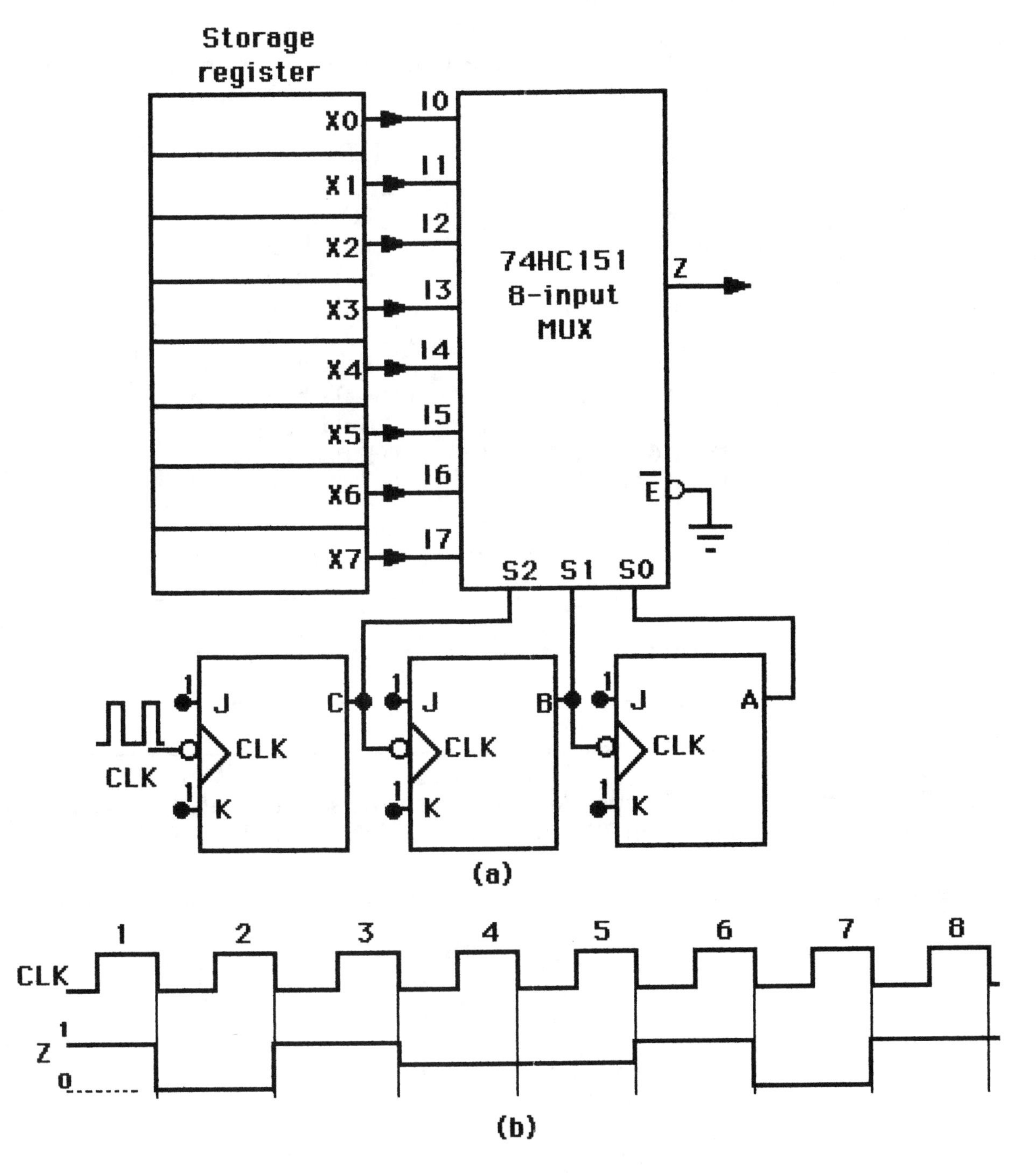

Figure 9.5: (a) Parallel-to-serial converter; (b) Output waveform Z.

9.12 The circuit of Figure 9.6 is built, tested, and the following observations were made. Find the probable cause for the malfunction.

Condition	LEDs
All doors closed	All LEDs off
All doors open	All LEDs flashing
Door 0 open	LED 0 flashing
Door 1 open	LED 2 flashing
Door 2 open	LED 1 flashing
Door 3 open	LED 3 flashing
Door 4 open	LED 4 flashing
Door 5 open	LED 6 flashing
Door 6 open	LED 5 flashing
Door 7 open	LED 7 flashing

9.13 What denotes the AND dependency between the select inputs and each of the data inputs 0 through 7 in the IEEE/ANSI symbol of Figure 9.7?

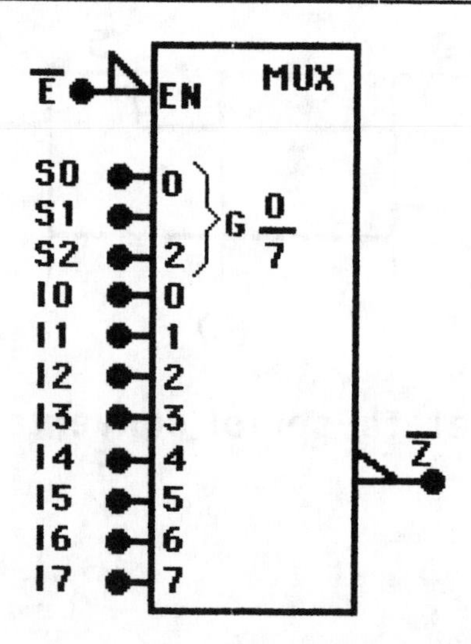

Figure 9.7: IEEE/ANSI symbol.

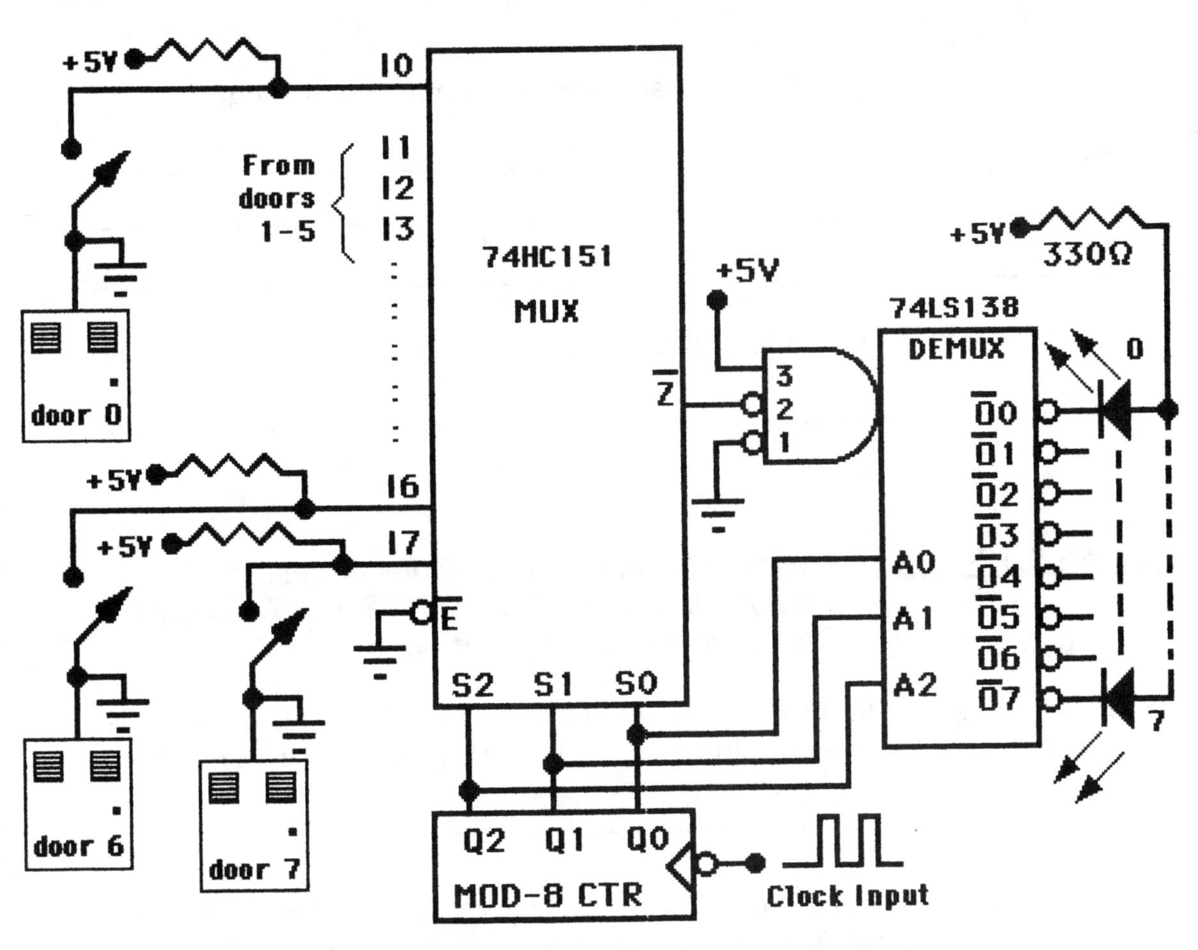

Figure 9.6: Security monitoring system.

9.14 Counter (Q_0 Q_1 Q_2) of Figure 9.8 counts up to binary 101_2 and then stops. Explain whether or not each of the following choices could cause the malfunction.

(a) Connection from counter output Q_2 to decoder input A_2 is open.

____

(b) Connection from Sensor 6 to I_6 of the multiplexer is open.

____

____

____

(c) Inverting amplifier for actuator #5 is not functioning.

(d) Output $\overline{O}5$ of the decoder is stuck HIGH.

SECTIONS 9.12-9.16 ***Magnitude Comparator/Code Converters/ Data Busing/The 74173-LS173-HC173 Tristate Register/Data Bus Operation.***

9.15 Refer to Figure 9.9. If the binary #A=11001101_2 and #B=11001110_2, what are the logic levels at the inputs $I_{A>B}$, $I_{A<B}$, $I_{A=B}$ of comparator Z2?

9.16 Show how to combine 74HC85 Comparator chips in order to compare two 9-bit binary numbers.

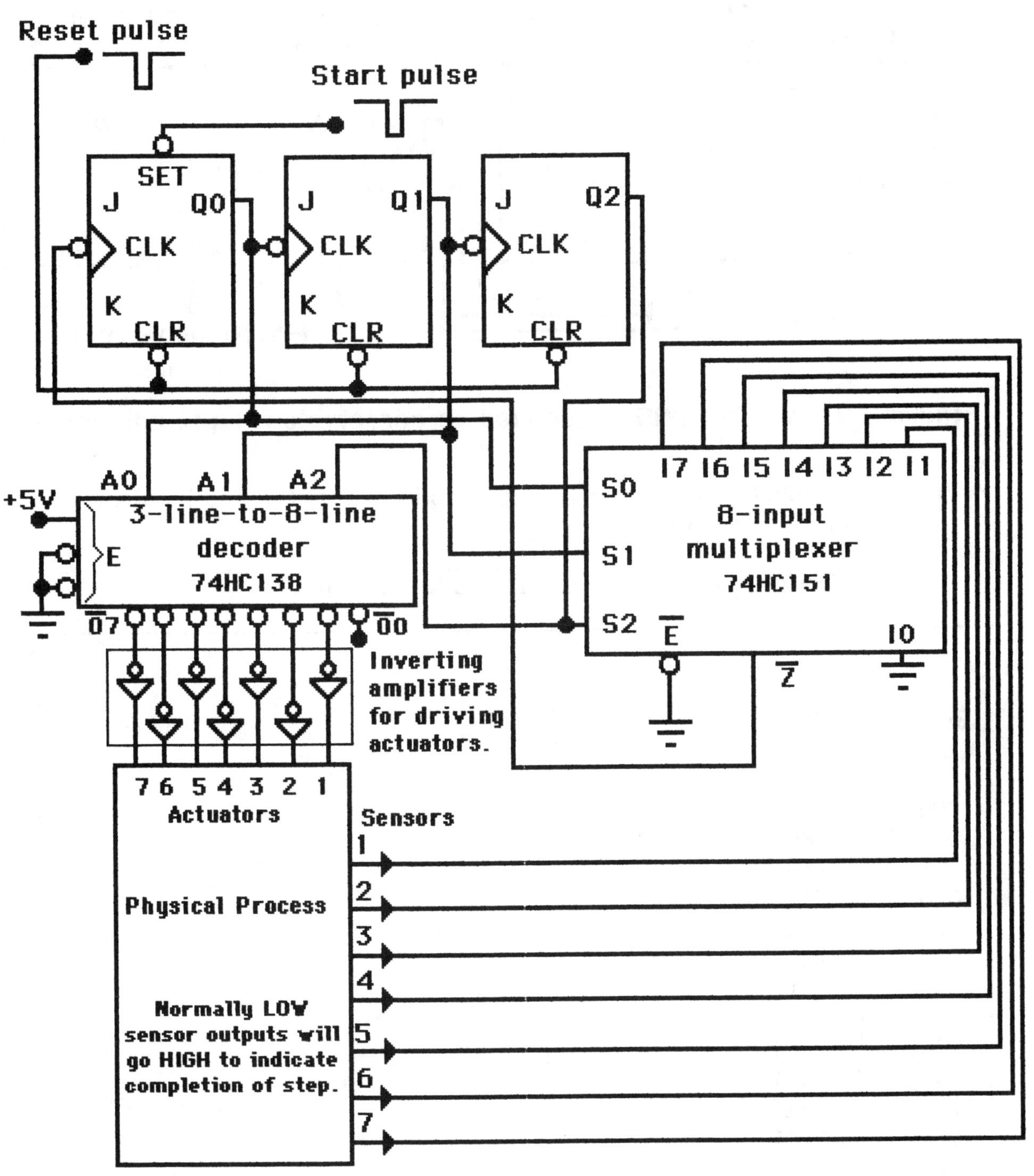

Figure 9.8: Seven-step control sequencer.

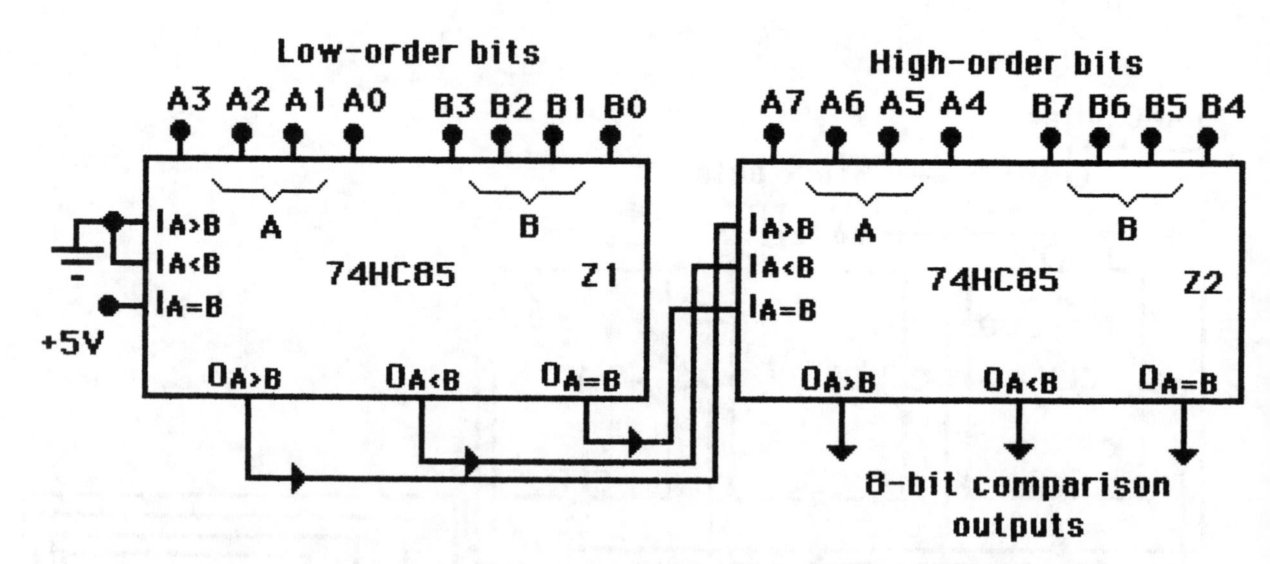

Figure 9.9: Two 74HC85s cascaded to perform an 8-bit comparison.

9.17 The table below shows a 4-bit gray code ($G_3G_2G_1G_0$) and its binary equivalencies ($B_3B_2B_1B_0$).

(a) Design a 4-bit Binary-to-Gray code converter.

Binary	Gray
0000	0000
0001	0001
0010	0011
0011	0010
0100	0110
0101	0111
0110	0101
0111	0100
1000	1100
1001	1101
1010	1111
1011	1110
1100	1010
1101	1011
1110	1001
1111	1000

(b) Design a 4-bit Gray-to-Binary code converter.

Binary	Gray
0000	0000
0001	0001
0010	0011
0011	0010
0100	0110
0101	0111
0110	0101
0111	0100
1000	1100
1001	1101
1010	1111
1011	1110
1100	1010
1101	1011
1110	1001
1111	1000

9.18 (a) What would happen to the circuit's operation of Figure 9.10 (next page), if $\overline{IE}_B$ is always LOW?

(b) What would happen to the circuit's operation of Figure 9.10 (next page), if $\overline{OE}_B$ is always LOW?

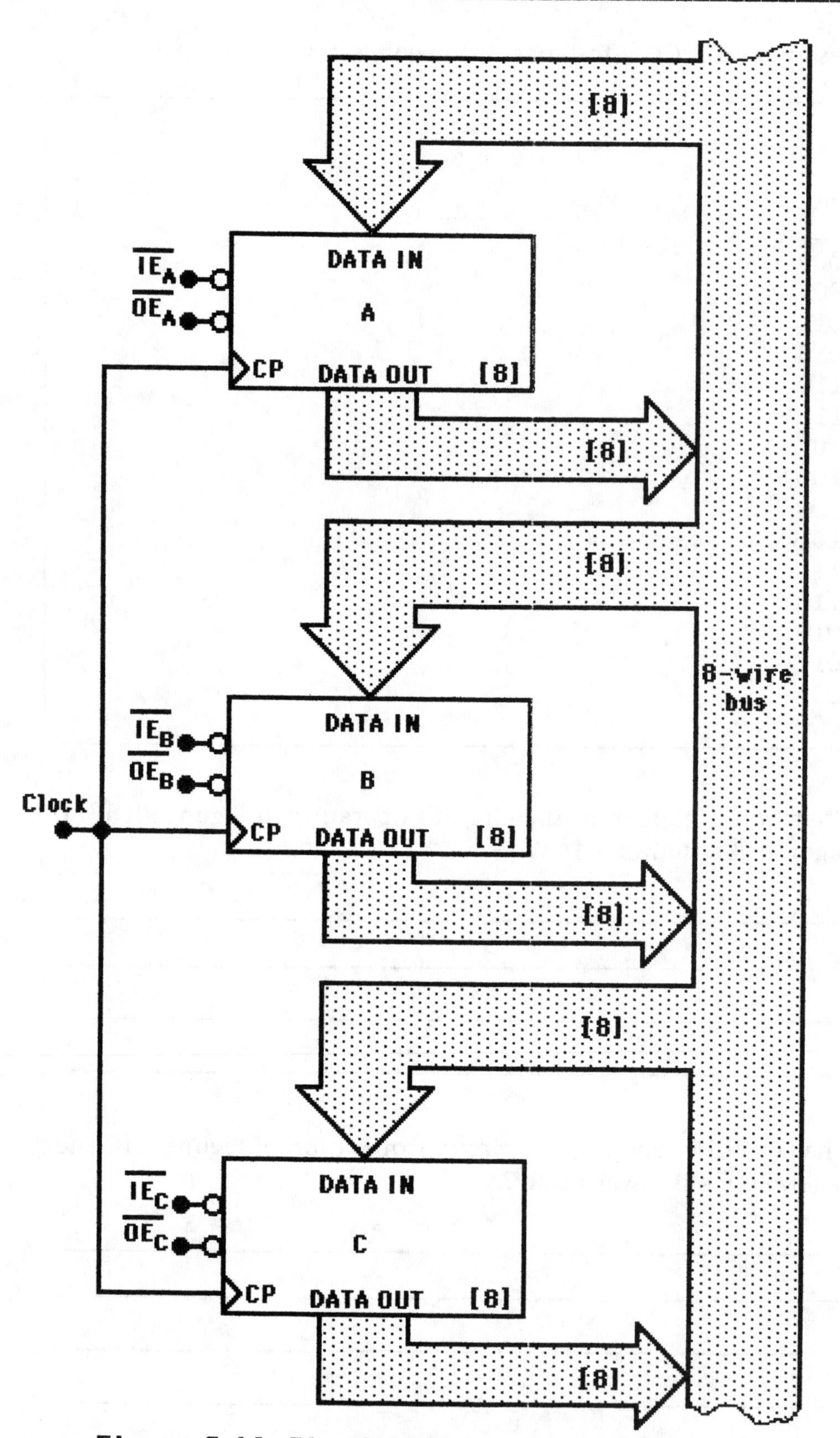

Figure 9.10: Simplified representation of bus arrangement.

TEST 9

1. A 32-input Multiplexer is to be used to perform parallel-to-serial data conversion. Which of the following counters would be required to provide the data select inputs?

   (a) MOD 4 (b) MOD 8 (c) MOD 16 (d) MOD 32

2. If two or more signals are active and are applied to the same point at the same time, it may result in what is called:

   (a) The Backplane effect (b) Time-share (c) Bus contention
   (d) A bidirectional data line

3. Another term used to describe a Demultiplexer is:

   (a) Data selector (b) Data distributor (c) Encoder (d) Code converter

4. This device may have only one input line active and may have more than one output line active at any given time.

   (a) Demultiplexer (b) Multiplexer (c) Encoder (d) Decoder

5. A 3-line-8-line decoder is often referred to as a:

   (a) Binary-to-octal decoder (b) BCD-to-decimal decoder
   (c) BCD-to-7-segment decoder (d) None of the above

6. When used inside an IEEE/ANSI symbol, HPRI/BCD indicates:

   (a) that the function of the IC is to convert the active input with the highest priority to its corresponding BCD code.
   (b) that the function of the IC is to convert the active input with the lowest priority to its corresponding BCD code.
   (c) that the function of the IC is to convert primarily BCD codes.
   (d) that this is a dual function IC: a high priority encoder and a BCD decoder.

7. A 74HC85 magnitude comparator is used to:

   (a) compare two input voltage levels and generate outputs to indicate which voltage has the greatest magnitude.
   (b) compare two input binary quantities and generate outputs to indicate which quantity has the greatest magnitude.
   (c) Logically AND two binary quantities and generate outputs only if they are equal in magnitude.
   (d) Logically exclusive-OR two binary quantities and generate outputs indicating whether the two binary numbers are equal to or different from each other.

8. When different sets of data are appearing on the same output line at different times, the data is:

   (a) frequency-shifted Demultiplexer.
   (b) frequency-shifted multiplexed.
   (c) time-division-Demultiplexer.
   (d) time-division-multiplexed

9. The Backplane is a connection common to all segments of a LED.

   (a) [TRUE] (b) [FALSE]

10. Reflective LCDs use ambient light for their operation.

   (a) [TRUE] (b) [FALSE]

10 INTERFACING WITH THE ANALOG WORLD

Objectives

Upon completion of this chapter, you will be able to:

- Understand the theory of operation and circuit limitations of several types of digital-to-analog converters (DACs).
- Read and understand the various DAC manufacturer specifications.
- Use different test procedures to troubleshoot DAC circuits.
- Compare the advantages and disadvantages among the digital-ramp analog-to-digital converter (ADC), successive-approximation ADC, and flash ADC.
- Analyze the process by which a computer in conjunction with an ADC digitizes an analog signal and then reconstructs that analog signal from the digital data.
- Describe the basic operation of a digital voltmeter.
- Understand the need for using sample-and-hold circuits in conjunction with ADCs.
- Describe the operation of an analog multiplexing system.
- Understand the features and basic operation of a digital storage oscilloscope.

Glossary of key terms covered in this chapter:

- **Acquisition Time:** The amount of time needed for the hold capacitor C_h of a sample-and-hold circuit to charge to the current value of the analog input voltage. The acquisition time depends on the value of C_h and the characteristics of the sample-and-hold circuit. [sec.10.15]

- **Actuator:** Electrically controlled device that controls a physical variable. [sec.10.1]

- **Analog-to-Digital Converter (ADC):** Circuit which converts an analog input to a corresponding digital output. [sec.10.1]

- **Bipolar DAC:** Digital-to-Analog Converter which accepts signed binary numbers as input and produces the corresponding positive or negative output value. [sec.10.2]

- **Conversion Time (t_c):** The interval between the end of the START pulse and the activation of the $\overline{EOC}$ output of an ADC. [sec.10.9]

- **Data Acquisition:** Process by which a computer acquires digitized analog data. [sec.10.10]

- **Digital-Ramp ADC:** Type of Analog-to-Decimal Converter in which an internal staircase waveform is generated and utilized for the purpose of accomplishing the conversion. The conversion time for this type of Analog-to-Decimal Converter varies depending on the value of the input analog signal. [sec.10.9]

- **Digital Storage Oscilloscope (DSO):** Oscilloscope that uses both DACs and ADCs to acquire, digitize, store, and display analog waveforms. [sec.10.17]

- **Digital-to-Analog Converter (DAC):** Circuit which converts a digital input to a corresponding analog output. [sec.10.1]

- **Digitize:** Process by which analog data is converted to digital data. [sec.10.6/10.10]

- **Dual-Slope ADC:** Type of analog-to-digital converter that linearly changes a capacitor from a current proportional to V_A for a fixed time interval, and then increments a counter as the capacitor is linearly discharged to zero. [sec.10.13]

- **Flash ADC:** Type of Analog-to-Decimal Converter which has the highest operating speed available. [sec.10.12]

- **Full-Scale Error:** Term used by some Digital-to-Analog Converter manufacturers to specify the accuracy of a Digital-to-Analog Converter. It's defined as the maximum deviation of a Digital-to-Analog Converter's output from its expected ideal value. [sec.10.4]

- **Full-Scale Output:** The maximum possible output value of a Digital-to-Analog Converter. [sec.10.2]

- **Linearity Error:** Term used by some Digital-to-Analog Converter manufacturers to specify the device's accuracy. It's defined as the maximum deviation in step-size from the ideal step-size. [sec.10.4]

- **Monotonicity:** A Digital-to-Analog Converter is said to be monotonic when its output either increases or stays the same as the input is increased. [sec.10.4]

- **Offset Error:** Under ideal conditions the output of a Digital-to-Analog Converter should be zero volts when the input is all 0s. In reality, there is a very small output voltage for this situation. This deviation from the ideal zero volts is called the offset error. [sec.10.4]

- **Percentage Resolution:** The ratio of the step-size to the full-scale value of a Digital-to-Analog Converter. Percentage Resolution can also be defined as the reciprocal of the maximum number of steps of a Digital-to-Analog Converter. [sec.10.2]

- **Quantization Error:** Also referred occasionally to as the Resolution, this is an inherent error of the device. [sec.10.9]

- **R/2R Ladder DAC:** Type of Digital-to-Analog Converter where its internal resistance values only span a range of 2 to 1. [sec.10.3]

- **Resolution:** The change that occurs in the analog output of a Digital-to-Analog Converter as a result of the change in the LSB of its digital input. [sec.10.2/10.4]

- **Sample-And-Hold Circuit:** Type of circuit which utilizes an unity-gain buffer amplifier in conjunction with a capacitor to accomplish a more stable analog-to-digital conversion process. [sec.10.15]

- **Settling Time:** The amount of time that takes the output of a Digital-to-Analog Converter to go from zero to its full-scale value as the input is changed from all 0s to all 1s. Settling time is sometimes referred to as the time needed for the DAC to settle within ± 1/2 step-size of its final value. [sec.10.4]

- **Staircase Test:** Process by which a Digital-to-Analog Converter's digital input is incremented and its output monitored to determine whether or not it exhibits a staircase format. The staircase waveform should be without any missing steps or any downward steps until it reaches its full-scale value. [sec.10.7]

- **Staircase Waveform:** Type of waveform generated at the output of a Digital-to-Analog Converter as its digital input signal is incrementally changed. [sec.10.2]

- **Static Accuracy Test:** When a fixed binary value is applied to the input of a Digital-to-Analog Converter and the analog output is accurately measured. The measured result should fall within the expected range specified by the Digital-to-Analog Converter's manufacturer. [sec.10.7]

- **Step-Size:** See Resolution. [sec.10.2]

- **Successive-Approximation ADC (SAC):** Type of Analog-to-Decimal Converter in which an internal parallel register and complex control logic are used to perform the conversion. The conversion time for this type of Analog-to-Decimal Converter is always the same regardless of the value of the input analog signal. [sec.10.11]

- **Tracking ADC:** See Up/Down Digital-Ramp ADC.

- **Transducer:** Device that converts a physical variable to an electrical variable. [sec.10.1]

- **Up/Down Digital-Ramp ADC:** Type of analog-to-digital converter that uses an up/down counter to step up or step down the voltage from a digital-to-analog converter until it intersects the analog input. [sec.10.13]

- **Voltage-Controlled Oscillator (VCO):** Circuit that produces an output signal with a frequency proportional to the voltage applied to its input. [sec.10.13]

- **Voltage-to-Frequency ADC:** Type of analog-to-digital converter that employs a voltage-controlled oscillator (VCO) to convert an analog voltage to a pulse frequency that is then counted to produce a digital output. [sec.10.13]

Problems

SECTIONS 10.1-10.2 *Interfacing With the Analog World/ Digital-to-Analog conversion.*

10.1 A 10-bit D/A converter has a F.S. value of 15V. What will be the output value if the binary input is 1010110111_2?

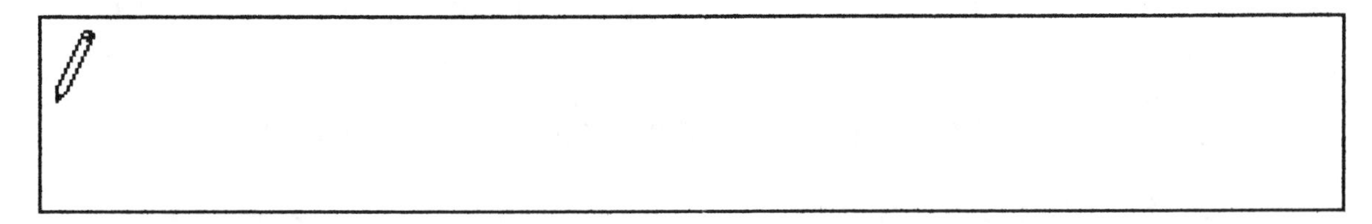

10.2 Determine the % Resolution for the DAC of problem 10.1.

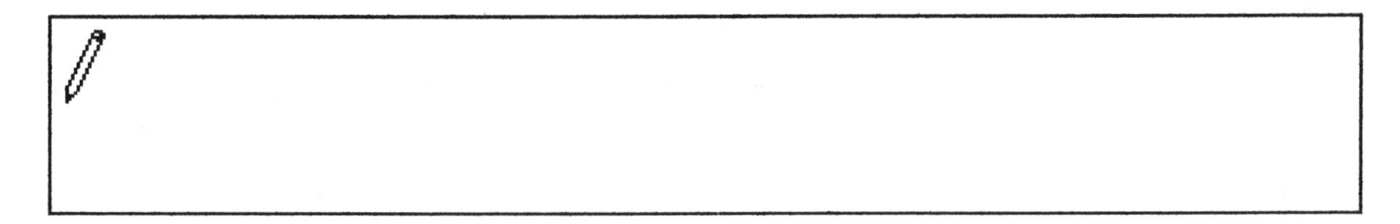

10.3 Assume that the DAC of Figure 10.1 (a) is a 4-bit DAC. Its output signal is shown in Figure 10.1 (b).

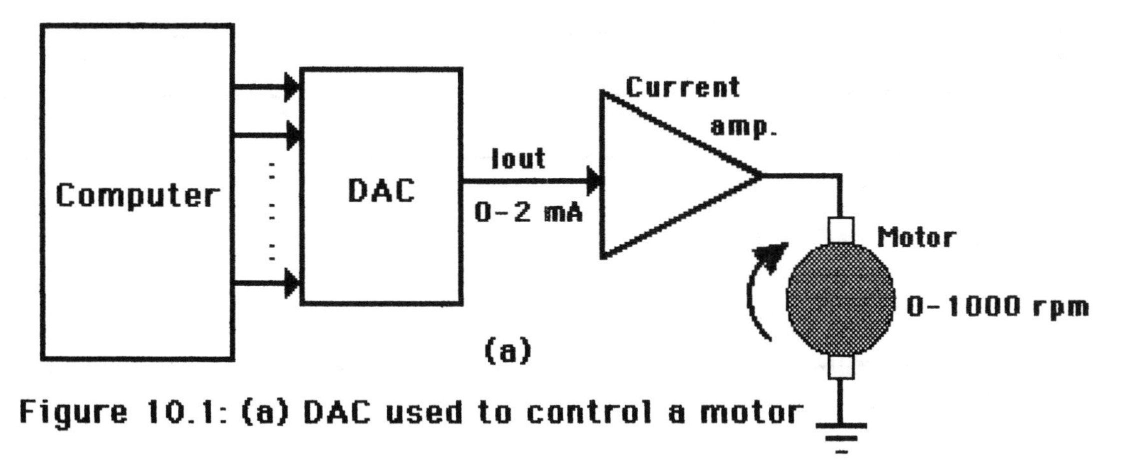

Figure 10.1: (a) DAC used to control a motor

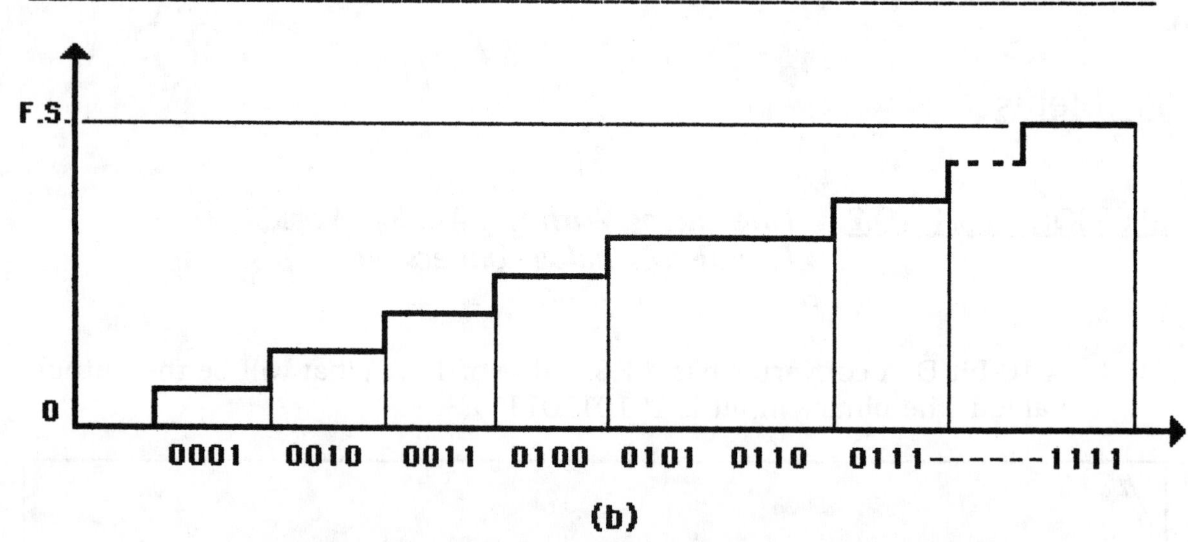

(b)

Figure 10.1: (b) Output signal of the DAC.

(a) Describe what is wrong with the DAC's output staircase waveform of Figure 10.1(b).

(b) What are the effects of that malfunction on the Motor's operation?

10.4 The weight of input D1 of Figure 10.2 is 2.24V, determine:

(a) The resolution of the BCD Digital-to-Analog Converter.

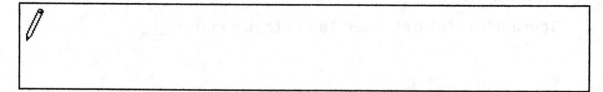

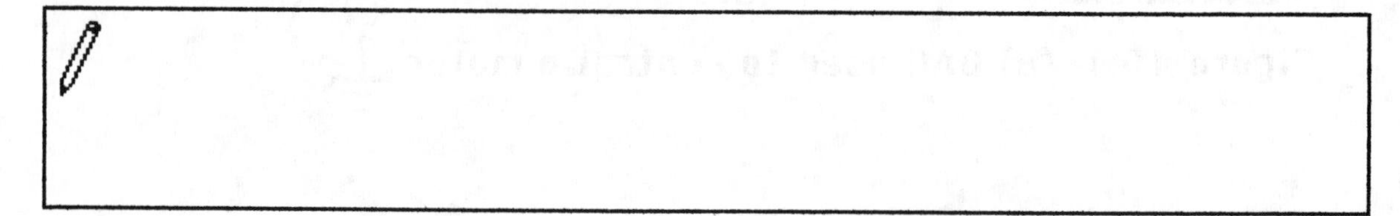

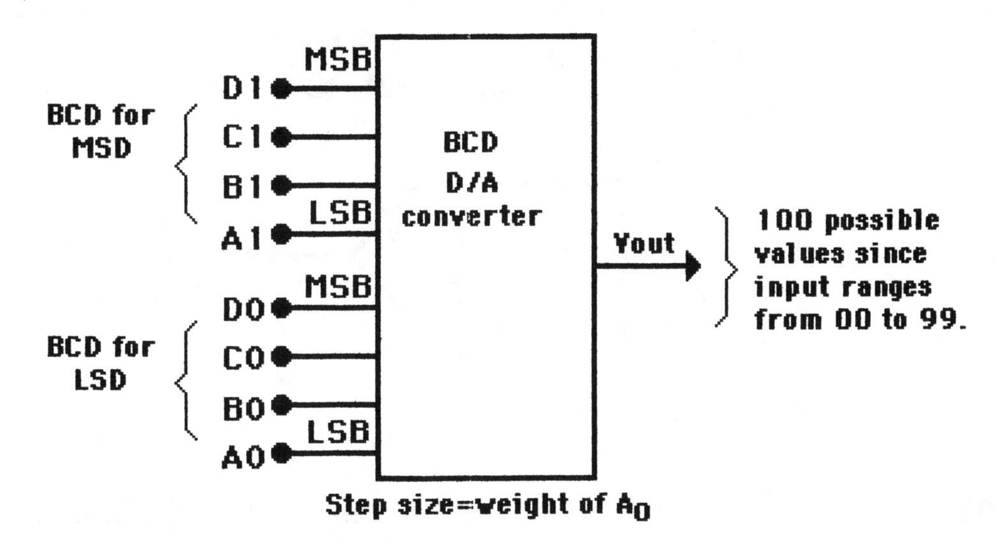

Figure 10.2: D/A converter using BCD input code.

(b) % Resolution of the BCD Digital-to-Analog Converter.

(c) The BCD input code when V_{out} is 2.156V.

SECTIONS 10.3-10.4,10.6   *D/A-Converter Circuitry/DAC Specifications DAC Applications*

10.5 Change the DAC circuit of Figure 10.3 so that its step-size is equal to -78.13mV.

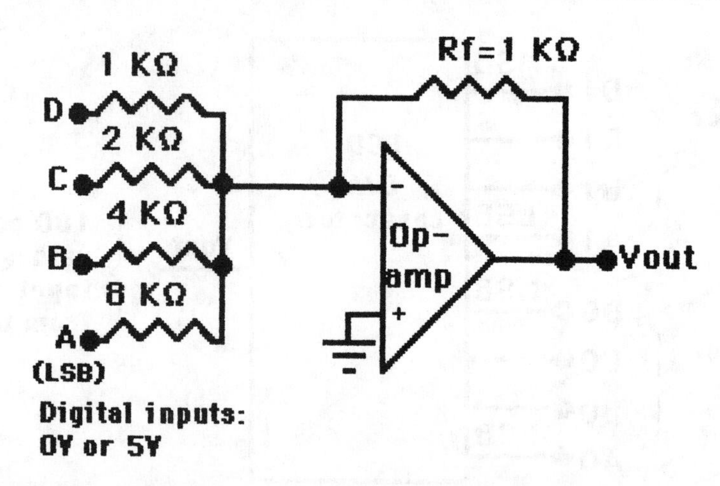

Figure 10.3: D/A converter using op-amp summing amplifier.

10.6 A 10-bit DAC has a full-scale output value of 1.590V and an accuracy of ± 0.25% full-scale. When the binary input is 1100110110_2, the output is equal to 1.283V. Determine whether or not this output value is within the DAC's specified accuracy?

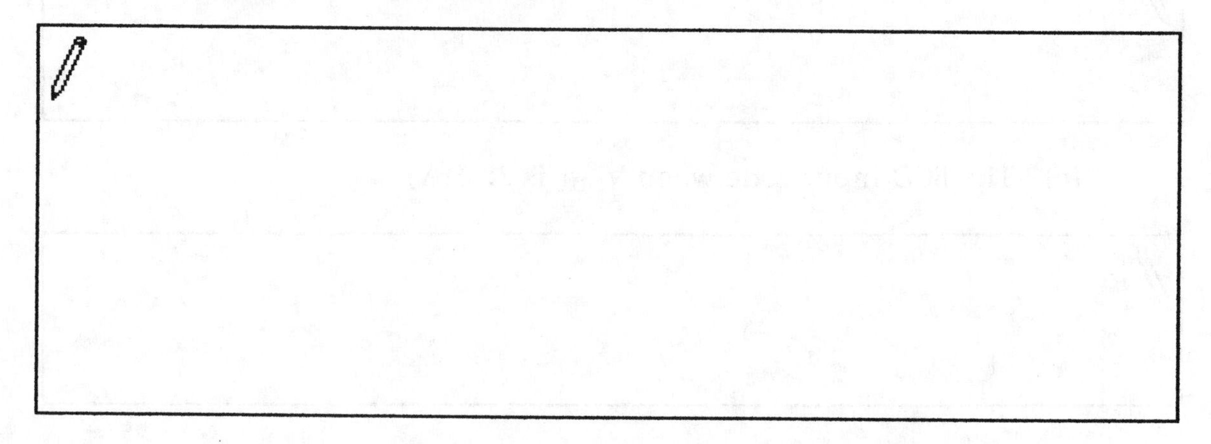

10.7 The input of a certain DAC is changed from all 0s to all 1s. The output response was recorded on a storage oscilloscope and is shown in Figure 10.4. Determine the approximate Settling time for the DAC.

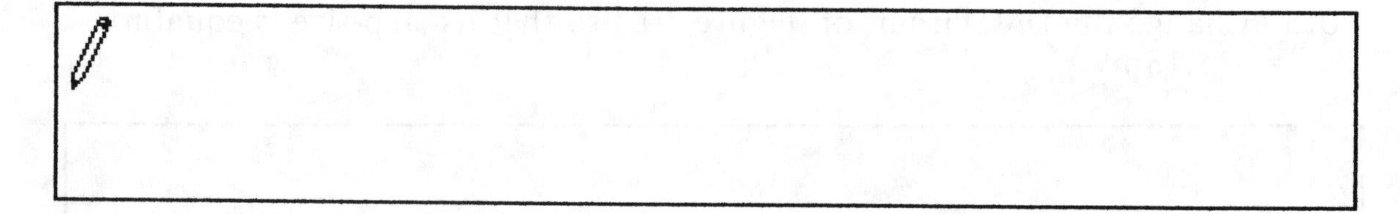

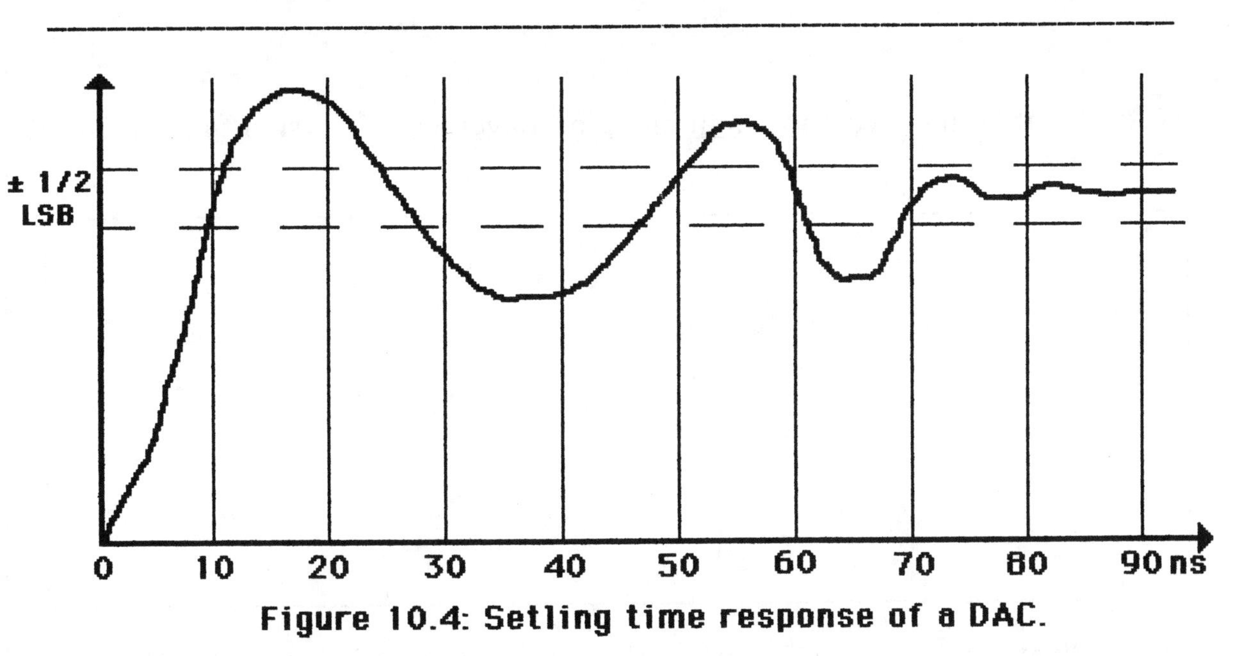

Figure 10.4: Setling time response of a DAC.

SECTIONS 10.7-10.9 *Troubleshooting DACs/ Analog-to-Digital Conversion/ Digital-Ramp ADC*

For the following five problems, refer to the ADC of Figure 10.5.

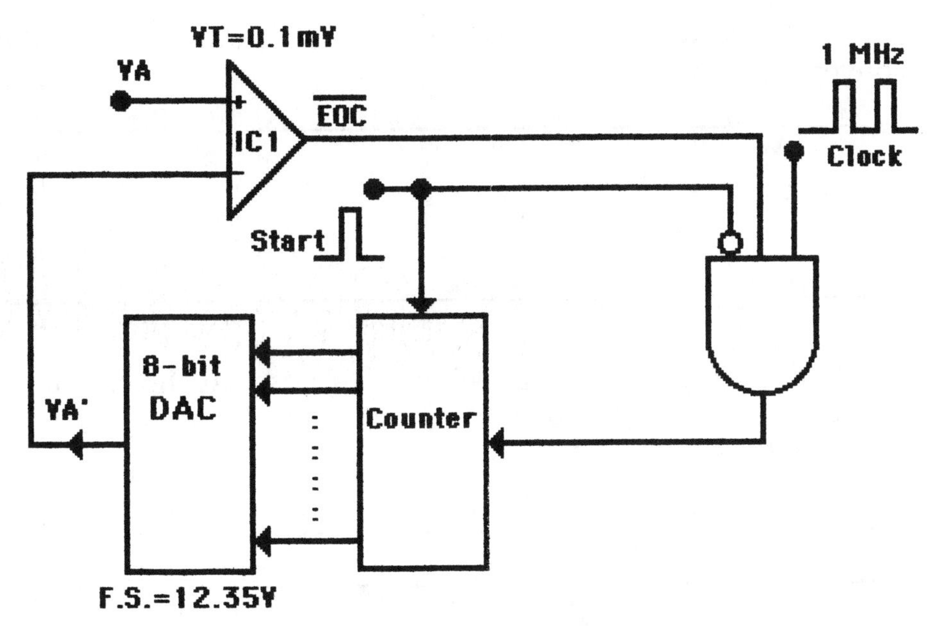

Figure 10.5: Digital-ramp A/D converter.

10.8 What is the maximum , minimum and average conversion time of this ADC?

10.9 What is the value of V_A, when the counter output is binary 10110001_2?

10.10 If V_A=8.572V, what is the DAC's input binary data at the end-of conversion time?

10.11 While testing the ADC, a technician finds that the counter never stops counting, regardless of the value of V_A. State one possible cause for the malfunction.

10.12 For each of the following circuit changes determine whether or not a decrease in the conversion time would occur.

(a) Increase the clock frequency.

____

(b) Replace the existing DAC with a 10-bit DAC.

____

(c) Use op-amps that have faster slew-rates.

____

(d) Replace the existing DAC with a 4-bit DAC.

____

SECTIONS 10.10-10.12 *Data Acquisition/Successive-Approximation Analog-to-Digital Converter/Flash ADCs.*

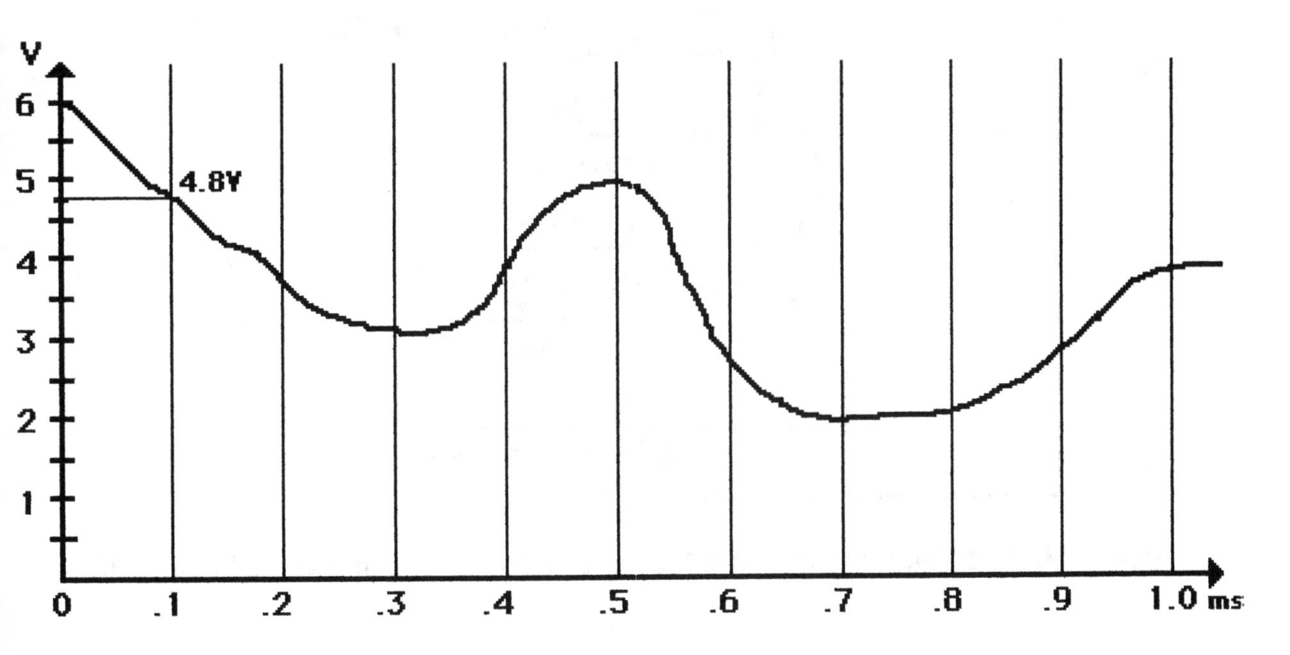

Figure 10.6: Analog signal to be digitized.

10.13 (a) Assume that the analog signal of Figure 10.6 is to be digitized by the ADC of Figure 10.5. Superimpose on Figure 10.6 the reconstructed signal using the data obtained during the digitalization process.

(b) If a highly accurate reproduction of the digitized analog signal of Figure 10.6 is desired, what type of ADC should be used?

10.14 Complete the timing diagram for the digital inputs D_0-D_7 and analog output $V_{A'}$ of the successive approximation ADC of Figure 10.7

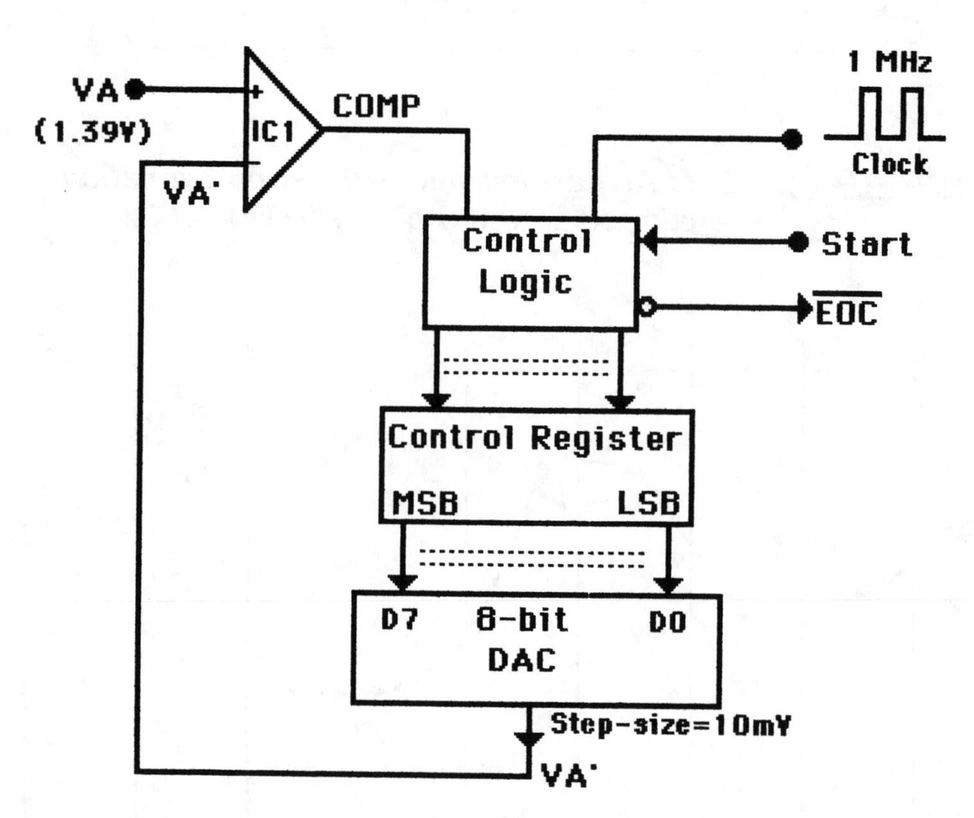

Figure 10.7: Block diagram of a successive-approximation ADC.

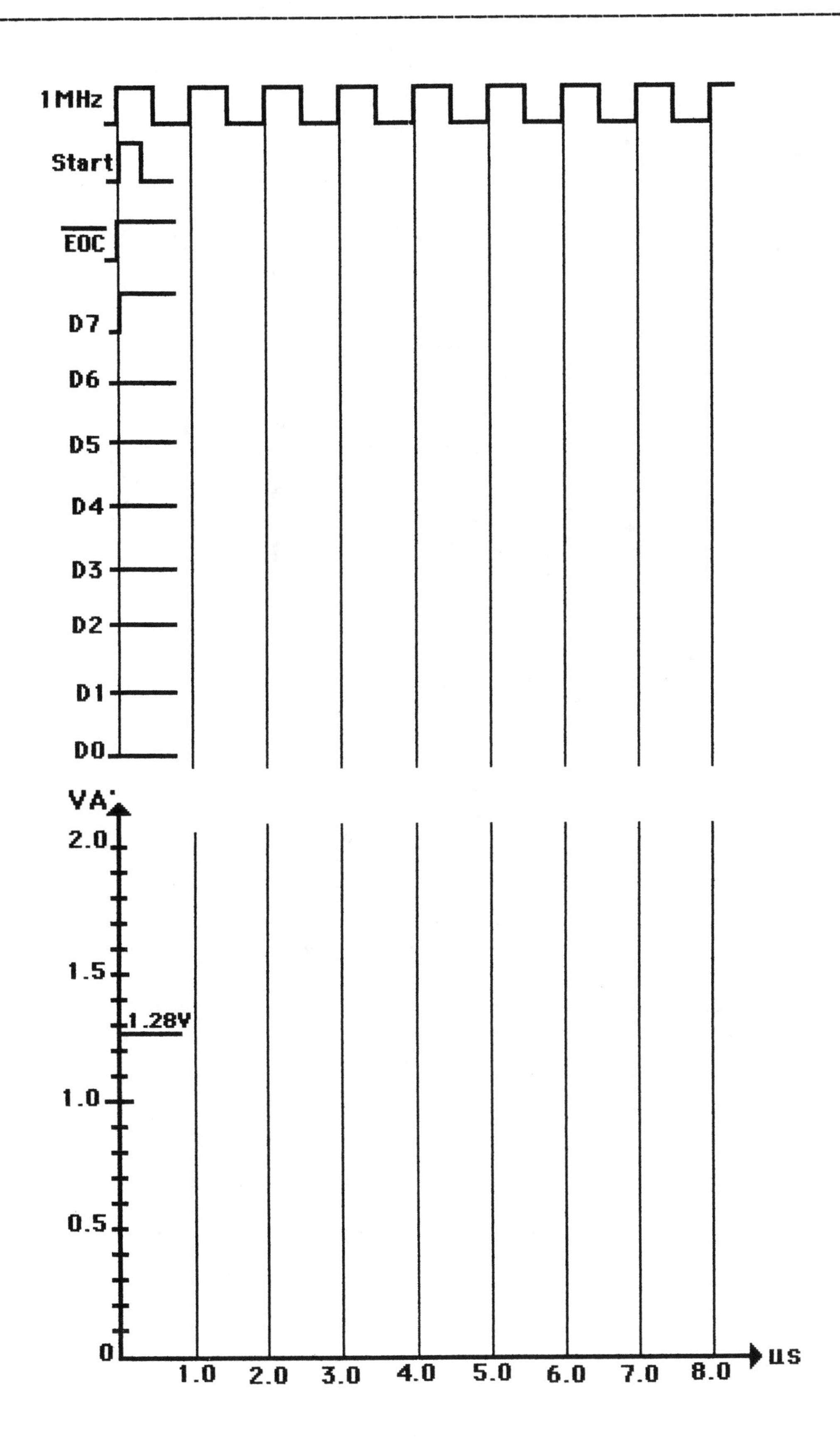
1MHz
Start
EOC
D7
D6
D5
D4
D3
D2
D1
D0
VA
2.0
1.5
1.28V
1.0
0.5
0
1.0
2.0
3.0
4.0
5.0
6.0
7.0
8.0
us

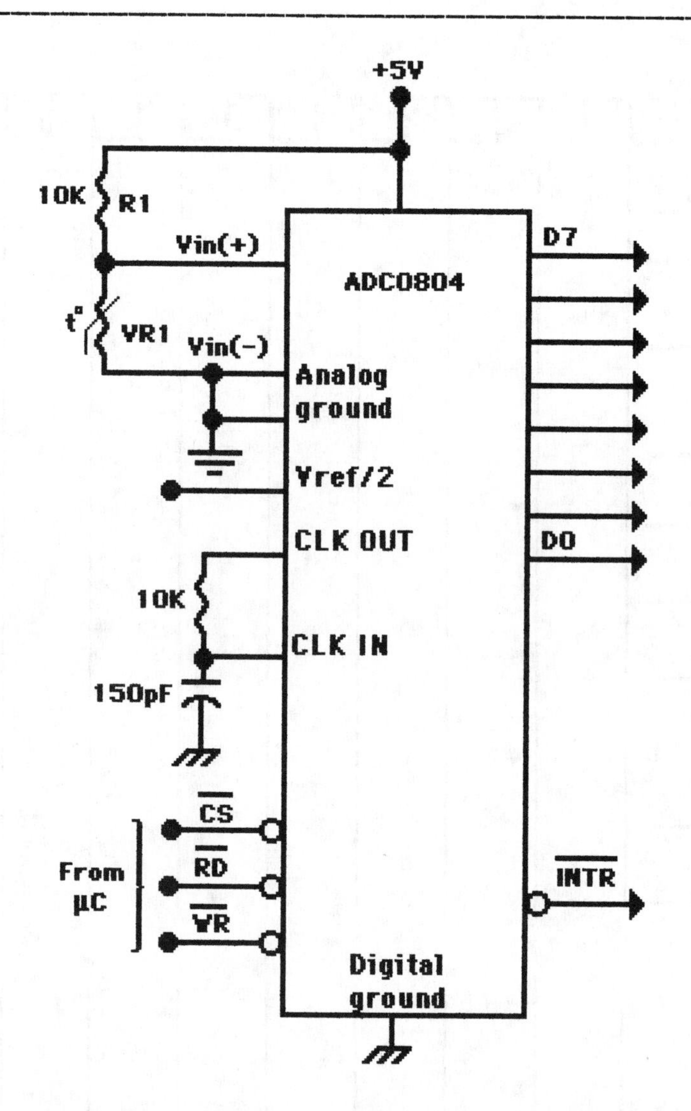

Figure 10.8: Digital thermometer

10.15 The circuit of Figure 10.8 is used to convert the ambient temperature, as sensed by the thermistor VR1, to an equivalent binary value D_7-D_0. The thermistor will detect a range of temperatures from 0˚- 60˚C. When the temperature is 60˚C, $V_{in}(+)$=5.0V.

(a) What is the binary number D_7-D_0 when the temperature is 60˚C?

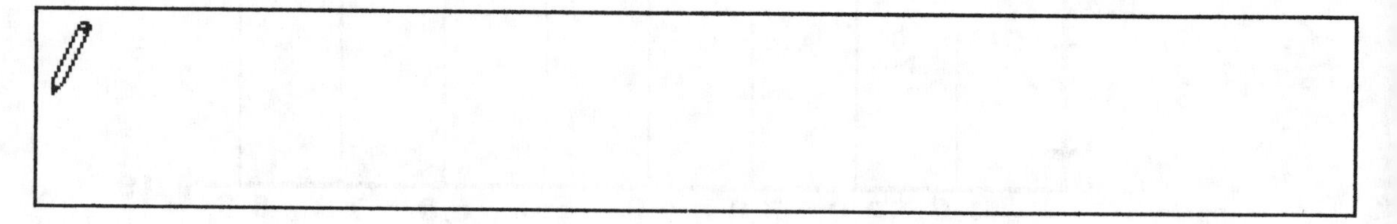

(b) After the circuit is designed it is determined that 30°C is the maximum ambient temperature. When the temperature is 30°C, $V_{in}(+)$=3.0V. Modify the circuit of Figure 10.8 in order for the maximum output binary number to be representative of the maximum ambient temperature?

____

SECTIONS 10.13-10.17 *Other A/D Conversion Methods/ Digital Voltmeter/Sample-and-Hold Circuits/ Multiplexing/Digital Storage Oscilloscope.*

10.16 Indicate which "Analog-to-Digital Conversion Method" is being described by each of the following statements:

(a) The basic operation of this converter involves the linear charging and discharging of a capacitor using constant currents.

____

(b) It produces an output frequency that is proportional to its input voltage.

____

(c) When a new conversion is to begin, the counter is not reset to zero, but begins counting up or down from its last value.

____

(d) Because of its slow conversion times, it is not used in any data acquisition applications.

____

⊕10.17 The DVM circuit of Figure 10.9 is tested and the results recorded in the accompanying table. Determine a possible cause for the discrepancy in the readings.

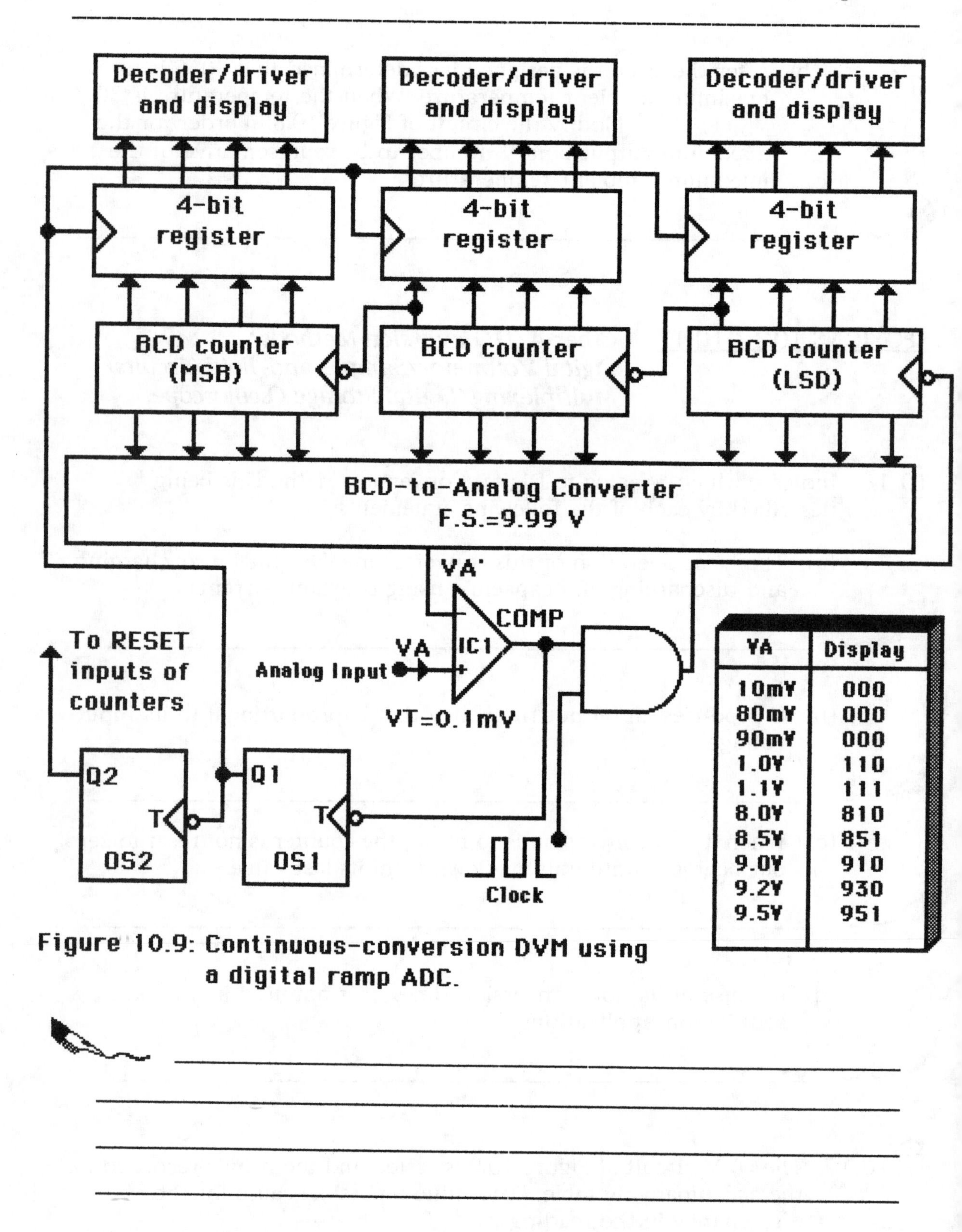

VA	Display
10mV	000
80mV	000
90mV	000
1.0V	110
1.1V	111
8.0V	810
8.5V	851
9.0V	910
9.2V	930
9.5V	951

Figure 10.9: Continuous-conversion DVM using a digital ramp ADC.

The circuit of Figure 10.10 uses a computer to input data to a DAC and to control which of the three transmission gates (IC1A-IC1C) is active. The output of each sample-and-hold circuit is connected to the input of the positioning controllers. The circuit of Figure 10.10 saves two DACs by using a multiplexing technique.

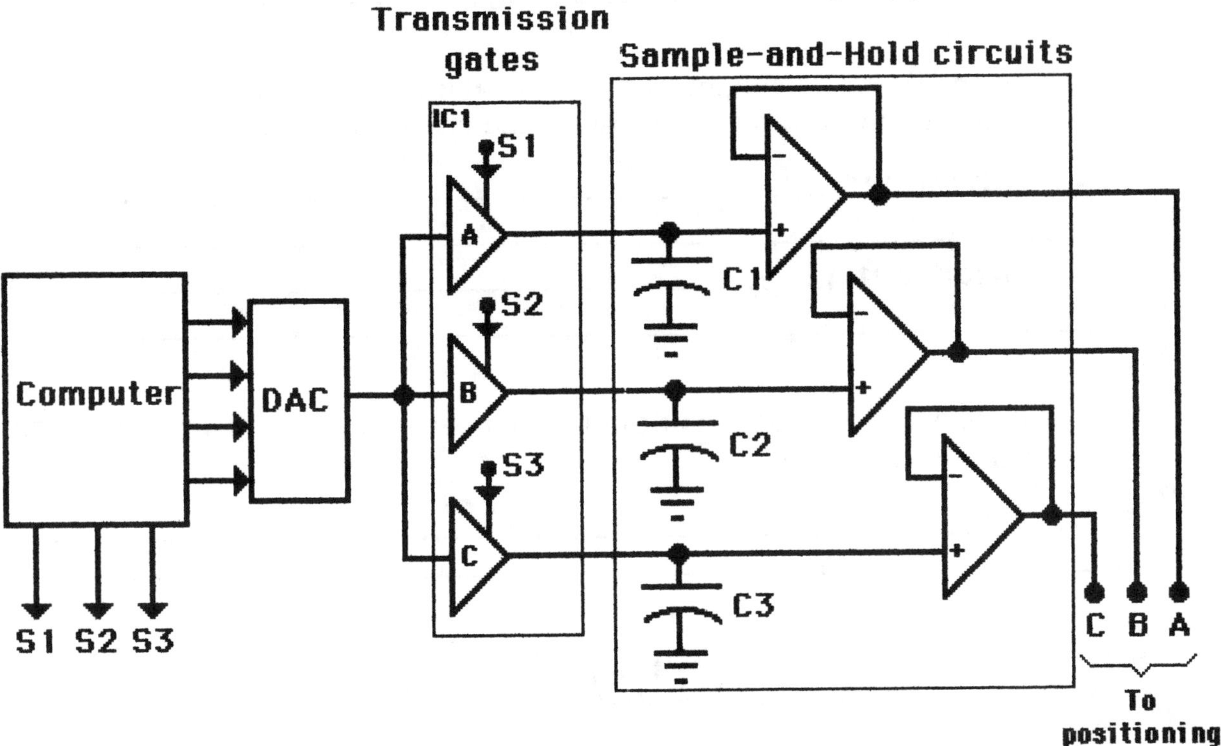

Figure 10.10: DAC used in a multiplexing application.

10.18 Assuming that the computer updates the positioning controllers continuously, describe the complete operation of the circuit including the proper sequencing of events.

10.19 What are the two major components in this circuit that limit the maximum speed at which data can be sent to each of the positioning controllers?

1. ______________________________
2. ______________________________

10.20 Cite the major advantage and disadvantage of using three DACs instead of multiplexing one DAC in the circuit of Figure 10.10?

Advantage: ______________________________

____

Disadvantage: ______________________________

____

10.21 Cite the basic sequence of operations performed in a Digital Storage Oscilloscope (DSO).

____

____

TEST 10

1. A certain 8-bit digital-to-analog converter has a full-scale output of 3mA and a full-scale error of ± 0.5% F.S. What is the range of possible outputs for an input of 10000001_2?

   (a) 1503-1533μA (b) 1518-1533μA (c) 994-1014μA (d) 1500-1530μA

2. A DAC has a 10-bit input and a 10.23V F.S. output. What is the DAC's step-size?

   (a) 100mV (b) 10mV (c) 10% (d) 9.9mV

3. A 12-bit DAC has a step-size of 10mV. Determine the % resolution.

   (a) ≈0.0244% (b) ≈0.1% (c) ≈1.0% (d) ≈2.5%

4. A Digital-to-Analog Converter is said to be monotonic when its output:

   (a) either increases or stays the same as the input is increased.
   (b) either decreases or stays the same as the input is increased.
   (c) either increases or stays the same as the input is decreased.
   (d) either decreases or stays the same as the input is decreased.

5. The amount of time that takes the output of a DAC to go from zero to its full-scale value as the input is changed from all 0s to all 1s is called:

   (a) Settling time (b) Acquisition time (c) Conversion time
   (d) Staircase time.

6. An ADC that employs a VCO to convert an analog voltage to a pulse frequency that is then counted to produce a digital output is called:

   (a) Voltage-to-frequency ADC (b) Dual-slope ADC (c) Flash ADC
   (d) Tracking ADC

7. The resolution of the ADC0804 successive-approximation ADC can be increased by:

   (a) Increasing the voltage going into the $V_{REF}/2$ input.
   (b) Decreasing the voltage going into the $V_{REF}/2$ input.
   (c) Increasing the clock frequency.
   (d) Leaving the $V_{REF}/2$ input open.

8. In a sample-and-hold circuit the acquisition time depends on:

   (a) The value of C_h and the characteristics of the S/H circuit.
   (b) The characteristics of the S/H circuit only.
   (c) The value of C_h only.
   (d) The gain of the amplifier used by the S/H circuit.

9. The weight of the LSB of a DAC is called the step-size.

   (a) [TRUE] (b) [FALSE]

10. The conversion time of a digital-ramp ADC is fixed regardless of the analog voltage V_A, while in a successive-approximation ADC the conversion time increases as the analog voltage V_A increases.

    (a) [TRUE] (b) [FALSE]

11 MEMORY DEVICES

Objectives

Upon completion of this chapter, you will be able to:

- Understand and correctly use the terminology associated with memory systems.
- Describe the difference between read/write memory and read-only memory.
- Discuss the difference between volatile and nonvolatile memory.
- Determine the capacity of a memory device from its inputs and outputs.
- Outline the steps that occur when the CPU reads from or writes to memory.
- Distinguish among the various types of ROMs and cite some common applications.
- Understand and describe the organization and operation of static and dynamic RAMs.
- Combine memory ICs to form larger memory capacities.
- Describe the architecture of the basic types of programmable logic devices and determine the fuse pattern for a given set of logic functions.
- Compare the relative advantages and disadvantages of EPROM, EEPROM, and flash memory.
- Combine memory ICs to form memory modules with larger word size and/or capacity.
- Use the test results on a RAM or ROM system to determine possible faults in the memory system.

Glossary of key terms covered in this chapter:

- **Access Time:** The time between the memory receiving a new input address and the output data becoming available in a read operation. [sec.11.1/11.6]
- **Address:** A number that uniquely identifies the location of a word in memory. [sec.11.1]
- **Address Bus:** A unidirectional bus that carries address bits from CPU to memory. [sec.11.3]
- **Address Multiplexing:** Used in dynamic RAMs to save IC pins, it involves latching the two halves of a complete address into the IC in separate steps. [sec.11.15]
- **Address Pointer Register:** Register that keeps track of where data are to be written and from where they are to be read during a FIFO operation. [sec.11.19]
- **Auxiliary Memory:** Nonvolatile memory used to store massive amounts of information external to the internal memory. [sec.11.1]
- **Bootstrap Program:** The program, stored in ROM, that a computer executes on power-up. [sec.11.9]
- **Burning:** The process of entering data into a ROM. [sec.11.3]
- **Burst Refresh:** When the normal memory operation is suspended, and each row of the DRAM is refreshed in succession until all rows have been refreshed. [sec.11.17]
- **Byte:** An 8-bit word. [sec.11.1]
- **Cache Memory:** Block memory that communicates directly with the CPU at high speed in order to achieve maximum system operation. [sec.11.19]
- **Capacity:** Amount of storage space in a memory expressed as number of bits or number of words. [sec.11.1]
- **$\overline{CAS}$-before-$\overline{RAS}$ Refresh:** A common method for refreshing a DRAM. In this method the $\overline{CAS}$ signal is driven LOW and held there as $\overline{RAS}$ is pulsed LOW. [sec.11.17]

- **Checksum:** A special data word stored in last ROM location. It is derived from the addition of all other data words in the ROM, and is used for error-checking purposes. [sec.11.21]

- **Column Address Strobe (CAS):** Signal used to latch the column address into a DRAM chip. [sec.11.15]

- **Control Bus:** A bus carrying control signals from CPU to memory. [sec.11.3]

- **Data Bus:** A Bidirectional bus that carries data between CPU and memory. [sec.11.3]

- **Data Rate Buffer:** An application of FIFOs where sequential data are written into the FIFO at one rate, and read out at a different rate. [sec.11.19]

- **Density:** Another term for "Capacity". [sec.11.1]

- **Distributed Refresh:** When the row refreshing of the DRAM is interspersed with the normal operations of the memory. [sec.11.17]

- **Downloading:** When the data to be burned into a PROM is obtained from a keyboard, from a disk drive, or transferred from a computer. [sec.11.7]

- **Dynamic Memory Devices:** Semiconductor memory devices in which the stored data will not remain permanently stored even with power applied. [sec.11.1]

- **Dynamic RAM (DRAM):** Type of semiconductor memory that stores data as capacitor charges that need to be refreshed periodically. [sec.11.14]

- **DRAM Controller:** IC used to handle refresh and address multiplexing operations needed by DRAM systems. [sec.11.17]

- **Electrical Erasable Programmable ROM (EEPROM):** A ROM that can be electrically programmed, erased, and reprogrammed. [sec.11.7]

- **Erasable PLDs:** PLD devices that are programmed and erased just like EEPROMs. [sec.11.10]

- **Erasable Programmable ROM (EPROM):** A ROM that can be electrically programmed by the user. It can be erased (usually with ultraviolet light) and reprogrammed as often as desired. [sec.11.7]

- **Fetch:** Another term for "Read operation." [sec.11.1]

- **Field-Programmable Array (FPLA):** Another name used to describe a Programmable Logic Array (PAL). [sec.11.10]

- **Firmware:** Computer programs stored in ROM. [sec.11.9]

- **First-In First-Out Memory (FIFO):** A semiconductor sequential access memory in which data words are read out in the same order that they were written in. [sec.11.19]

- **Flash Memory:** A nonvolatile memory with the EEPROM's in-circuit electrically erasability, but with densities and costs much closer to EPROMs, while retaining the high-speed read access of both. [sec.11.8]

- **Fusible-link:** Fuse links that make up a PROM. The user burns these fuse links to produce the desired stored memory data. [sec.11.7]

- **Internal Memory:** Semiconductor memory that stores instructions and data the CPU is currently working on. [sec.11.1]

- **Mask-Programmed ROM (MROM):** A ROM that is programmed by the manufacturer according to the customer's specifications. It cannot be erased or reprogrammed. [sec.11.7]

- **Mass Storage:** Another term used for "Auxiliary Memory." [sec.11.1]

- **Memory Cell:** A device that stores a single bit. [sec.11.1]

- **Memory Word:** A group of bits in memory that represents instructions or data of some type. [sec.11.1]

- **Nonvolatile Memory:** Memory that will keep storing its information without the need for electrical power. [sec 11.3]

- **Polarity Fuse:** Feature present on many PLDs that gives the user the option to invert any of the device outputs. [sec.11.10]

- **Power-Down Storage:** A special memory function that allows data to be saved in a nonvolatile memory when the system power is shut down. [sec.11.19]

- **Power-Up Self-Test:** A program stored in ROM and executed by the CPU on power-up to test RAM and/or ROM portions of the computer circuitry. [sec.11.20]

- **Programmable Array Logic (PAL):** A type of Programmable Logic Device that contains an array of AND gates and OR gates as with PROMs, but in the PAL the inputs to the AND gates are programmable while the inputs to the OR gates are hard-wired. [sec.11.10]

- **Programmable Logic Array (PLA):** A PLA combines the characteristics of the PROM and the PAL by providing both a programmable OR array and a programmable AND array. [sec.11.10]

- **Programmable Logic Device (PLD):** An IC that contains a large number of interconnected logic functions. The user can program the IC for a specific function by selectively breaking the appropriate interconnections. [sec.11.10]

- **Programmable ROM (PROM):** A ROM that can be electrically programmed by the user. It cannot be erased and reprogrammed. [sec.11.7]

- **Programming:** The process of entering data into a ROM. [sec.11.3]

- **Random Access Memory (RAM):** Memory in which the access time is the same for any location. [sec.11.1]

- **$\overline{RAS}$-Only Refresh:** A common method for refreshing a DRAM. A row address is strobed with $\overline{RAS}$ while $\overline{CAS}$ and $R/\overline{W}$ remain HIGH. [sec.11.17]

- **Read-Only Memory (ROM):** Memory devices that are designed for applications where the ratio of read operations to write operations is very high. [sec.11.1]

- **Read Operation:** A word in a specific memory location is sensed and possibly transferred to another device. [sec.11.1]

- **Read/Write Input:** Input that controls which memory operation is to take place - read (R) or write (W). [sec.11.2]

- **Read/Write Memory (RWM):** Any memory that can be read from and written into with equal ease. [sec.11.1]

- **Refresh Counter:** During the refresh cycle of a DRAM, this counter supplies row addresses to the DRAM address inputs. [sec.11.17]

- **Refreshing:** The process of recharging the cells of a dynamic memory. [sec.11.14]

- **Row Address Strobe (RAS):** Signal used to latch row address into dynamic RAM chip. [sec.11.15]

- **Sequential Access Memory (SAM):** Memory in which the access time will vary depending on where the data is stored. [sec.11.1]

- **Static Memory Devices:** Semiconductor memory devices in which the stored data will remain permanently stored as long as power is applied. [sec.11.1]

- **Static RAM (SRAM):** Semiconductor RAM that stores information in flip- flop cells that do not have to be periodically refreshed. [sec.11.13]

- **Store:** Another term for "Write operation." [sec.11.1]

- **Volatile Memory:** Requires electrical power to keep information stored. [sec.11.1]

- **Working Memory:** Another term for "Internal Memory." [sec.11.1]

- **Write Operation:** A new word is placed into a specific memory location. [sec.11.1]

Problems

SECTIONS 11.1-11.2 *Memory Terminology/General Memory Operation.*

11.1 Complete each of the following statements by filling in the blank:

a) The ____________ of a certain memory device is 1Kx8.
b) The term used to describe an 8-bit word is a __________.
c) Any device which is capable of storing a single bit can be called a ________ ______.
d) A __________ memory is a memory that requires the application of electrical power in order to store information.
e) The amount of time required to perform a read operation is called the _______ time.
f) A __________ memory is a semiconductor memory in which the stored data will remain permanently stored as long as power is applied. Data stored in a ________ memory, on the other hand, does not remain stored even with power applied, unless it is periodically refreshed.
g) __________ memory stores instructions and data the CPU is currently working on.

11.2 Describe the step-by-step procedure necessary to read the contents of memory location 11001_2 of Figure 11.1.

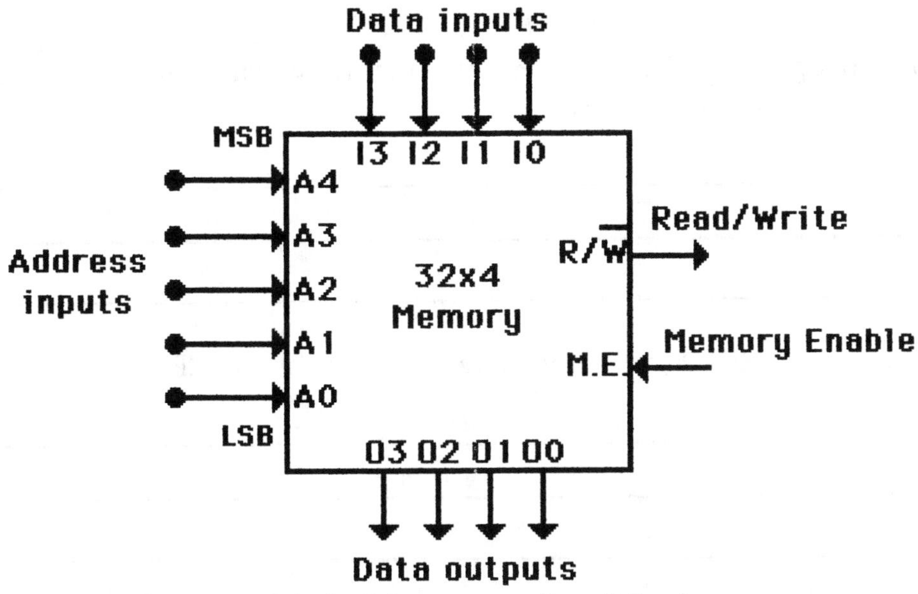

Figure 11.1: Diagram of a 32x4 memory.

11.3 A certain memory device has 10 address lines. How many memory locations does it have?

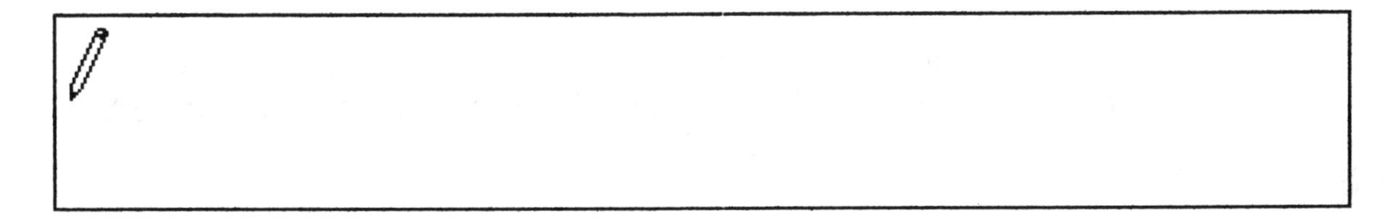

SECTION 11.3 *CPU-Memory Connections*

11.4 The following statements are all wrong. Rewrite each of them in order to make them true.

a) The bus which carries data between the CPU and the memory ICs is unidirectional.

b) During a WRITE operation data may flow from a memory IC into the CPU via the Data bus.

c) During a READ operation data flows out of the CPU via the Data bus.

SECTIONS 11.4-11.6 *Read-Only Memories/ROM Architecture/ ROM Timing.*

11.5 For the following signal conditions ($A_0=1$, $A_1=1$, $A_2=1$, $A_3=0$, CS=1), which one of the ROM's register of Figure 11.2 is sending data to the Output buffers?

11.6 The 16x8 ROM circuit of Figure 11.2 is tested and the technician finds that everything works as expected except that the contents of Register 14 cannot be accessed. Determine what malfunction could cause the circuit to behave this way.

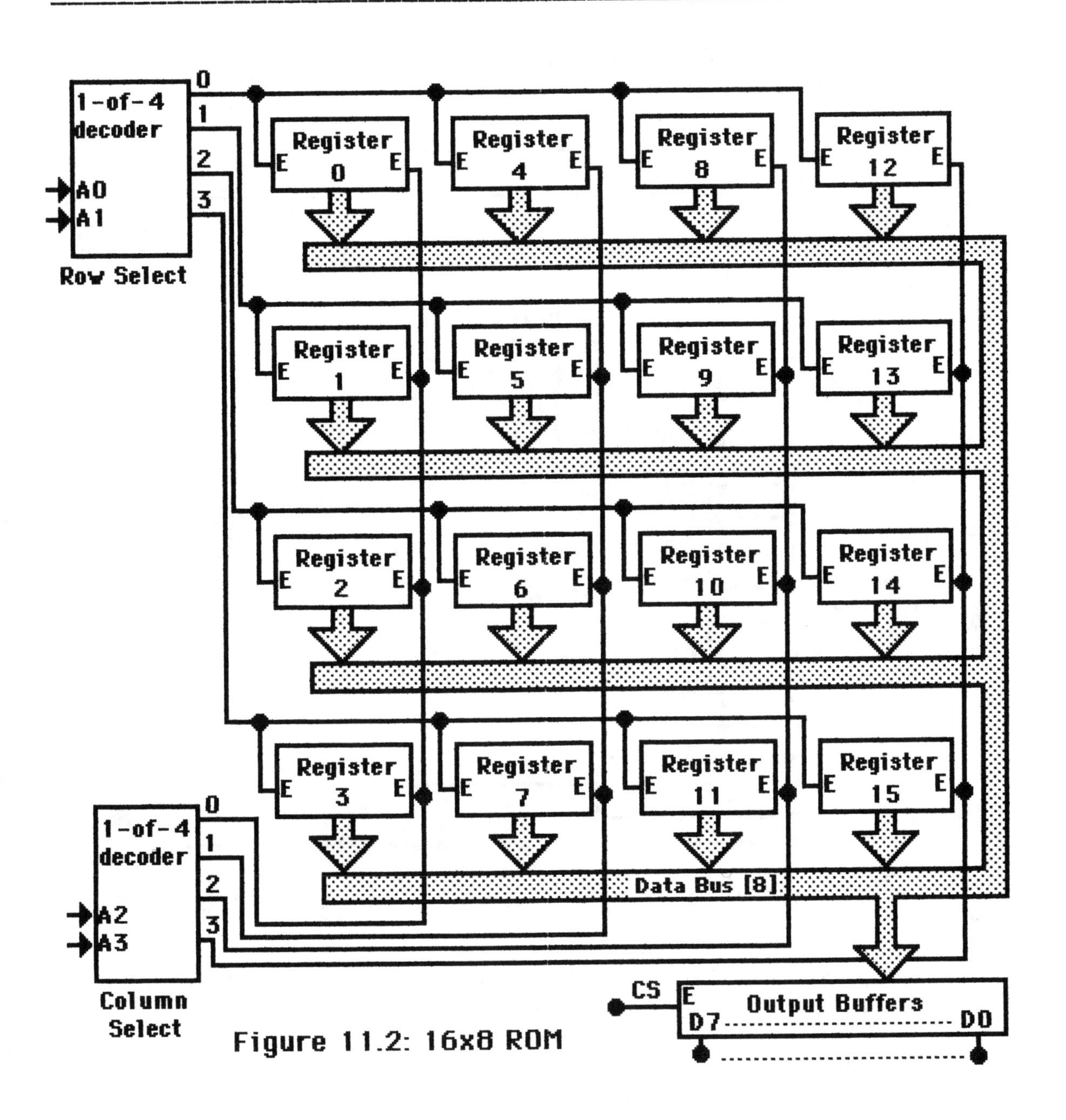

Figure 11.2: 16x8 ROM

11.7 a) Determine the capacity of the memory of Figure 11.3.

b) Is this a ROM or a RAM? Why?

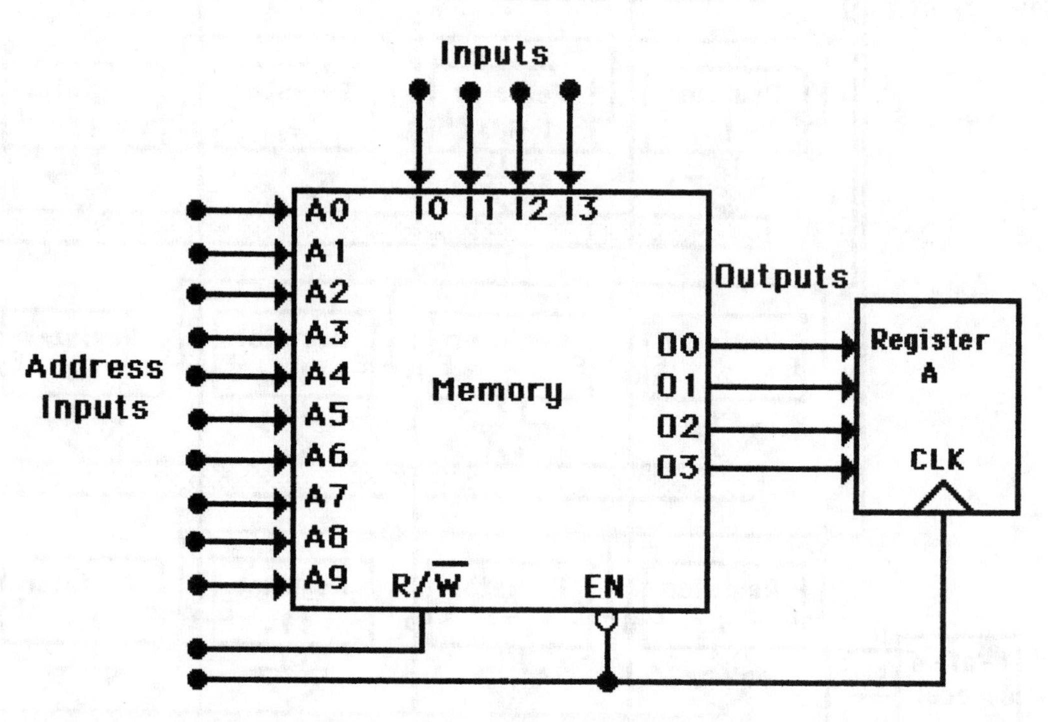

Figure 11.3: Diagram of a memory connected to a register.

11.8 The memory of Figure 11.3 has a t_{OE} requirement of 150ns. In this circuit, memory location 1111100000_2 is loaded with data 1111_2. While the signals on the address bus are kept stable at 1111100000_2, the Enable signal goes LOW for 100ns. Predict what data would be clocked into Register A.

SECTIONS 11.7-11.10 *Types of ROMs/Flash Memory/ROM Applications/ Programmable Logic Devices.*

11.9 Determine the Boolean expressions for the programmed output functions (O_0-O_3) of the PROM shown in Figure 11.4.

11.10 Program the unprogrammed PAL of Figure 11.5 by removing the appropriate Xs, so that its output logic functions are equivalent to those of the PROM of Figure 11.4.

11.11 True or False:

a) EPROMs are examples of volatile memories. [T] , [F]
b) EEPROMs and Flash memories can be easily erased and/or programmed without removing them from the circuit. [T] , [F]
c) MROMs should be used in circuit designs where few units are required to be built. [T] , [F]
d) Fusible-link PROMs operate in the same manner as MROMs. [T] , [F]
e) A PLA combines the characteristics of the PROM and the PAL. [T] , [F]
f) Polarity fuses gives the designer the option of inverting any of the PLD outputs. [T] , [F]
g) PALs are often referred to as FPLAs. [T] , [F]

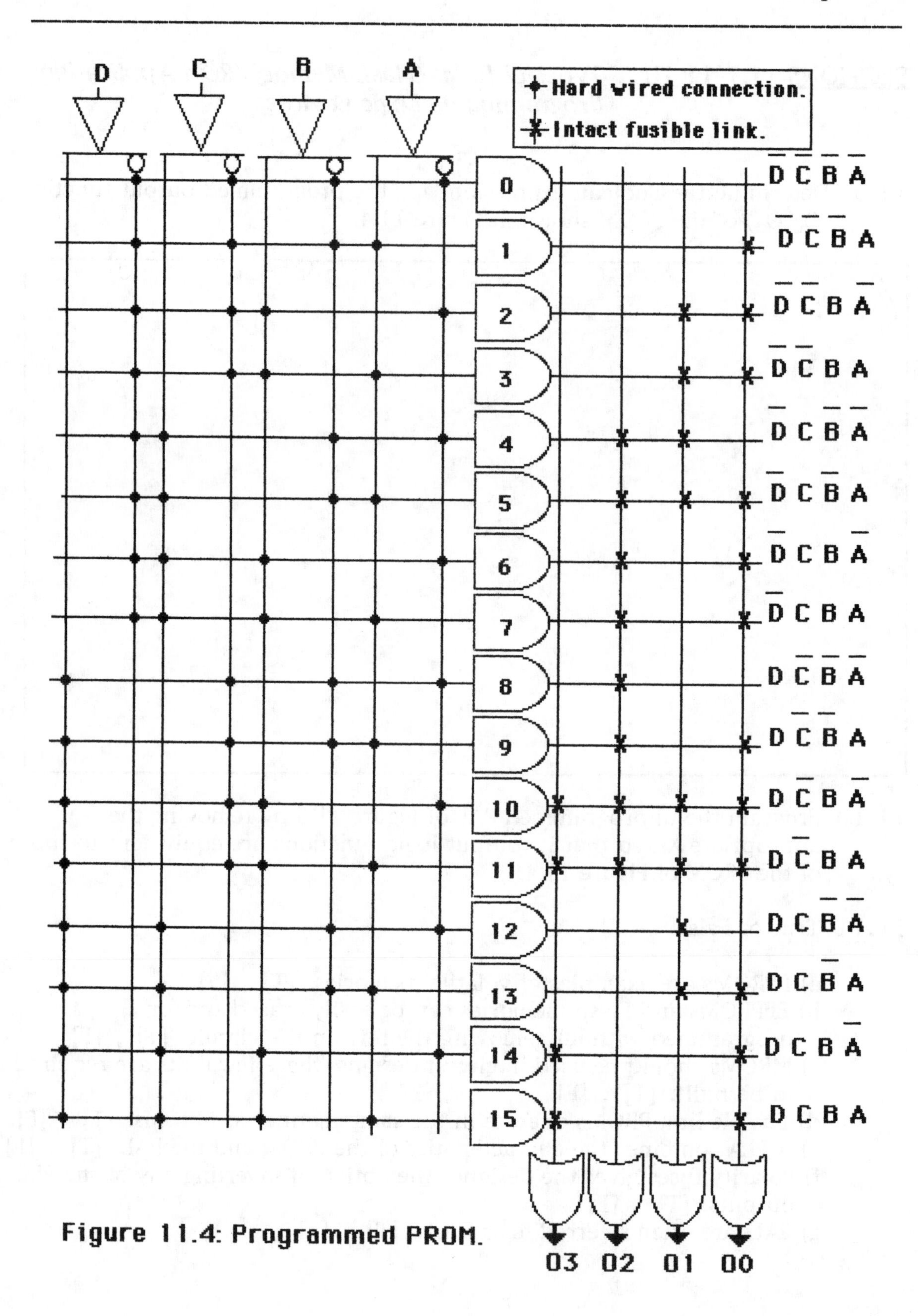

Figure 11.4: Programmed PROM.

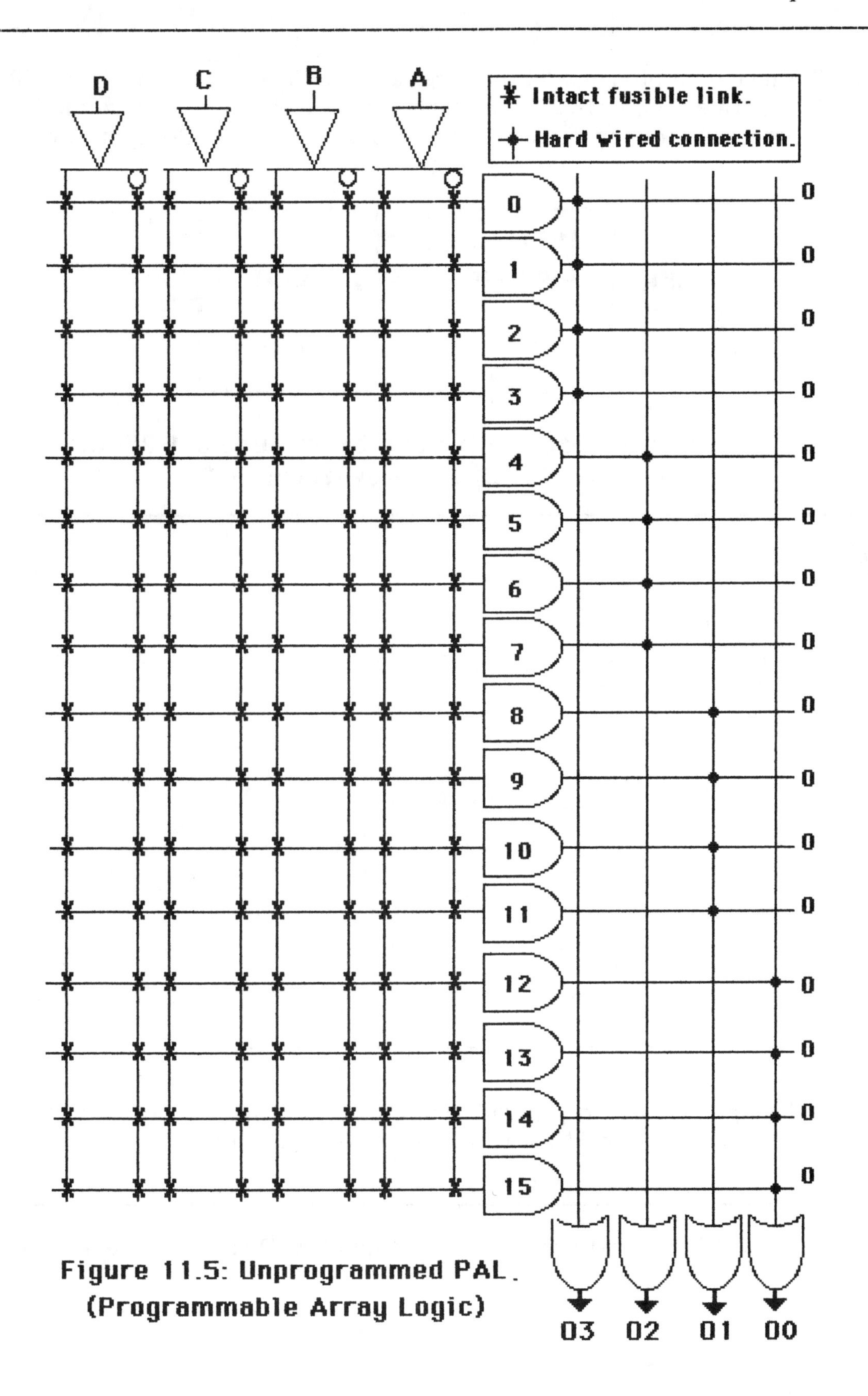

Figure 11.5: Unprogrammed PAL.
(Programmable Array Logic)

11.12 The circuit of Figure 11.6 utilizes a MOD-16 counter and a 16x1 PROM to generate a 1.25 KHz, 37.5% Duty-Cycle waveform at X. Show the PROM programming table necessary to achieve the desired output waveform X.

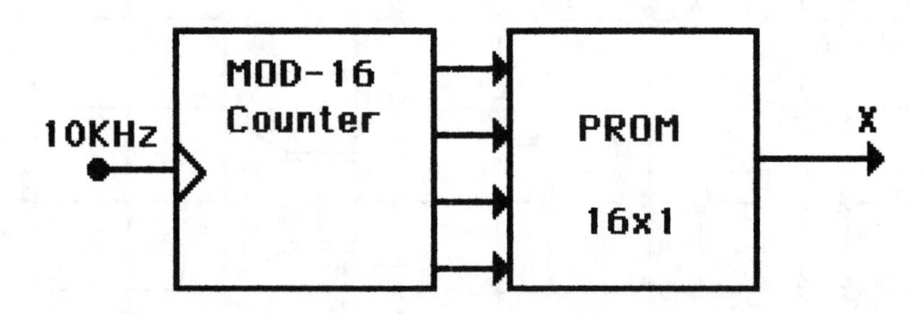

Figure 11.6: Circuit used to produce a 1.25KHz, 37.5% d.c. waveform at X.

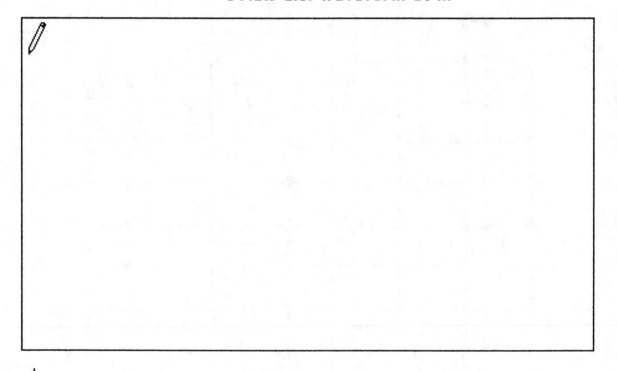

11.13 (a) How will the operation of the PAL of Figure 11.5 be affected if the output of AND gate 0 is shorted to ground?

______

(b) Repeat (a) but with the output of AND gate 0 shorted to Vcc.

______

11.14 A technician executes the following operational steps to a 28F256A CMOS flash memory:

I) Writes the code 20_{16} to the command register.
II) Again, he writes the code 20_{16} to the command register.

a) How was the flash memory affected by the two-step procedure?

____

b) What data would you expect to find in memory location 0300_{16} after the two step procedure?

____

SECTIONS 11.13-11.17 ***Static RAM (SRAM)/Dynamic RAM (DRAM)/ Dynamic RAM Structure and Operation/ DRAM Read/Write Cycles/DRAM Refreshing***

11.15 Fill in the missing word/s in order to make the statement true:

a) SRAM memory cells are essentially ____________ that will stay in a given state indefinitely, provided that power to the circuit is not interrupted.
b) If the t_{RC} of a SRAM is 50ns, the CPU can read ____ million words per second.
c) Dynamic RAMs require periodic recharging of the memory cells; this is called ______________the memory.
d) In order to reduce the number of pins on high-capacity DRAM chips, manufacturers utilize address ______________.
e) During a read cycle the $\overline{RAS}$ signal is activated __________ the $\overline{CAS}$ signal.
f) $\overline{RAS}$-only refresh method is performed by strobing in a row address with _____ while _____ and _____ remain HIGH.

SECTIONS 11.18-11.19 *Expanding Word Size and Capacity/ Special Memory Functions*

11.16 Design a 4Kx8 memory module using 1Kx4 RAM chips.

11.17 How many 2125A RAM ICs would be necessary to build the 4Kx8 memory module of problem 11.16?

11.18 While testing the circuit of Figure 11.7, the technician finds that regardless of the address on the Address bus, the data on the Data bus are always in the Hi-Z state. Determine a possible cause for the malfunction.

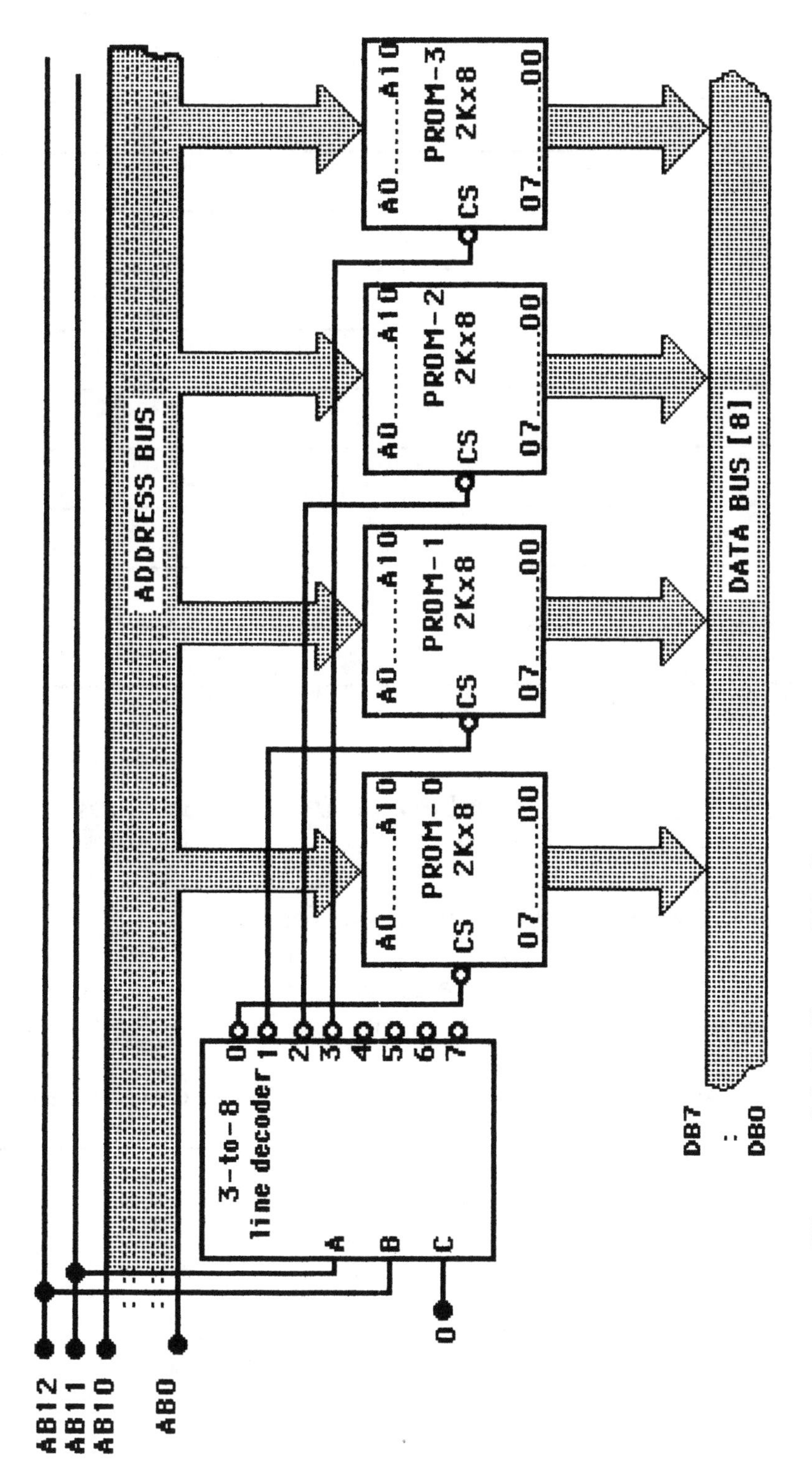

Figure 11.7: Four 2Kx8 arranged to form a total capacity of 8Kx8.

11.19 Give three different ways to prevent the loss of memory data during a system power failure.

SECTIONS 11.20-11.21 *Troubleshooting RAM Systems/Testing ROM.*

11.20 Modify the circuit of Figure 11.8 as follows:

1) Disconnect A15 from the OR gate and $\overline{E3}$ from Vcc.
2) Connect A15 to $\overline{E3}$ and the unused input of the OR gate to Ground.

Determine the complete new range of addresses for each RAM module?

11.21 The operation of the circuit of Figure 11.8 is checked and it is determined that the CPU is placing 'valid' data on the data bus to be written into memory. After data are written into a certain memory location, the technician is able to verify the data by reading them back from that memory location. However, the contents of other memory locations have been changed with completely different and unpredictable data. Determine what could cause the malfunction and how you would troubleshoot the circuit in order to verify your prediction.

11.22 The table below represents the contents of a 16x8 ROM. Determine the checksum word that should be stored in memory location 1111_2.

Address	Data
0000	00110110
0001	11001100
0010	11000001
0011	00111001
0100	11111111
0101	00000000
0110	01110011
0111	10011001
1000	00010001
1001	01011111
1010	11000011
1011	10010010
1100	00111101
1101	11100000
1110	10000000
1111	[ ]

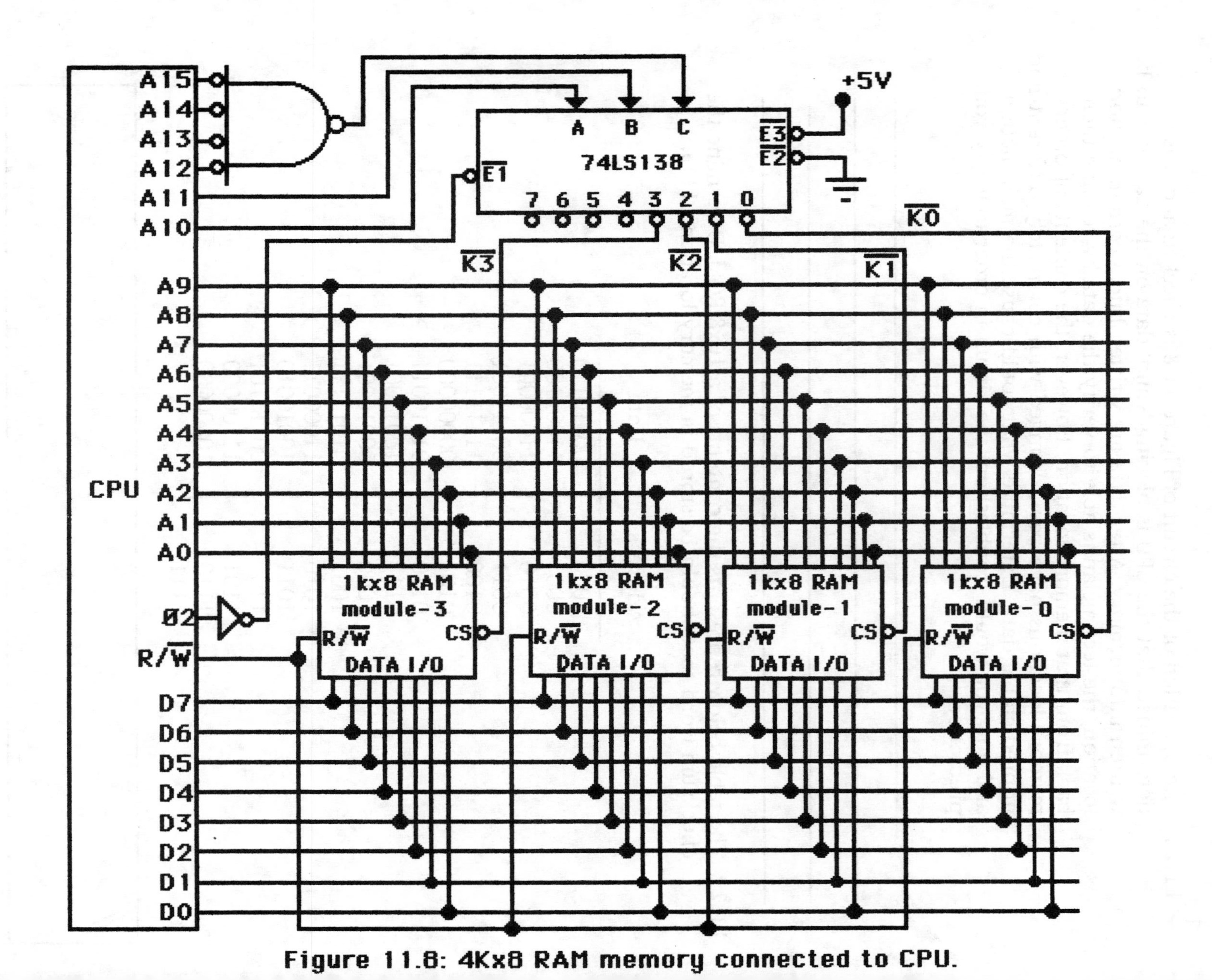

Figure 11.8: 4Kx8 RAM memory connected to CPU.

TEST 11

1. How many address lines would be required for a 2Kx8 memory?

   (a) 9 (b) 10 (c) 11 (d) 12

2. This kind of memory is an in-circuit, electrically erasable, byte-by-byte memory.

   (a) EPROM (b) MROM (c) Flash (d) EEPROM

3. How many bits of data can be stored in a memory with a capacity of 128K?

   (a) 131,072 (b) 262,144 (c) 1,048,576 (d) 128,000

4. A ROM that stores programming instructions is often referred to as:

   (a) hardware (b) software (c) firmware (d) None of the above

5. Which of the following has a programmable AND array and a hard-wired OR array?

   (a) PROM (b) PAL (c) PLA (d) FPLA

6. Which type of memory is associated with the "Command Register"?

   (a) PROM (b) EPROM (c) EEPROM (d) Flash

7. This type of memory can store data as long as power is applied to the chip.

   (a) DRAM (b) EEPROM (c) SRAM (d) FIFO

8. How many 1Kx4 memory chips are needed to build an 8Kx16 memory module?

   (a) 64 (b) 32 (c) 16 (d) 8

9. The most common method for refreshing a DRAM is the "$\overline{CAS}$-before-$\overline{RAS}$."

   (a) [TRUE] (b) [FALSE]

10. A ROM IC chip normally has an $R/\overline{W}$ control line.

    (a) [TRUE] (b) [FALSE]

12 APPLICATIONS OF A PROGRAMMABLE LOGIC DEVICE

Objectives

Upon completion of this chapter, you will be able to:

- identify applications of programmable logic devices.
- Understand the hardware architecture of a typical programmable logic device.
- Use software tools capable of programming a large variety of programmable devices.
- Completely implement a logic design using programmable logic devices.

Glossary of key terms covered in this chapter:

- **ASICS (Application Specific Integrated Circuits):** Often used to describe FPGAs as well as other non-programmable devices that are developed for a specific application. [sec.12.5]
- **CULP (Universal Compiler for Programmable Logic:** One of the most popular high-level compilers available today for the development of PLDs. [sec.12.4]
- **Downloading:** When the programming file is sent over a cable to the programming device. [sec.12.2]
- **FMUX (Feedback Multiplexer):** A multiplexer within an Output Logic Macro Cell (OLMC) that selects the logic signal that is fed back to the input matrix of a GAL16V8A. [sec.12.1]
- **FPGA (Field Programmable Gate Arrays):** An array of PLD-like cells arranged on a single IC chip. [sec.12.5]
- **Fuse Plot:** A file that is like a map that shows which fuses in a programmable device are to be fused open and which ones are to remain intact. [sec.12.2]
- **JEDEC:** Joint Electronic Device Engineering Council. [sec.12.2]
- **JEDEC File:** A standardized file which is loaded into any JEDEC-compatible PLD programmer that is capable of programming the desired type of PLD. [sec.12.2]
- **OLMC (Output Logic Macro Cells):** Vital output logic circuitry of a GAL 16V8A device. Products (outputs of AND gates) are applied as inputs to the OLMC and then these products are ORed together within the OLMC to generate the sum-of-products. [sec.12.1]
- **OMUX (Output Multiplexer):** Two input multiplexer within an Output Logic Macro Cell (OLMC). [sec.12.1]
- **PTMUX (Product Term Multiplexer):** Multiplexer within an Output Logic Macro Cell (OLMC). [sec.12.1]
- **Simulator:** A computer program that calculates the correct output logic states based on the description of the logic circuit and the current inputs. [sec.12.4]

- **SOP (Sum-of-Products):** An OLMC will OR together eight different products in order to generate the SOP. [sec.12.1]

- **TSMUX (Tri-State Multiplexer):** Four input tri-state multiplexer within an Output Logic Macro Cell (OLMC). [sec.12.1]

- **ZIF (Zero Insertion Force):** A special IC socket where the IC chip is dropped in the socket and then the contacts are manually clamped to the IC pins. [sec.12.2]

Problems

SECTIONS 12.1-12.2 *The GAL16V8A/Programming PLDs*

12.1 Complete the following statements by filling in the blank spaces.

(a) The major components of the GAL16V8A devices are the input term _______, the ______ gates, and the _________ _________ _________ _________.

(b) Within each OLMC the products are ______ together to generate the SOP.

(c) The GAL16V8A has three different modes of operation. They are the _______ mode, the ________ mode, and the __________ mode.

12.2 What must be the conditions of signals SYN, AC0, and AC1 in order to program the GAL16V8A in *complex mode*?

____

12.3 Refer to the Simple Combinational Logic Implementation of Figure 12.1. Determine the SOP expression present at output Z of the OLMC 19.

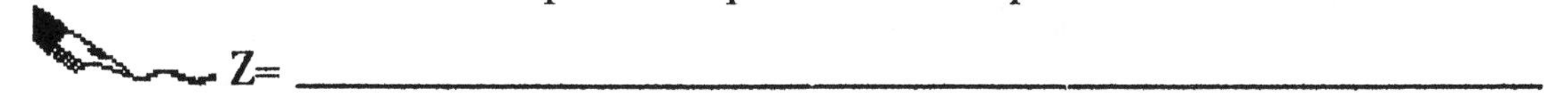

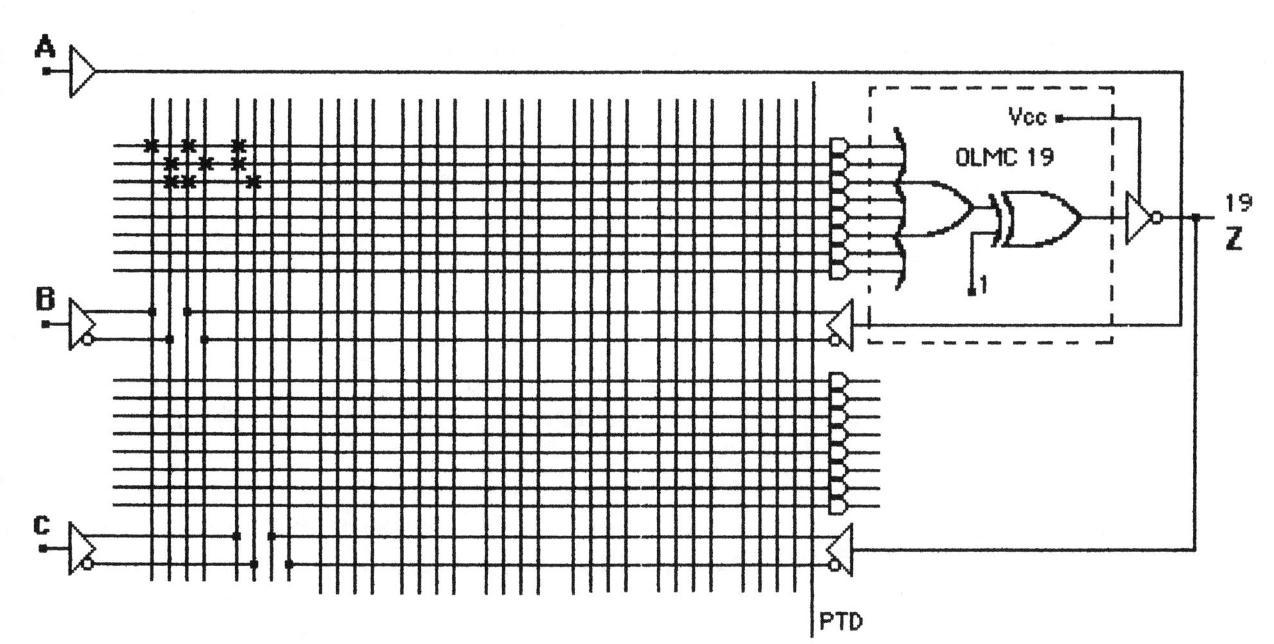

Figure 12.1: Simple Combinational Logic Implementation.

12.4 Complete the table below by determining the proper logic levels for the signals *SYN*, *AC0* and *AC1*.

Mode	Configuration	SYN	AC0	AC1
Simple mode	*IN*			
	OUT			
Complex mode				
Registered mode	*Registered*			
	Combinational			

12.5 State the five essential pieces of equipment that are necessary to design and build circuits using PLDs.

1. ____________________
2. ____________________
3. ____________________
4. ____________________
5. ____________________

12.6 Complete the following statements:

(a) A __________ __________ is a file that is like a map that shows which fuses in a programmable device are to be fused open and which ones are to remain intact.

(b) ____________________ is a term used to describe the action of sending a programming file over a cable to the programming device.

(c) A __________ is standardized file which is loaded into any __________-compatible PLD programmer that is capable of programming the desired type of PLD.

SECTIONS 12.3-12.4 *Development Software/Universal Compiler for Programmable Logic (CUPL).*

12.7 Which are the two main software package categories that automatically create the JEDEC file for a specified device?

1. ______________________________
2. ______________________________

12.8 True or False:

(a) Low-level development systems will accept the Boolean equations in an ASCII input file.
(b) High-level compilers allow the entry of a logic circuit schematic diagram by using a CAD software package.
(c) High-level compilers allow you to enter your design in the form of truth tables or state tables.

12.9 Translate each of the following logic functions into CUPL format statements:

(a) AB ------> ______________
(b) $A+B$ ------> ______________
(c) $\overline{A}$ ------> ______________
(d) $A \oplus B$ ------> ______________
(e) $\overline{A \oplus B}$ ------> ______________

12.10 A compiler generates a documentation file that shows the reduced expression: X = !A # !C & D # A $ C # !B & D. Express the simplified expression using the conventional logic format.

12.11 Write the source file that would program a GAL16V8A to function as a 1-of-4 decoder.

Source File

12.12 Write the equations in CUPL format that will implement the synchronous counter of problem 7.14(a) *(See page 125)*.

Source File

TEST 12

1. Which configuration of the GAL16V8A is capable of getting eight product terms into a SOP expression to generate a combinational output.

   (a) Simple mode/dedicated input.
   (b) Simple mode/dedicated combinational output.
   (c) Complex mode
   (d) Registered output.

2. A certain design uses some basic gates, a D latch, and has a tri-state output. Which mode of operation should the GAL16V8A be programmed for?

   (a) Simple
   (b) Complex
   (c) Registered
   (d) Any of the above

3. The ! sign is used by the CUPL in order to denote which operation?

   (a) XOR (b) OR (c) NOT (d) AND

4. The XOR gates present at the output of some of the OLMCs are used:

   (a) as a SYN signal.
   (b) to program the output polarity.
   (c) to allow the I/O pin to be used a an input.
   (d) as a feedback signal to the input matrix.

5. The following equation when expressed in words means:
   The D input of the QA flip-flop is connected to the complement of the QA output.

   (a) $QA\$D = \overline{QA}$ (b) QA#D = !QA (c) $QA.D = \overline{QA}$ (d) QA.D = !QA

6. The following are used to enclose any comments that might make the input file more understandable:

   (a) "/!" and "!/" (b) "/*" and "*/" (c) "/#" and "#/" (d) "/$" and "$/"

7. The following device calculates the correct output logic states based on the description of the logic circuit and the current inputs:

   (a) Simulator (b) OLMC (c) JEDEC (d) Fuse plot

8. The source file is where:

   (a) Compiling errors are corrected.
   (b) Compiling errors are identified.
   (c) The relationship between inputs and outputs are specified.
   (d) (a) and (c)

9. The GAL16V8A has JK flip-flops in some of the OLMCs.

   (a) True (b) False

10. Low-level development systems as well as high-level logic compilers allow you to enter your design in the form of truth tables or state tables.

    (a) True (b) False

13 INTRODUCTION TO THE MICROPROCESSOR AND MICROCOMPUTER

Objectives

Upon completion of this chapter, you will be able to:

- Describe the function and operation of each one of the five basic elements of any computer organization.
- Understand that a computer continually repeats a sequence of fetch and execute operations.
- Understand the difference between a microprocessor and microcomputer.
- Analyze the fetch and execute cycles during the execution of a machine-language program.
- Understand the operational role of the different types of buses and their signals in a microcomputer.
- Cite the major functions performed by the microcomputer.
- Describe the different types of computer words.
- Upgrade your knowledge by learning more details and advanced concepts of microprocessor-based systems.

Glossary of key terms covered in this chapter:

- **Accumulator:** A register in the Arithmetic Logic Unit (ALU). [sec.13.7]
- **Address Bus:** Unidirectional lines that carry the address code from the CPU to memory and I/O devices. [sec.13.9]
- **Arithmetic Logic Unit (ALU):** A digital circuit used in computers to perform various arithmetic and logic operations. [sec.13.4/13.16]
- **BASIC:** A High-Level language (Beginner's All-purpose Symbolic Instruction Code). [sec.13.8]
- **Byte:** A group of 8 bits. [sec.13.6]
- **Central Processing Unit (CPU):** That part of a computer which is composed of the Arithmetic Logic Unit (ALU) and the Control Unit. [sec.13.4]
- **Conditional Branch:** See Conditional Jump. [sec.13.3]
- **Conditional Jump:** Computer decision-making instruction which causes the computer to JUMP to some place in memory regardless of any other condition. [sec.13.3]
- **Control Bus:** A set of signal lines that are used to synchronize the activities of the CPU and the separate μC elements. [sec.13.9]
- **Control Unit:** That part of a computer which provides decoding of program instructions and the necessary timing and control signals for the execution of such instructions. [sec.13.4]
- **Data Bus:** Bidirectional lines that carry data between the CPU and memory, or between the CPU and I/O devices. [sec.13.9]
- **Execute Cycle:** The period during which a computer's Control Unit performs the operation specified by the fetched op code. [sec.13.8]
- **Fetch Cycle:** The period during which a computer's Control Unit obtains instruction codes from memory. [sec.13.8]
- **High-Level Languages:** Computer programming languages which utilize the English language in order to facilitate the writing of a computer program. [sec.13.8]

- **High-Order Byte:** The 8 higher-order bits of a 16-bit data word. [sec.13.7]
- **Input Unit:** That part of a computer which facilitates the feeding of information into the computer's memory unit or ALU. [sec.13.4]
- **Interfacing:** The joining of dissimilar devices in such a way that they are able to function in a compatible and coordinated manner. [sec.13.4]
- **Low-Order Byte:** The 8 lower-order bits of a 16-bit data word. [sec.13.7]
- **Machine Cycle:** Each read or write operation performed by the CPU is often referred to as a "machine cycle." [sec.13.9]
- **Machine-Language:** Computer programming language in which groups of 1s and 0s are used to represent instructions. Machine-Language is also the only language a computer understands. [sec.13.8]
- **Memory:** Place within the architecture of a computer where programming instructions reside. [sec.13.3]
- **Memory Unit:** Part of a computer which stores instructions and data received from the Input unit, as well as results from the Arithmetic Logic Unit. [sec.13.4]
- **Microprocessor (MPU):** The LSI computer chip which contains the Central Processing Unit (CPU). [sec.13.5]
- **Mnemonic:** An abbreviation which represents the op code of a computer instruction. [sec.13.8]
- **Multibyte Instruction:** Computer instruction which in order to be represented requires more than one one-byte instruction. [sec.13.7]
- **Op Code:** That part of a computer instruction that defines what type of operation the computer is to execute on specified data. [sec.13.7]
- **Operand:** Data that are operated on by the computer as it executes a program. [sec.13.3]
- **Operand Address:** Address in memory where operand is currently stored or is to be stored. [sec.13.3/13.7]
- **Operational Code:** See Op-code.

- **Output Unit:** That part of a computer which receives data from the memory unit or ALU and presents it to the outside world. [sec.13.4]
- **Peripherals:** Devices that make up the input and output units of a computer. [sec.13.4]
- **Program:** A sequence of binary-coded instructions designed to accomplish a particular task by a computer. [sec.13.1]
- **Program Counter (PC):** CPU register where the address of the next instruction to be fetched is stored. [sec.13.8]
- **RAM:** One or more LSI memory chips which are used to store programs and data that will change often during the course of computer operation. [sec.13.5]
- **ROM:** One or more LSI memory chips which are used to store programs and data that do not change during the course of computer operation. [sec.13.5]
- **Word:** A group of binary bits which is the primary unit of information in a computer. [sec.13.6]
- **Word Size:** The number of bits that make up a word. [sec.13.6]

Problems

SECTIONS 13.4-13.6 *Basic Computer System Organization/ Basic μC Elements/Computer Words.*

13.1 Match the appropriate statement preceded by a letter with its corresponding term preceded by a number:

1 - CPU	2 - Peripherals
3 - RAM and/or ROM	4 - Output Unit
5 - Byte	6 - Control Unit
7 - ALU	8 - Memory Unit
9 - Magnetic-Strip	10 - Interfacing
11 - Word.	

a) Where all of the arithmetic and logic operations are performed. [ ]
b) One of the computer units where instructions are stored. [ ]
c) Device that could be used to enter data into the computer's ALU. [ ]
d) This is where the computer sends its results to be viewed by the outside world. [ ]
e) Directs the operation of all the other units in a computer. [ ]
f) Where the ALU and Control Unit reside. [ ]
g) A group of 8-bits. [ ]
h) Devices that make up the input and output units. [ ]
i) Principal unit of information in a computer. [ ]

SECTIONS 13.7-13.8 *Instruction Words/Executing a Machine-Language Program.*

13.2 How many bytes of code are needed to instruct an 8-bit μC to CLA?.

____

13.3 In an 8-bit microcomputer, what does a single-byte instruction represent?

____

13.4 In a three-byte instruction, what do the last two bytes of the instruction represent?

____

13.5 What is the only language that a computer understands?

____

13.6 Complete the following statements by filling the blank spaces:

a) LDA is the mnemonic which instructs the computer to ________ __________.

b) STA $0300 instructs the computer to take the contents of the ________ and store them in memory location ________.

c) The Program Counter is a counter within the ________ Unit.

d) The Computer is always in one of two kinds of operating cycles: ________ cycle or _______ cycle.

13.7 For the following 8085 program, determine what is stored in address location $0400 at the end of the program's execution.

Address	Memory word	Mnemonic
0000	3A	LDA $0500
0001	00	
0002	05	
0003	C3	JMP $0300
0004	00	
0005	03	
:	:	
0300	32	STA $0400
0301	00	
0302	04	
0303	76	HLT
:	:	
0400	[ ]	
:	:	
0500	FF	

SECTIONS 13.9 *Typical μC Structure*

13.8 What determines the direction of data flow on the Data bus?

____

13.9 How many T states (clock cycles) are needed to execute the instruction STA $0400 if we are using a 8085μP based computer?

____

13.10 How many READ and WRITE operations must a computer perform as it executes the program of problem 13.7?

TEST 13

1. The first byte of a three byte instruction is called:

   (a) Operand address (b) Data word (c) Op code (d) LOW-order byte

2. Which two units are combined into the central processing unit CPU?

   (a) Input and output units. (b) Memory and ALU.
   (c) Control and Memory units. (d) Control unit and ALU.

3. The terms read and write always refer to operations performed by the:

   (a) Memory unit (b) ALU (c) CPU (d) Modem

4. The only language a computer understands is:

   (a) BASIC (b) PASCAL (c) Machine (d) High-Level languages

5. Each read or write operation performed by the CPU is referred to as:

   (a) Machine cycle (b) Delay cycle (c) Execute cycle (d) Fetch cycle

6. Which of the following buses is Bidirectional?

   (a) Address bus (b) Control bus (c) I/O bus (d) Data bus

7. The logical operation A+B is performed by:

   (a) Memory (b) I/O device (c) Control unit (d) None of the above

8. RAM and ROM do not have interrupting capability.

   (a) [TRUE] (b) [FALSE]

9. The ALU is a register in the Control Unit.

   (a) [TRUE] (b) [FALSE]

10. Instruction CLRA instructs the computer to clear all internal registers.

    (a) [TRUE] (b) [FALSE]

1 INTRODUCTORY CONCEPTS

SECTION 1.1 *Numerical Representations*

1.1 (a) ANALOG is the continuous representation of a quantity, while DIGITAL is the discrete representation of a quantity.

(b) Systems in which both analog and digital quantities are manipulated are called HYBRID.

SECTION 1.3 *Digital Number Systems*

1.2 (a) $10110_2 = (\underline{1} \times 2^4) + (\underline{0} \times 2^3) + (\underline{1} \times 2^2) + (\underline{1} \times 2^1) + (\underline{0} \times 2^0) = 22_{10}$

(b) $11101_2 = (\underline{1} \times 2^4) + (\underline{1} \times 2^3) + (\underline{1} \times 2^2) + (\underline{0} \times 2^1) + (\underline{1} \times 2^0) = 29_{10}$

(c) $11011110110010_2 = (\underline{1} \times 2^{13}) + (\underline{1} \times 2^{12}) + (\underline{0} \times 2^{11}) + (\underline{1} \times 2^{10}) +$
$+ (\underline{1} \times 2^9) + (\underline{1} \times 2^8) + (\underline{1} \times 2^7) + (\underline{0} \times 2^6) + (\underline{1} \times 2^5) + (\underline{1} \times 2^4) +$
$+ (\underline{0} \times 2^3) + (\underline{0} \times 2^2) + (\underline{1} \times 2^1) + (\underline{0} \times 2^0) = 14258_{10}$

(d) $1101.1101_2 = (\underline{1} \times 2^3) + (\underline{1} \times 2^2) + (\underline{0} \times 2^1) + (\underline{1} \times 2^0) + (\underline{1} \times 2^{-4}) +$
$+ (\underline{1} \times 2^{-3}) + (\underline{0} \times 2^{-2}) + (\underline{1} \times 2^{-1}) = 13.8125_{10}$

(e) $0.111011101_2 = (\underline{1} \times 2^{-1}) + (\underline{1} \times 2^{-2}) + (\underline{1} \times 2^{-3}) + (\underline{0} \times 2^{-4}) + (\underline{1} \times 2^{-5}) +$
$+ (\underline{1} \times 2^{-6}) + (\underline{1} \times 2^{-7}) + (\underline{0} \times 2^{-8}) + (\underline{1} \times 2^{-9}) = 0.93164_{10}$

1.3 $2^N - 1 = 2^{12} - 1 = 4095_{10} = 111111111111_2$

1.4 $2^9 = 512_{10}$ and $2^{10} = 1024_{10}$. Therefore, we need 10 binary bits.

1.5 Counts 0100_2, 1010_2 and 1110_2 are missing from the expected descending sequence.

SECTION 1.6 *Parallel and Serial Transmission*

1.6 00110_2

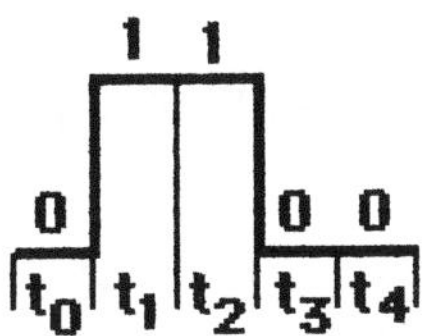

In this example the LSB (t_0) is the first bit to be transmitted.

1.7 Since 5 bits have to be transmitted and each bit takes 10ms to transmit, the complete serial transmission will take 50ms.

1.8 In the circuit of Figure 1.1(b) all bits are transmitted simultaneously (parallel transmission), compared with one bit at a time for the circuit of Figure 1.1(a) (serial transmission). Thus, the circuit of Figure 1.1(b) is five times faster than the circuit of Figure 1.1(a).

1.9 00110_2

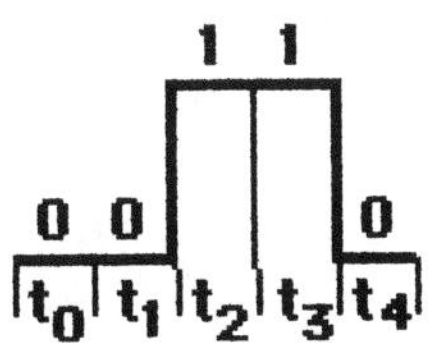

In this example the MSB (t_0) is the first bit to be transmitted.

1.10 In parallel transmissions all bits are transmitted at once. Therefore, the number of lines required between the transmitter and the receiver must be equal to the number of bits being transmitted. Thus, ten lines are necessary to transmit 1100011010_2 using the parallel transmition method.

SECTION 1.8 *Digital Computers*

1.11 1 - Input Unit 2 - Memory Unit 3 - Arithmetic Logic Unit
4 - Output Unit 5 - Control Unit

1.12 Supercomputer.

2 NUMBER SYSTEMS AND CODES

SECTION 2.1 *Binary-to-Decimal Conversions*

2.1 (a) $1101_2 = (\underline{1} \times 2^3) + (\underline{1} \times 2^2) + (\underline{0} \times 2^1) + (\underline{1} \times 2^0) = 13_{10}$

(b) $1010_2 = (\underline{1} \times 2^3) + (\underline{0} \times 2^2) + (\underline{1} \times 2^1) + (\underline{0} \times 2^0) = 10_{10}$

(c) $1100011010_2 = (\underline{1} \times 2^9) + (\underline{1} \times 2^8) + (\underline{0} \times 2^7) + (\underline{0} \times 2^6) + (\underline{0} \times 2^5) + (\underline{1} \times 2^4) + (\underline{1} \times 2^3) + (\underline{0} \times 2^2) + (\underline{1} \times 2^1) + (\underline{0} \times 2^0) = 794_{10}$

(d) $1000001101_2 = (\underline{1} \times 2^9) + (\underline{0} \times 2^8) + (\underline{0} \times 2^7) + (\underline{0} \times 2^6) + (\underline{0} \times 2^5) + (\underline{0} \times 2^4) + (\underline{1} \times 2^3) + (\underline{1} \times 2^2) + (\underline{0} \times 2^1) + (\underline{1} \times 2^0) = 525_{10}$

(e) $11101_2 = (\underline{1} \times 2^4) + (\underline{1} \times 2^3) + (\underline{1} \times 2^2) + (\underline{0} \times 2^1) + (\underline{1} \times 2^0) = 29_{10}$

SECTION 2.2 *Decimal-to-Binary Conversions*

Converting decimal numbers to their binary equivalent can be done by using two different methods:

One way

By adding together the weights of the various positions in the binary number which contain a 1.

2.2 (a) $51_{10} = 32+16+0+0+2+1$
$= (1 \times 2^5) + (1 \times 2^4) + (0 \times 2^3) + (0 \times 2^2) + (1 \times 2^1) + (1 \times 2^0)$
$= 110011_2$

Thus, $51_{10} = 110011_2$

(b) $13_{10} = 8+4+0+1$
$= (1 \times 2^3) + (1 \times 2^2) + (0 \times 2^1) + (1 \times 2^0)$
$= 1101_2$

Thus, $13_{10} = 1101_2$

Another way

By using the Repeated Division method.

(c) $137_{10}/2 = 68_{10}$ + remainder of 1 (LSB)
$68_{10}/2 = 34_{10}$ + remainder of 0
$34_{10}/2 = 17_{10}$ + remainder of 0
$17_{10}/2 = 8_{10}$ + remainder of 1
$8_{10}/2 = 4_{10}$ + remainder of 0
$4_{10}/2 = 2_{10}$ + remainder of 0
$2_{10}/2 = 1_{10}$ + remainder of 0
$1_{10}/2 = 0_{10}$ + remainder of 1 (MSB)

Thus, $137_{10} = 10001001_2$

(d) $567_{10}/2 = 283_{10}$ + remainder of 1 (LSB)
$283_{10}/2 = 141_{10}$ + remainder of 1
$141_{10}/2 = 70_{10}$ + remainder of 1
$70_{10}/2 = 35_{10}$ + remainder of 0
$35_{10}/2 = 17_{10}$ + remainder of 1
$17_{10}/2 = 8_{10}$ + remainder of 1
$8_{10}/2 = 4_{10}$ + remainder of 0
$4_{10}/2 = 2_{10}$ + remainder of 0
$2_{10}/2 = 1_{10}$ + remainder of 0
$1_{10}/2 = 0_{10}$ + remainder of 1 (MSB)

Thus, $567_{10} = 1000110111_2$

2.3 $2^N-1 = 1023_{10}$; N = 10
Therefore, we need 10 binary bits.

2.4 $2^N-1 = X$; $2^7-1 = X$; $128-1 = X$
Therefore, $X = 127_{10}$

SECTION 2.3 *Octal Number System*

2.5 (a) $217_8 = (2 \times 8^2)+(1 \times 8^1)+(7 \times 8^0) = 143_{10}$

(b) $55_8 = (5 \times 8^1)+(5 \times 8^0) = 45_{10}$

(c) $5076_8 = (5 \times 8^3)+(0 \times 8^2)+(7 \times 8^1)+(6 \times 8^0) = 2622_{10}$

(d) $511_8 = (5 \times 8^2)+(1 \times 8^1)+(1 \times 8^0) = 329_{10}$

(e) $100_8 = (1 \times 8^2)+(0 \times 8^1)+(0 \times 8^0) = 64_{10}$

(f) 898_8 = Illegal octal number. Octal numbers cannot have digits greater than 7_{10}.

2.6 (a) $323_{10}/8 = 40$ + remainder of 3 (LSD)
$40_{10}/8 = 5$ + remainder of 0
$5_{10}/8 = 0$ + remainder of 5 (MSD)

Thus, $323_{10} = 503_8$

(b) $123_{10}/8 = 15$ + remainder of 3 (LSD)
$15_{10}/8 = 1$ + remainder of 7
$1_{10}/8 = 0$ + remainder of 1 (MSD)

Thus, $123_{10} = 173_8$

(c) $898_{10}/8 = 112$ + remainder of 2 (LSD)
$112_{10}/8 = 14$ + remainder of 0
$14_{10}/8 = 1$ + remainder of 6
$1_{10}/8 = 0$ + remainder of 1 (MSD)

Thus, $898_{10} = 1602_8$

(d) $32536_{10}/8 = 4067$ + remainder of 0 (LSD)
$4067_{10}/8 = 508$ + remainder of 3
$508_{10}/8 = 63$ + remainder of 4
$63_{10}/8 = 7$ + remainder of 7
$7_{10}/8 = 0$ + remainder of 7 (MSD)

Thus, $32536_{10} = 77430_8$

(e) $245_{10}/8 = 30$ + remainder of 5 (LSD)
$30_{10}/8 = 3$ + remainder of 6
$3_{10}/8 = 0$ + remainder of 3 MSD)

Thus, $245_{10} = 365_8$

2.7 To represent any octal digit it requires three binary bits.

(a) 217_8 = 010 001 111

Thus, $217_8 = 010001111_2$

(b) 55_8 = 101 101

Thus, $55_8 = 101101_2$

(c) 5076_8 = 101 000 111 110

Thus, $5076_8 = 101000111110_2$

(d) 511_8 = 101 001 001

Thus, $511_8 = 101001001_2$

(e) 100_8 = 001 000 000

Thus, $100_8 = 001000000_2$

(f) 898_8 = 1000 1001 1000

898 is an illegal octal number. Only three bits can be used to represent any octal digit. Obviously, this limits the range of octal digits from 0_{10} to 7_{10}. Therefore, any number which may contain digits greater than 7_{10} (8_{10} and 9_{10}) is an illegal octal number.

2.8 To convert a binary number to its octal equivalent, the bits which make the binary number must be grouped into groups of three bits, starting at the LSB and moving towards the MSB.

(a) 1101_2 = 001 101 = 15_8

(b) 1010_2 = 001 010 = 12_8

(c) 1100011010_2 = 001 100 011 010 = 1432_8

(d) 1000001101_2 = 001 000 001 101 = 1015_8

(e) 11101_2 = 011 101 = 35_8

SECTION 2.4 *Hexadecimal Number System*

2.9 (a) $FF_{16} = (15 \times 16^1)+(15 \times 16^0) = 255_{10}$

(b) $AD3_{16} = (10 \times 16^2)+(13 \times 16^1)+(3 \times 16^0) = 2771_{10}$

(c) $589_{16} = (5 \times 16^2)+(8 \times 16^1)+(9 \times 16^0) = 1417_{10}$

(d) $3AFD_{16} = (3 \times 16^3)+(10 \times 16^2)+(15 \times 16^1)+(13 \times 16^0) = 15101_{10}$

(e) $FEED_{16} = (15 \times 16^3)+(14 \times 16^2)+(14 \times 16^1)+(13 \times 16^0 = 65261_{10}$

2.10 (a)
$323_{10}/16 = 20$ + remainder of 3 (LSD)
$20_{10}/16 = 1$ + remainder of 4
$1_{10}/16 = 0$ + remainder of 1 (MSD)

Thus, $323_{10} = 143_{16}$

(b)
$123_{10}/16 = 7$ + remainder of 11 (LSD)
$7_{10}/16 = 0$ + remainder of 7 (MSD)

Thus, $123_{10} = 7B_{16}$

(c)
$898_{10}/16 = 56$ + remainder of 2 (LSD)
$56_{10}/16 = 3$ + remainder of 8
$3_{10}/16 = 0$ + remainder of 3 (MSD)

Thus, $898_{10} = 382_{16}$

(d)
$32536_{10}/16 = 2033$ + remainder of 8 (LSD)
$2033_{10}/16 = 127$ + remainder of 1
$127_{10}/16 = 7$ + remainder of 15
$7_{10}/16 = 0$ + remainder of 7 (MSD)

Thus, $32536_{10} = 7F18_{16}$

(e)
$245_{10}/16 = 15$ + remainder of 5 (LSD)
$15_{10}/16 = 0$ + remainder of 15 (MSD)

Thus, $245_{10} = F5_{16}$

2.11 To represent any hexadecimal digit it requires four binary bits.

(a) FF_{16} = 1111 1111

Thus, $FF_{16} = 11111111_2$

(b) $AD3_{16}$ = 1010 1101 0011

Thus, $AD3_{16} = 101011010011_2$

(c) 589_{16} = 0101 1000 1001

Thus, $589_{16} = 010110001001_2$

(d) $3AFD_{16}$ = 0011 1010 1111 1101

Thus, $3AFD_{16} = 0011101011111101_2$

(e) $FEED_{16}$ = 1111 1110 1110 1101

Thus, $FEED_{16} = 1111111011101101_2$

2.12 To convert a binary number to its hexadecimal equivalent, we must divide the binary number into groups of four bits, starting at the LSB and moving towards the MSB.

(a) $1101_2 = D_{16}$

(b) $1010_2 = A_{16}$

(c) $1100011010_2 = 001100011010_2 = 31A_{16}$

When binary zeros are added to the left of the MSB, the binary number is not affected.

(d) $1000001101_2 = 001000001101_2 = 20_{16}$

(e) $11101_2 = 00011101_2 = 1D_{16}$

2.13 (a) $1230_5 = (1 \times 5^3) + (2 \times 5^2) + (3 \times 5^1) + (0 \times 5^0) = 190_{10}$

(b) $777_{10}/5$ = 155 + remainder of 2 (LSD)
$155_{10}/5$ = 31 + remainder of 0
$31_{10}/5$ = 6 + remainder of 1
$6_{10}/5$ = 1 + remainder of 1
$1_{10}/5$ = 0 + remainder of 1 (MSD)

Thus, $777_{10} = 11102_5$

2.14 (a) $333_4 = (3 \times 4^2) + (3 \times 4^1) + (3 \times 4^0) = 63_{10}$

(b) $777_{10}/4 = 194$ + remainder of 1 (LSD)
$194_{10}/4 = 48$ + remainder of 2
$48_{10}/4 = 12$ + remainder of 0
$12_{10}/4 = 3$ + remainder of 0
$3_{10}/4 = 0$ + remainder of 3 (MSD)

Thus, $777_{10} = 30021_4$

2.15 (a) 96_{16}, 97_{16}, 98_{16}, 99_{16}, $9A_{16}$, $9B_{16}$, $9C_{16}$, $9D_{16}$, $9E_{16}$, $9F_{16}$, $A0_{16}$, $A1_{16}$, $A2_{16}$, $A3_{16}$, $A4_{16}$, $A5_{16}$, $A6_{16}$, $A7_{16}$, $A8_{16}$, $A9_{16}$, AA_{16}, AB_{16}, AC_{16}, AD_{16}, AE_{16}, AF_{16}.

SECTION 2.5 *BCD Code*

2.16 (a) $63_{10} = 0110\ 0011_{BCD}$

(b) $105_{10} = 0001\ 0000\ 0101_{BCD}$

(c) $757_{10} = 0111\ 0101\ 0111_{BCD}$

(d) $999_{10} = 1001\ 1001\ 1001_{BCD}$

(e) $36543_{10} = 0011\ 0110\ 0101\ 0100\ 0011_{BCD}$

2.17 (a) $0111\ 0100\ 0010\ 1001_{BCD} = 7429_{10}$

(b) $1010\ 0010\ 0111_{BCD} = ?27_{10}$
|___|
|_______ Illegal BCD code

1010_2, 1011_2, 1100_2, 1101_2, 1110_2, and 1111_2 are illegal BCD codes.

(c) $0101\ 0011\ 1001_{BCD} = 539_{10}$

SECTION 2.6 *Putting it All Together*

2.18 First we must convert the decimal number 1000_{10} to binary. Therefore, $1000_{10} = 1111101000_2$. Now, in order to convert from binary to octal the binary number has to be divided into groups of three bits, starting with the LSB. Thus, $1111101000_2 = 1\ 111\ 101\ 000_2$. Next, each individual group of three bits is evaluated. Hence, $1\ 111\ 101\ 000_2 = 1750_8$.

In order to convert from binary to hexadecimal the binary number has to be divided into groups of four bits, starting with the LSB.
Thus, $1111101000_2 = 11\ 1110\ 1000_2$. Next, each individual group of four bits is evaluated. Hence, $11\ 1110\ 1000_2 = 3E8_{16}$.

Finally, in order to convert from decimal to BCD each individual decimal digit has to be expressed by using four binary bits.
Thus, $1000_{10} = 0001\ 0000\ 0000\ 0000_{BCD}$.

SECTION 2.7 *Gray Code*

2.19 (a) $1_{10} = 0001$ (b) $2_{10} = 0011$ (c) $3_{10} = 0010$ (d) $7_{10} = 0100$
(e) $13_{10} = 1011$

SECTION 2.8 *Alphanumerical Codes*

2.20 (a) FLIP-FLOP

	7-bit ASCII code
F	100 0110
L	100 1100
I	100 1001
P	101 0000
-	010 1101
F	100 0110
L	100 1100
O	100 1111
P	101 0000

(b) PRINT (5+10/2)

	7-bit ASCII code
P	101 0000
R	101 0010
I	100 1001
N	100 1110
T	101 0100
	010 0000
(	010 1000
5	011 0101
+	010 1011
1	011 0001
0	011 0000
/	010 1111
2	011 0010
)	010 1001

SECTION 2.9 *Parity Method For Error Detection*

2.21 (a) $\underline{1}00110001_2$
|________ Even parity bit.

(b) 0100001$_2$
|________ Even parity bit.

2.22 (a) FLIP-FLOP 7-bit ASCII code

F	0100 0110
L	0100 1100
I	0100 1001
P	1101 0000
-	1010 1101
F	0100 0110
L	0100 1100
O	0100 1111
P	1101 0000

(Bits underlined are the Odd parity bits)

(b) PRINT (5+10/2) 7-bit ASCII code

P	1101 0000
R	0101 0010
I	0100 1001
N	1100 1110
T	0101 0100
	0010 0000
(	1010 1000
5	1011 0101
+	1010 1011
1	0011 0001
0	1011 0000
/	0010 1111
2	0011 0010
)	0010 1001

(Bits underlined are the Odd parity bits)

2.23 (a)

FLIP-FLOP	Message Transmitted	Message Stored
F	0100 0110 = 46_{16}	1100 0110 = $C6_{16}$
L	0100 1100 = $4C_{16}$	1100 1100 = CC_{16}
I	0100 1001 = 49_{16}	1100 1001 = $C9_{16}$
P	1101 0000 = $D0_{16}$	0101 0000 = 50_{16}
-	1010 1101 = AD_{16}	0010 1101 = $2D_{16}$
F	0100 0110 = 46_{16}	1100 0110 = $C6_{16}$
L	0100 1100 = $4C_{16}$	1100 1100 = CC_{16}
O	0100 1111 = $4F_{16}$	1100 1111 = CF_{16}
P	1101 0000 = $D0_{16}$	0101 0000 = 50_{16}

By comparing the "Message Transmitted" with the "Message Stored" in the computer's memory, it can be observed that the codes received have the wrong parity bit. In fact, all of the codes received have had an even rather than an odd parity bit added to the original 7-bit ASCII code. Thus, it appears that the transmitter added an even parity bit instead of an odd parity bit to each and all of the ASCII codes transmitted.

In cases such as this, the parity checker (special circuit) in the receiver, that in this case would have been checking for odd parity, would have detected an error for each ASCII code received. Therefore, in this particular example a problem exists with the parity generator (special circuit) in the transmitter. It may be malfunctioning or simply programmed for even rather than odd parity generation.

3 LOGIC GATES AND BOOLEAN ALGEBRA

SECTION 3.1 *Boolean Constants and Variables*

3.1 (a) Logical Addition, can also be referred to as OR Addition.
(b) Inversion or logical complementation is often called the NOT operation.
(c) The AND operation can also be referred to as the Logical Multiplication.

SECTIONS 3.2-3.5 *Truth Tables/OR, AND and NOT Operations*

3.2 (a) The output of an OR gate is LOW only when all of its inputs are LOW. If one or more inputs is HIGH, then the output is HIGH (see Figure P3.2(a)).

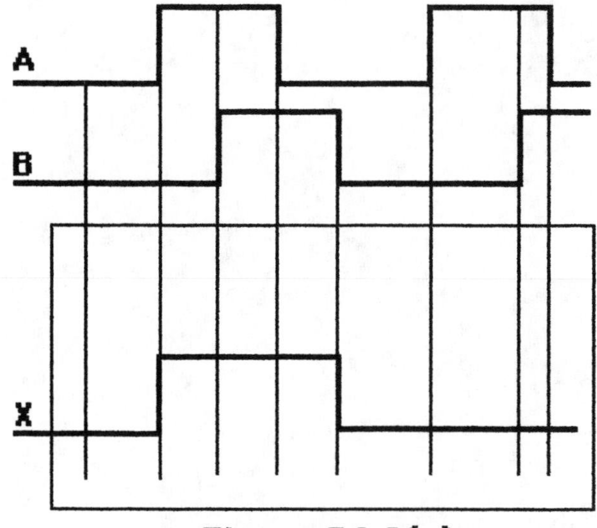

Figure P3.2(a)

(b) The output of a 2-input AND gate is HIGH only when both of its inputs are HIGH. If either or both inputs are LOW, then the output is LOW (see Figure P3.2(b)).

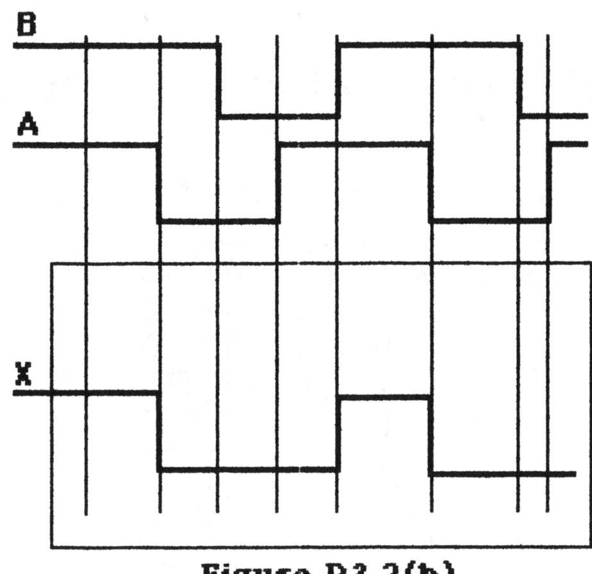

Figure P3.2(b)

3.3 There are four possible input combinations in this circuit. The practical thing to do in order to determine when $(\overline{A}+B) = 0$, is to generate the truth table for the circuit.

	A	B	$X = (\overline{A}+B)$
Case 1	0	0	1 = (1+0)
Case 2	0	1	1 = (1+1)
Case 3	1	0	0 = (0+0)
Case 4	1	1	1 = (0+1)

Thus, Case 3 is the only combination of A and B which causes X = 0.

3.4 (a) False. The output of a 3-input AND gate is HIGH only when all of its inputs are HIGH.

(b) False. The output of a 3-input OR gate is LOW only when all of its inputs are LOW.

(c) True.

SECTIONS 3.6-3.7 *Describing Logic Circuits Algebraically/ Evaluating Logic-Circuit Outputs*

3.5 $X=\overline{(\overline{A+B})C}$

3.6 A, B, C and D are the four variables in the expression. Therefore, there are 16 (2^4) combinations.

A	B	C	D	$Y=\overline{(\overline{A}BC)+D}$
0	0	0	0	$1=\overline{(\overline{0}00)+0}$
0	0	0	1	$0=\overline{(\overline{0}00)+1}$
0	0	1	0	$1=\overline{(\overline{0}01)+0}$
0	0	1	1	$0=\overline{(\overline{0}01)+1}$
0	1	0	0	$1=\overline{(\overline{0}10)+0}$
0	1	0	1	$0=\overline{(\overline{0}10)+1}$
0	1	1	0	$1=\overline{(\overline{0}11)+0}$
0	1	1	1	$0=\overline{(\overline{0}11)+1}$
1	0	0	0	$1=\overline{(\overline{1}00)+0}$
1	0	0	1	$0=\overline{(\overline{1}00)+1}$
1	0	1	0	$1=\overline{(\overline{1}01)+0}$
1	0	1	1	$0=\overline{(\overline{1}01)+1}$
1	1	0	0	$1=\overline{(\overline{1}10)+0}$
1	1	0	1	$0=\overline{(\overline{1}10)+1}$
1	1	1	0	$0=\overline{(\overline{1}11)+0}$
1	1	1	1	$0=\overline{(\overline{1}11)+1}$

3.7 $Y=AB+C\overline{B}+\overline{A}BC$ (<u>Conditions</u>: A=0, B=1 and C=1)

$Y=0\,1+1\overline{1}+\overline{0}11$

$Y=0\,1+10+111$

$Y=0+0+1$

$Y=1$

SECTIONS 3.8-3.9 *Implementing Circuits From Boolean Expressions/NOR gates and NAND Gates*

3.8 (a) $X = \overline{(\overline{A+B})(\overline{B}C)}$

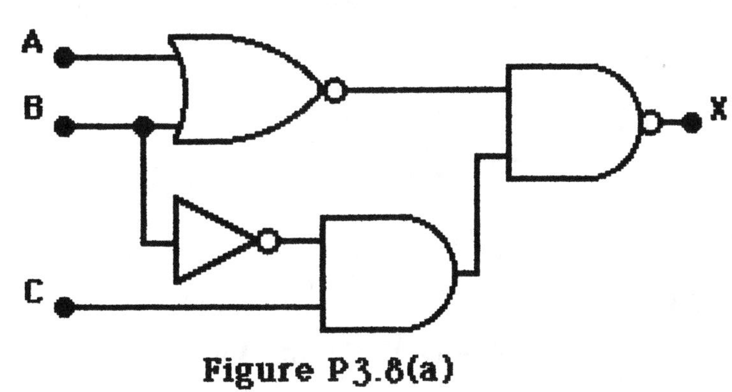

Figure P3.8(a)

(b) $X = \overline{\overline{ABC}(A+D)}$

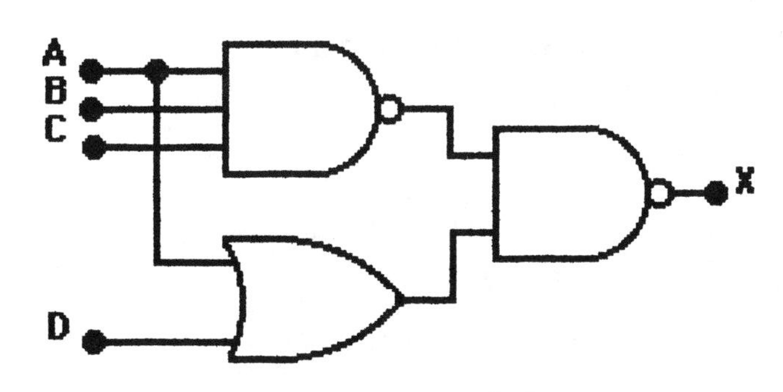

Figure P3.8(b)

SECTIONS 3.10-3.11 *Boolean Theorems/DeMorgans's Theorems*

3.9 (a) $X = \overline{A}\overline{B}\overline{C} + \overline{B}\overline{C}\overline{C}$

]----> Theorem #3

$X = \overline{A}\overline{B}\overline{C} + \overline{B}\overline{C}$

$X = \overline{B}\overline{C}\,(\overline{A} + 1)$

]------> Theorem #6

$X = \overline{B}\overline{C}(1)$

]----------> Theorem #2

$X = \overline{B}\overline{C}$

(b) $X = \overline{A}BC + A\overline{B}C + \overline{A}\overline{B}C$

$X = \overline{B}C(A + \overline{A}) + \overline{A}BC$

]----> Theorem #8

$X = \overline{B}C(1) + \overline{A}BC$

]---------> Theorem #2

$X = \overline{B}C + \overline{A}BC$

$X = C + (\overline{B} + \overline{A}B)$

]----------> Theorem #15

$X = C + (\overline{B} + \overline{A})$

(c) $X = \overline{KL(\overline{M+N})KL}$

]--------> Theorem #17

$X = \overline{KL} + (\overline{\overline{M+N}}) + \overline{KL}$

$X = \overline{KL} + (M+N) + \overline{KL}$

]-----> Theorem #17

$X = \overline{K} + \overline{L} + M + N + \overline{K} + \overline{L}$

]-----> Theorem #7

$X = \overline{K} + \overline{L} + M + N$

(d) $X = \overline{\overline{AB}(A+B)}$

]---------------> Theorem #17

$X = \overline{\overline{AB}} + (\overline{A+B})$

]---------------> Theorem #16

$X = AB + \overline{A}\,\overline{B}$ (EX-NOR gate)

SECTION 3.12 *Universality of NAND Gates and NOR Gates*

3.10 The circuit that implements X = (A + B)(C + D) uses one 2-input AND gate and two 2-input OR gates. It requires two NOR gates to make an OR gate, and three NOR gates to make an AND gate. After the double inversions are cancelled, the final circuit will have a total of three 2-input NOR gates as shown in Figure P3.10.

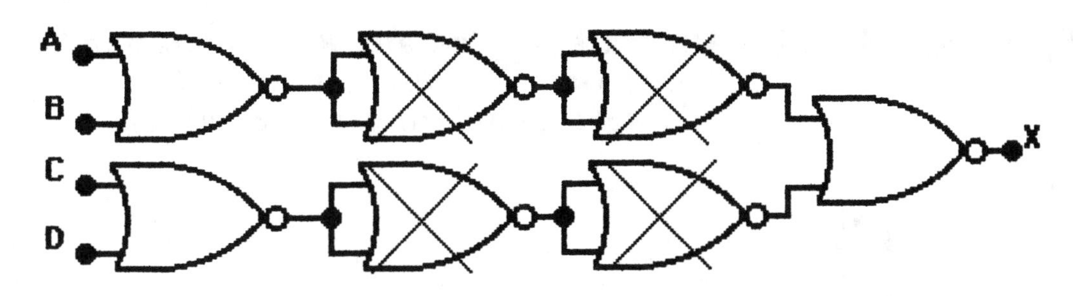

Figure P3.10: X=(A+B)(C+D)

3.11 Expression $X=((A+\overline{B})(B+C)B)$ should first of all, be simplified as much as possible.

$X=((A+\overline{B})(B+C)B)$

]----------> Theorem #13b

$X=(AB+AC+\overline{B}B+\overline{B}C)B$

]--------> Theorem #13a

$X=ABB+ACB+\overline{B}BB+\overline{B}CB$

]-------> Theorem #4

$X=ABB+ACB$

]-----------------------> Theorem #3

$X=AB(1+CB)$

]-----------------------> Theorem #6

$X=AB$

Thus, it requires two 2-input NAND gates, as shown in Figure P3.11, to implement the expression X = AB.

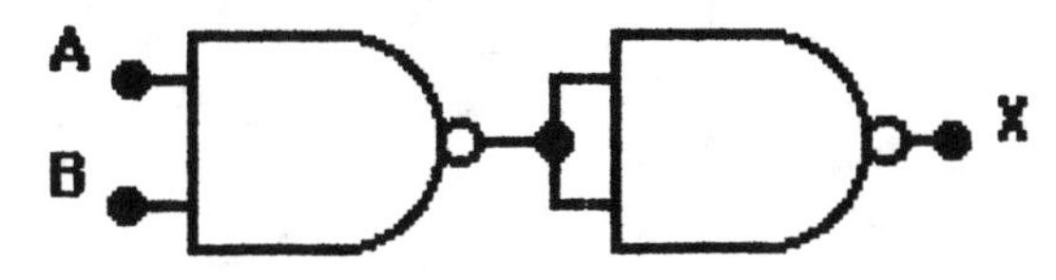

Figure P3.11: X=AB

3.12

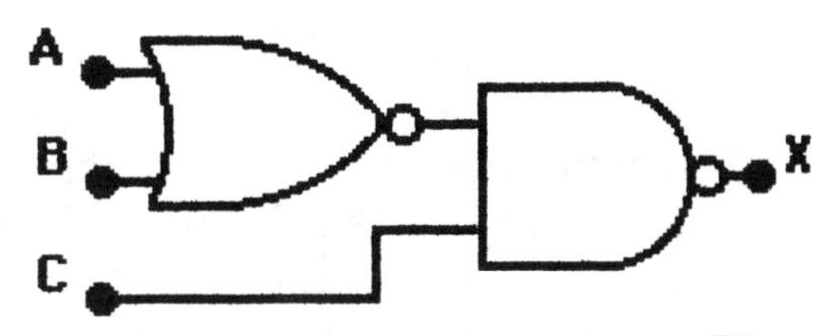

Figure P3.12: $X=A+B+\overline{C}$

SECTIONS 3.13-3.15 ***Alternate Logic-Gate Representation/ Which Gate Representation to Use/ New IEEE Standard Logic Symbols.***

3.13

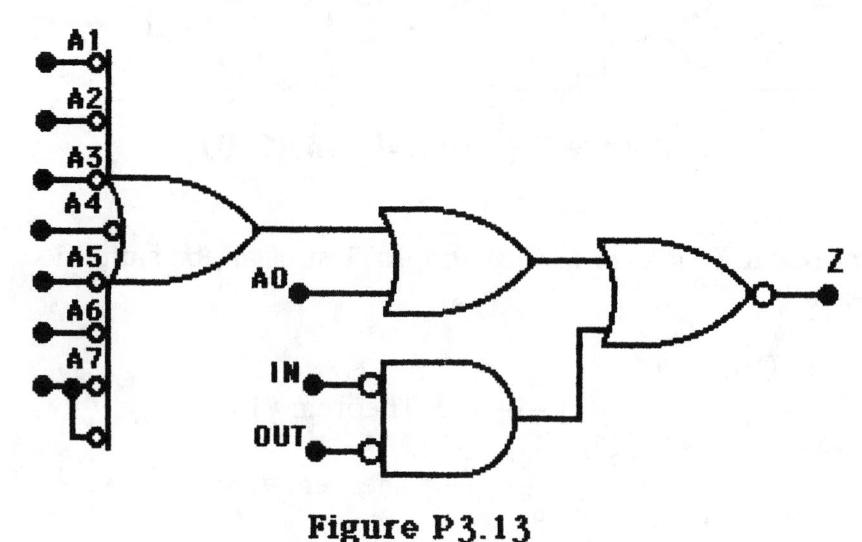

Figure P3.13

3.14

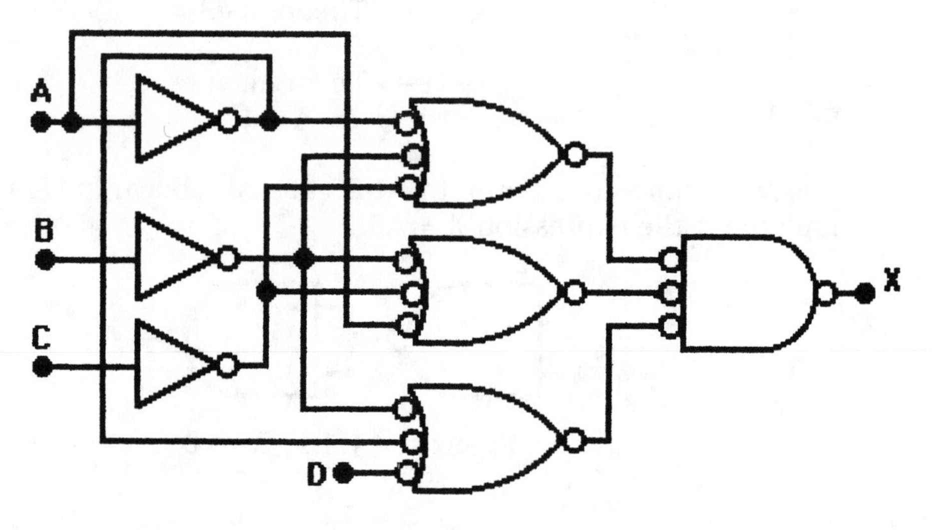

Figure P3.14

3.15 The term unasserted means Not-Active. For the circuit of Figure 3.6, the LED turns ON when the output of the 2-input NOR gate is LOW. Therefore, we must determine the necessary state conditions at the inputs A and B that will cause output X to be HIGH (unasserted).
Hence, signal $\overline{\text{LIGHT}}$ will be HIGH if A=0 and B=1, or if, A=1 and B=0.

3.16

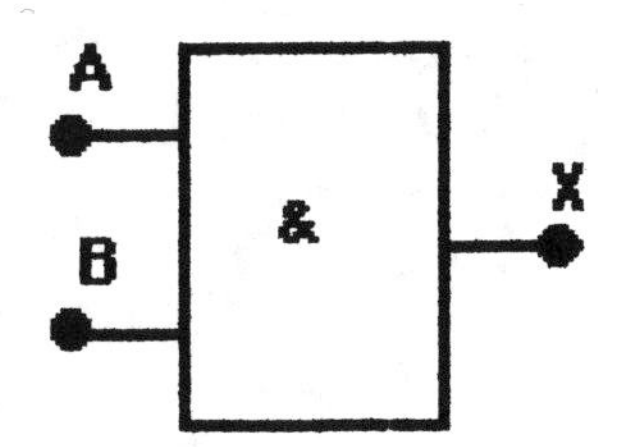

Figure P3.16: IEEE/ANSI symbol for an AND gate.

3.17

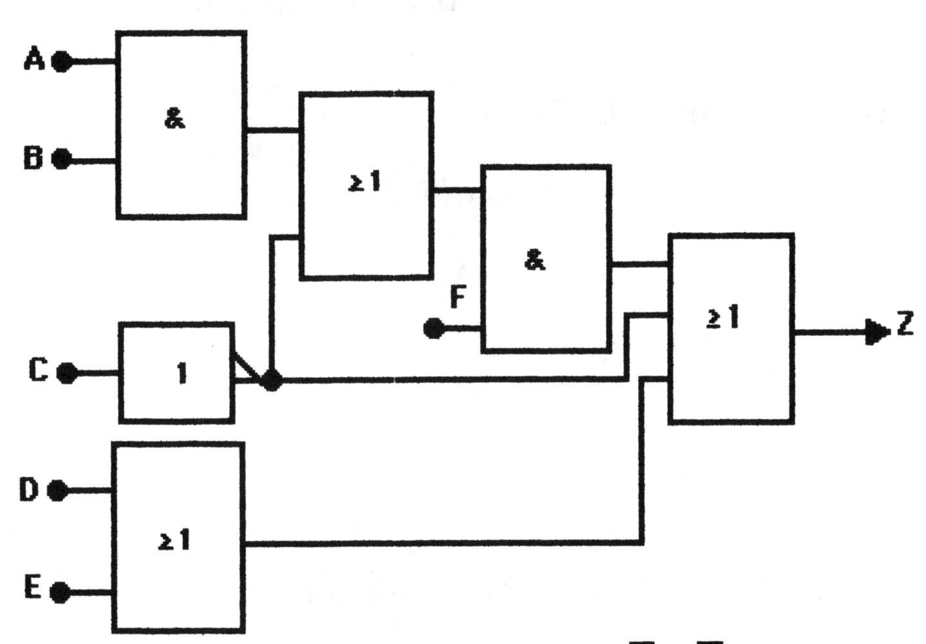

Figure P3.17: $Z=(AB+\overline{C})F+\overline{C}+D+E$

3.18 (a) $Z=(AB+\overline{C})F+\overline{C}+D+E$

]-------------> Theorem #13a

$Z=ABF+\overline{C}F+\overline{C}+D+E$

$Z=ABF+\overline{C}(F+1)+D+E$

]-------------> Theorem #6

$Z=ABF+\overline{C}+D+E$

(b) It requires three logic circuits to implement the simplified expression.

(c) The three logic circuits are: a 4-input OR gate, a 3-input AND gate and an INVERTER.

4 COMBINATIONAL LOGIC CIRCUITS

SECTIONS 4.1-4.3 ***Sum-of-Products Form/Simplifying Logic Circuits/Algebraic Simplification***

4.1 (a) $X = ABC + A\overline{B}C + \overline{A}BC + (\overline{\overline{A} + B})C$

$X = ABC + A\overline{B}C + \overline{A}BC + (\overline{A} \cdot \overline{B})C$

$X = ABC + A\overline{B}C + \overline{A}BC + \overline{A}\overline{B}C$

$X = AC(B + \overline{B}) + \overline{A}C(B + \overline{B})$

$X = AC + \overline{A}C$

$X = C$

(b) $K = \overline{X}\overline{Y}Z + \overline{X}YZ + X\overline{Y}\overline{Z} + X\overline{Y}Z + (\overline{\overline{X} + \overline{Y} + \overline{Z}})$

$K = \overline{X}\overline{Y}Z + \overline{X}YZ + X\overline{Y}\overline{Z} + X\overline{Y}Z + XYZ$

$K = \overline{X}Z(\overline{Y} + Y) + X\overline{Y}(\overline{Z} + Z) + XYZ$

$K = \overline{X}Z + X\overline{Y} + XYZ$

$K = X(\overline{Y} + YZ) + \overline{X}Z$

$K = X(\overline{Y} + Z) + \overline{X}Z$

$K = X\overline{Y} + XZ + \overline{X}Z$

$K = Z(X + \overline{X}) + X\overline{Y}$

$K = Z + X\overline{Y}$

(c) $W=(K+N+M)(K+\overline{N}+M)+(\overline{K}+N+M)(\overline{K}+\overline{N}+M)$

$W=(KK+K\overline{N}+KM+NK+N\overline{N}+NM+MK+M\overline{N}+MM)+$

$(\overline{K}\overline{K}+\overline{K}\overline{N}+\overline{K}M+N\overline{K}+N\overline{N}+NM+M\overline{K}+M\overline{N}+MM)$

$W=(K+K\overline{N}+KM+NK+NM+MK+M\overline{N}+M)+$

$(\overline{K}+\overline{K}\overline{N}+\overline{K}M+N\overline{K}+NM+M\overline{K}+M\overline{N}+M)$

$W=(K(1+\overline{N}+M+N+M)+NM+M\overline{N}+M)+(\overline{K}(1+\overline{N}+M+N+M)+NM+M\overline{N}+M)$

$W=(K+NM+M\overline{N}+M)+(\overline{K}+NM+M\overline{N}+M)$

$W=M(1+\overline{N}+N)+K+M(1+\overline{N}+N)+\overline{K}$

$W=M+K+M+\overline{K}$

$W=M+K+\overline{K}$

$W=1+M$

$W=1$

4.2 (a) DRIV=A; $\overline{BELTP}=\overline{BELTD}=B$; PASS=C; IGN=E; $\overline{ALARM}=X$

$X=\overline{(\overline{\overline{A\overline{B}}}+\overline{\overline{\overline{B}C}})E}$

(b) $X=\overline{\overline{A\overline{B}}}\cdot\overline{\overline{\overline{B}C}}+\overline{E}$

$X=\overline{(A\overline{B})}(\overline{\overline{B}C})+\overline{E}$

$X=(\overline{A}+B)(B+\overline{C})+\overline{E}$

$X=\overline{A}B+\overline{A}\overline{C}+BB+B\overline{C}+\overline{E}$

$X=B(\overline{A}+1+\overline{C})+\overline{A}\overline{C}+\overline{E}$

$X=B+\overline{A}\overline{C}+\overline{E}$

(c)

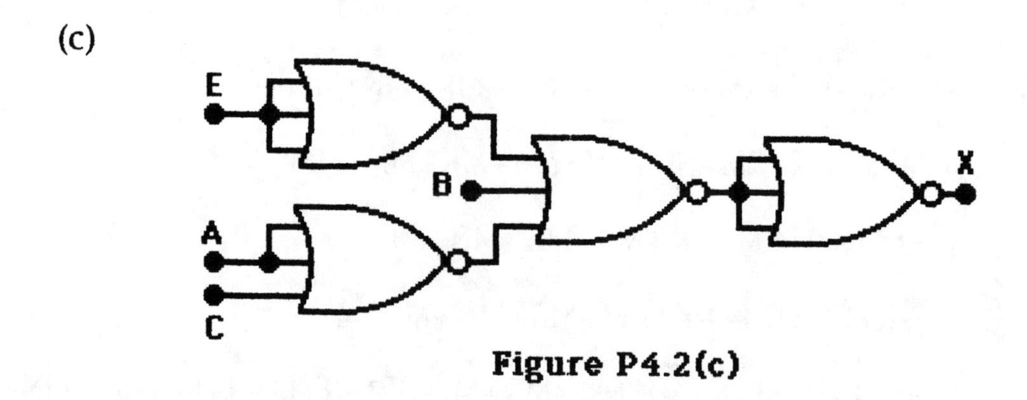

Figure P4.2(c)

SECTION 4.4 *Designing Combinational Logic Circuits*

4.3 Step #1: Set up the truth table for the problem.

A	B	C	D	X
0	0	0	0	1
0	0	0	1	1
0	0	1	0	1
0	0	1	1	1
0	1	0	0	0
0	1	0	1	0
0	1	1	0	0
0	1	1	1	1
1	0	0	0	0
1	0	0	1	0
1	0	1	0	0
1	0	1	1	1
1	1	0	0	0
1	1	0	1	0
1	1	1	0	0
1	1	1	1	1

Step #2: Write the AND term for each case where output X=1.

A	B	C	D	X	
0	0	0	0	1	$\overline{A}\,\overline{B}\,\overline{C}\,\overline{D}$
0	0	0	1	1	$\overline{A}\,\overline{B}\,\overline{C}\,D$
0	0	1	0	1	$\overline{A}\,\overline{B}\,C\,\overline{D}$
0	0	1	1	1	$\overline{A}\,\overline{B}\,C\,D$
0	1	1	1	1	$\overline{A}\,B\,C\,D$
1	0	1	1	1	$A\,\overline{B}\,C\,D$
1	1	1	1	1	$A\,B\,C\,D$

Step #3: Write the Sum-of-Products expression for all the cases where output X = 1.

$$X=\overline{A}\,\overline{B}\,\overline{C}\,\overline{D}+\overline{A}\,\overline{B}\,\overline{C}\,D+\overline{A}\,\overline{B}\,C\,\overline{D}+\overline{A}\,\overline{B}\,C\,D+\overline{A}\,B\,C\,D+A\,\overline{B}\,C\,D+A\,B\,C\,D$$

Step #4: Simplify the Sum-of-Products expression as much as possible using Boolean algebra.

$$X=\overline{A}\,\overline{B}\,\overline{C}(\overline{D}+D)+\overline{A}\,\overline{B}\,C(\overline{D}+D)+A\,C\,D(\overline{B}+B)+\overline{A}\,B\,C\,D$$

$$X=\overline{A}\,\overline{B}\,\overline{C}+\overline{A}\,\overline{B}\,C+A\,C\,D+\overline{A}\,B\,C\,D+[\overline{A}\,\overline{B}\,C\,D]^{*}$$

* *This term can be added since it exists in the original Boolean expression. This technique is often used in order make the simplification process easier.*

$$X=\overline{A}\,\overline{B}\,\overline{C}+\overline{A}\,\overline{B}\,C+A\,C\,D+\overline{A}\,C\,D(\overline{B}+B)$$

$$X=\overline{A}\,\overline{B}\,\overline{C}+\overline{A}\,\overline{B}\,C+A\,C\,D+\overline{A}\,C\,D$$

$$X=\overline{A}\,\overline{B}(\overline{C}+C)+C\,D(\overline{A}+A)$$

$$X=\overline{A}\,\overline{B}+C\,D$$

Step #5: Implement the simplified expression using logic gates (either circuit (a) or (b) of Figure P4.3 is acceptable).

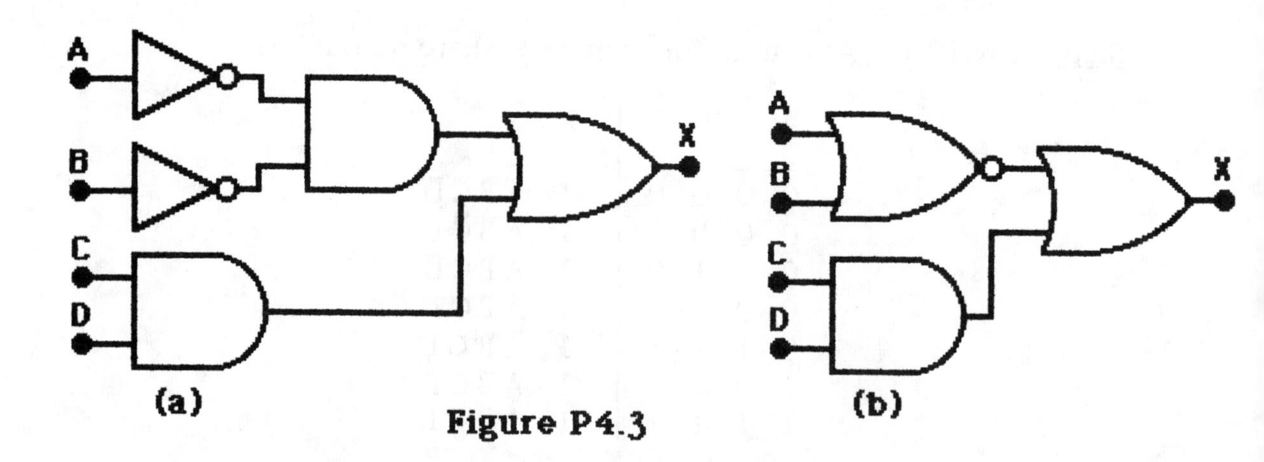

Figure P4.3

4.4 In this problem we have two different sources of signals. First, we have a 3-bit binary number coming from the external fuel tanks, which monitors the fuel pressure. Second, we have a 3-bit binary number coming from the computer, which represents the desired pressure in the fuel tanks. Consequently, we have to design a logic circuit which compares the two sets of 3-bit binary numbers and determines when they are equal. Once the circuit determines the equality of the two binary numbers, a green indicator light is turned ON. Let's follow the procedure step-by-step of the previous problem. However, before we do that, let's call $X_2\ X_1\ X_0$ the binary number coming from the transducer, and $Y_2\ Y_1\ Y_0$ the binary number from the on board computer and finally Z, the green indicator light.

Since we are only concerned with those situations when the two numbers are equal, only 8 combinations of $Y_2Y_1Y_0X_2X_1X_0$ have to be considered:

Step #1: Set up the truth table for the problem.

Y_2	Y_1	Y_0	X_2	X_1	X_0	Z	
0	0	0	0	0	0	1	(both numbers equal to 0_{10})
0	0	1	0	0	1	1	(both numbers equal to 1_{10})
0	1	0	0	1	0	1	(both numbers equal to 2_{10})
0	1	1	0	1	1	1	(both numbers equal to 3_{10})
1	0	0	1	0	0	1	(both numbers equal to 4_{10})
1	0	1	1	0	1	1	(both numbers equal to 5_{10})
1	1	0	1	1	0	1	(both numbers equal to 6_{10})
1	1	1	1	1	1	1	(both numbers equal to 7_{10})

<u>Step #2</u> : Write the AND term for each case where output Z=1.

<u>Step #3</u> : Write the Sum-of-Products expression for all the cases where output Z=1.

$$Z = \overline{Y}_2\overline{Y}_1\overline{Y}_0\overline{X}_2\overline{X}_1\overline{X}_0 + \overline{Y}_2\overline{Y}_1 Y_0\overline{X}_2\overline{X}_1 X_0 + \overline{Y}_2 Y_1\overline{Y}_0\overline{X}_2 X_1\overline{X}_0 + \overline{Y}_2 Y_1 Y_0\overline{X}_2 X_1 X_0 +$$

$$+ Y_2\overline{Y}_1\overline{Y}_0 X_2\overline{X}_1\overline{X}_0 + Y_2\overline{Y}_1 Y_0 X_2\overline{X}_1 X_0 + Y_2 Y_1\overline{Y}_0 X_2 X_1\overline{X}_0 + Y_2 Y_1 Y_0 X_2 X_1 X_0$$

<u>Step #4</u>: Simplify the Sum-of-Products expression as much as possible using Boolean algebra.

$$Z = \overline{X}_0\overline{Y}_0(\overline{X}_1\overline{Y}_1\overline{X}_2\overline{Y}_2 + X_1 Y_1\overline{X}_2\overline{Y}_2 + \overline{X}_1\overline{Y}_1 X_2 Y_2 + X_1 Y_1 X_2 Y_2) +$$

$$+ X_0 Y_0(\overline{X}_1\overline{Y}_1\overline{X}_2\overline{Y}_2 + X_1 Y_1\overline{X}_2\overline{Y}_2 + \overline{X}_1\overline{Y}_1 X_2 Y_2 + X_1 Y_1 X_2 Y_2)$$

$$Z = \overline{X}_0\overline{Y}_0(\overline{X}_2\overline{Y}_2(\overline{X}_1\overline{Y}_1 + X_1 Y_1) + X_2 Y_2(\overline{X}_1\overline{Y}_1 + X_1 Y_1)) +$$

$$+ X_0 Y_0(\overline{X}_2\overline{Y}_2(\overline{X}_1\overline{Y}_1 + X_1 Y_1) + X_2 Y_2(\overline{X}_1\overline{Y}_1 + X_1 Y_1))$$

Let $K = (\overline{X}_1\overline{Y}_1 + X_1 Y_1) = (\overline{X_1 \oplus Y_1})$

$$Z = \overline{X}_0\overline{Y}_0(\overline{X}_2\overline{Y}_2(K) + X_2 Y_2(K)) + X_0 Y_0(\overline{X}_2\overline{Y}_2(K) + X_2 Y_2(K))$$

$$Z = \overline{X}_0\overline{Y}_0 K(\overline{X}_2\overline{Y}_2 + X_2 Y_2) + X_0 Y_0 K(\overline{X}_2\overline{Y}_2 + X_2 Y_2)$$

Let $Q = (\overline{X}_2\overline{Y}_2 + X_2 Y_2) = (\overline{X_2 \oplus Y_2})$

$$Z = \overline{X}_0\overline{Y}_0 KQ + X_0 Y_0 KQ$$

$$Z = KQ(\overline{X}_0\overline{Y}_0 + X_0 Y_0)$$

$$Z = (\overline{X_1 \oplus Y_1}) \cdot (\overline{X_2 \oplus Y_2}) \cdot (\overline{X_0 \oplus Y_0})$$

STEP #5: Implement the simplified expression using logic gates.

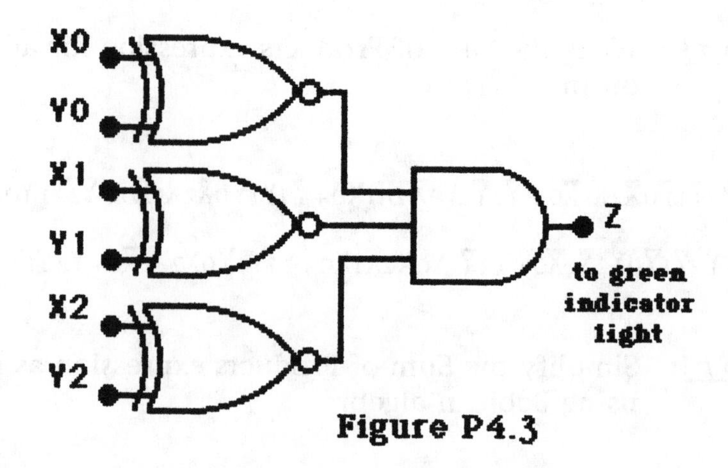

Figure P4.3

4.5 The LSB represents 100 p.s.i.. When the pressure of 600 p.s.i. is reached inside of the tanks, the transducer should have its output $X_2\,X_1\,X_0$ equal to 110_2. Thus, a circuit has to be designed that detects when the binary number $X_2\,X_1\,X_0$ exceeds 110_2.

X_2	X_1	X_0	W	
0	0	0	0	
0	0	1	0	
0	1	0	0	
0	1	1	0	
1	0	0	0	
1	0	1	0	
1	1	0	0	
1	1	1	1	$X_2\,X_1\,X_0$

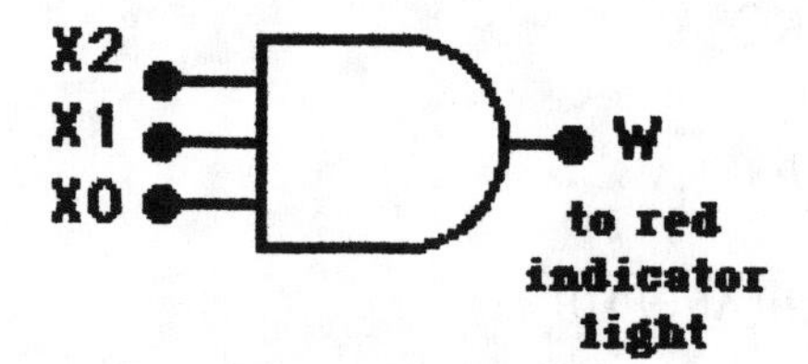

Figure P4.5

4.6 First set up the truth table, then write the S-of-P expression, then simplify it using Boolean algebra and finally implement the logic circuit.

A	B	C	X	
0	0	0	0	
0	0	1	0	
0	1	0	0	
0	1	1	0	
1	0	0	0	
1	0	1	0	
1	1	0	1	$A B \overline{C}$
1	1	1	0	

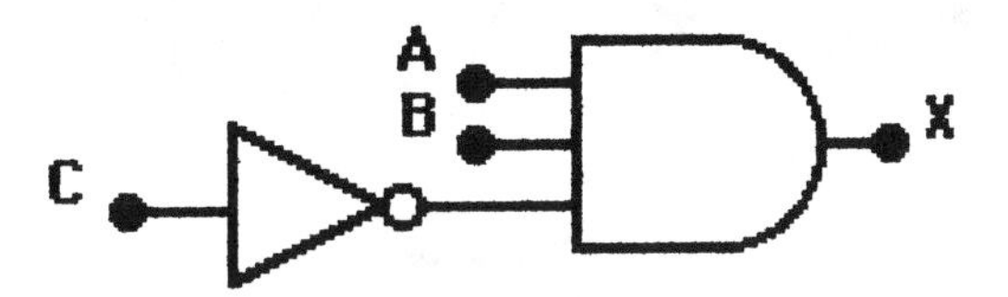

Figure P4.6

4.7 First set up the truth table, then write the S-of-P expression, then simplify it using Boolean algebra and finally implement the logic circuit.

A	B	C	D	Z	
0	0	0	0	0	
0	0	0	1	0	
0	0	1	0	0	
0	0	1	1	0	
0	1	0	0	1	$\overline{A} B \overline{C} \overline{D}$
0	1	0	1	0	
0	1	1	0	0	
0	1	1	1	0	
1	0	0	0	1	$A \overline{B} \overline{C} \overline{D}$
1	0	0	1	0	
1	0	1	0	0	
1	0	1	1	0	
1	1	0	0	1	$A B \overline{C} \overline{D}$
1	1	0	1	0	
1	1	1	0	0	
1	1	1	1	0	

$Z=\overline{A}\,B\,\overline{C}\,\overline{D}+A\,\overline{B}\,\overline{C}\,\overline{D}+A\,B\,\overline{C}\,\overline{D}$

$Z=\overline{C}\,\overline{D}(\overline{A}\,B+A\,\overline{B}+A\,B)$

$Z=\overline{C}\,\overline{D}(\overline{A}\,B+A\,(\overline{B}+B))$

$Z=\overline{C}\,\overline{D}(\overline{A}\,B+A)$

$Z=\overline{C}\,\overline{D}(A+B)$

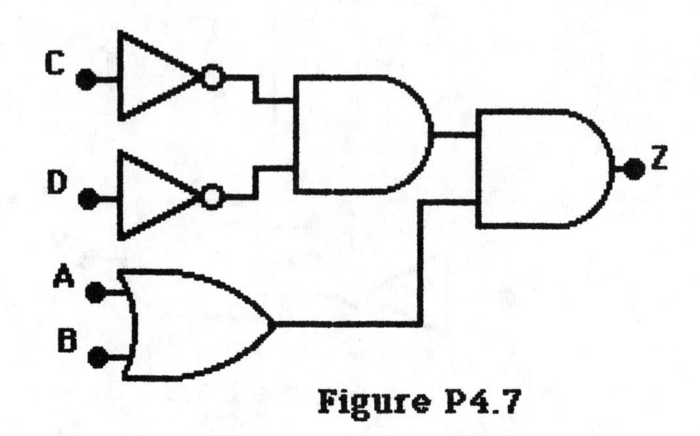

Figure P4.7

4.8 Let's first set up the truth table.

P	M	W	E	Z	
0	0	0	0	0	$\overline{P}\,\overline{M}\,\overline{W}\,\overline{E}$
0	0	0	1	1	
0	0	1	0	1	
0	0	1	1	1	
0	1	0	0	1	
0	1	0	1	1	
0	1	1	0	1	
0	1	1	1	1	
1	0	0	0	1	
1	0	0	1	1	
1	0	1	0	1	
1	0	1	1	1	
1	1	0	0	1	
1	1	0	1	1	
1	1	1	0	1	
1	1	1	1	1	

Note : In this truth table Z=1 for all cases except when P=M=W=E=0. There are two easy ways to simplify the S-of-P expression obtained from this truth table:

Method # 1. Use Karnaugh mapping for all the cases where Z = 1. (You may skip this method until it's covered in section 4.5 of your text book)

Method # 2. By writing the S-of-P expression for $\overline{Z}$, and then inverting it.

Method # 1

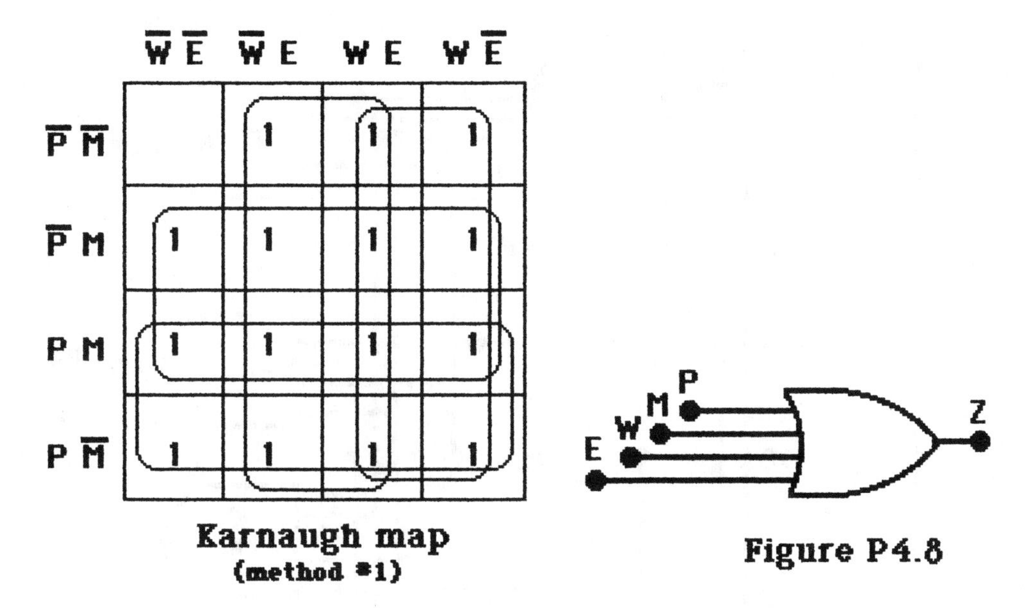

Karnaugh map
(method #1)

Figure P4.8

Thus, by using method #1, from the Karnaugh map we get: Z = P+M+E+W.

Method # 2

The Sum-of Products expression for $\overline{Z} = \overline{P}\,\overline{M}\,\overline{W}\,\overline{E}$

After we invert Z, and simplify by using DeMorgan's theorems we have $Z = \overline{\overline{P}\,\overline{M}\,\overline{W}\,\overline{E}} = P + M + W + E$. Either method yields the same circuit of Figure P4.8.

4.9 An 8-bit microcomputer has an 8-bit microprocessor with 16 address bus lines and 8 data bus lines. Since we are interested in decoding the two most significant hexadecimal digits of the address bus, we have to consider address lines A_8-A_{15}. Keep in mind that signals DCL, SBB, and ECS are active HIGH, while signal $\overline{ION}$ is active LOW.

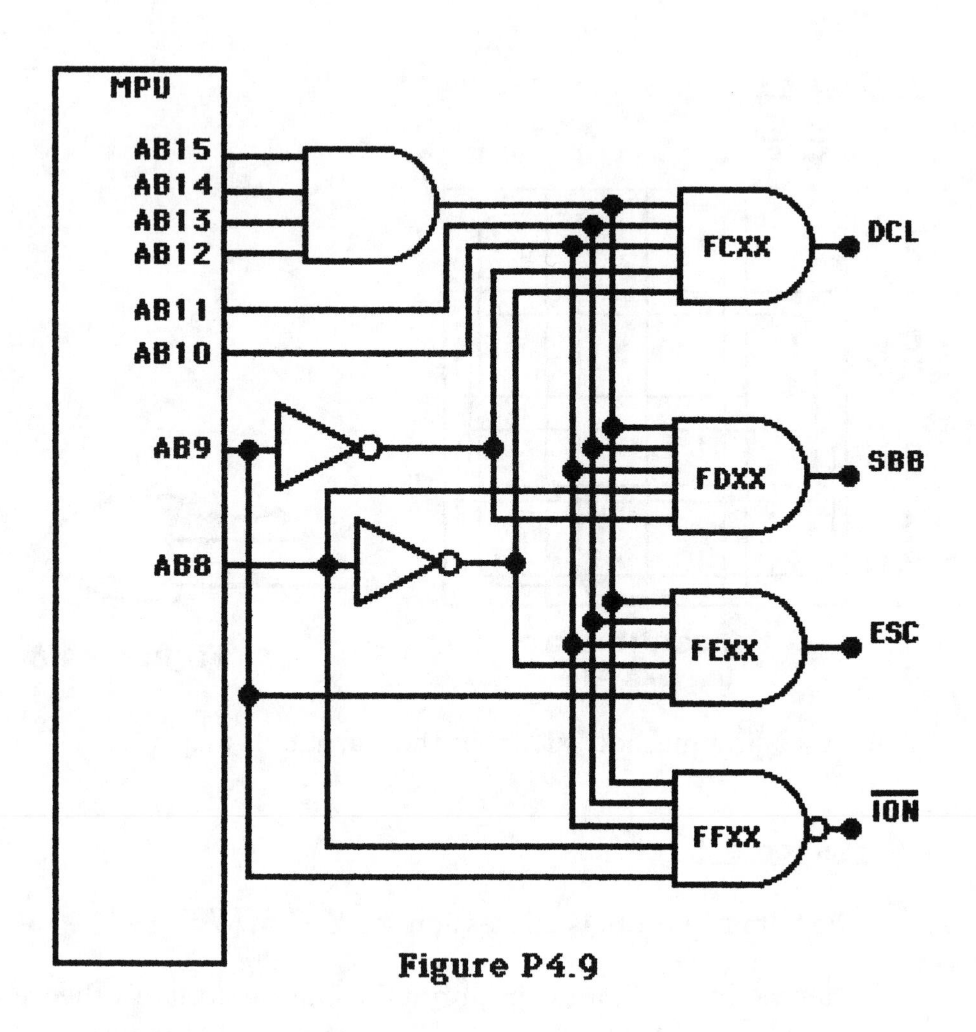

Figure P4.9

SECTIONS 4.5-4.6 *The Karnaugh map Method/Exclusive-OR and Exclusive-NOR Circuits*

4.10 (a) $X=\overline{A}\,\overline{B}\,\overline{C}\,\overline{D}+\overline{A}\,\overline{B}C\overline{D}+A\overline{B}C\overline{D}+\overline{A}B\overline{C}D+\overline{A}BCD+A\overline{B}\,\overline{C}\,\overline{D}$

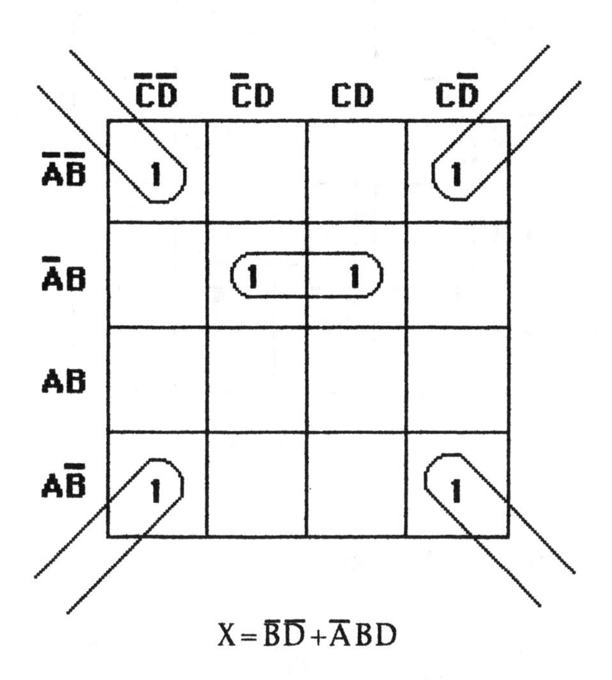

$X=\overline{B}\,\overline{D}+\overline{A}BD$

(b) $Z=\overline{K}\,\overline{N}M+\overline{K}\,\overline{N}\,\overline{M}+\overline{K}N\overline{M}$

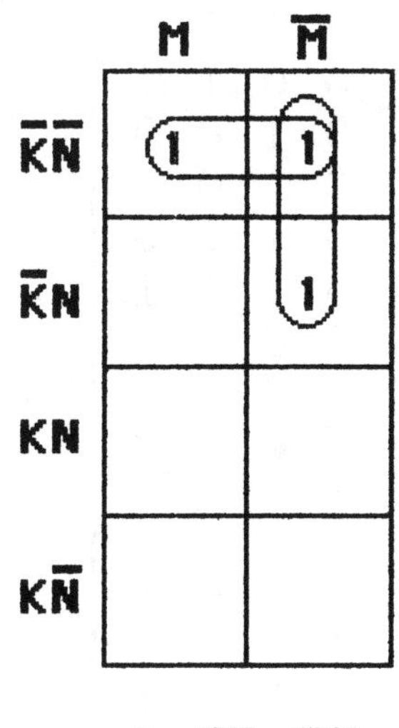

$Z=\overline{K}\,\overline{N}+\overline{K}\,\overline{M}$

(c) $Y = \overline{P}Q\overline{R}S + \overline{P}QRS + PQ\overline{R}S + PQRS$

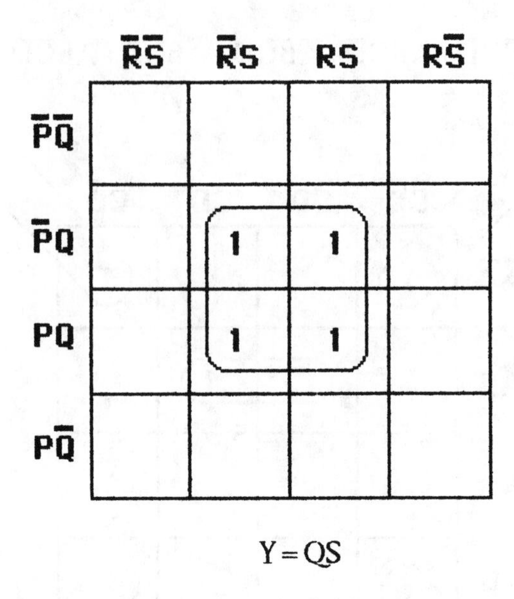

$Y = QS$

4.11 (a) First set up the truth table and write the AND term for each case where Z=1.

A	B	C	D	Z	
0	0	0	0	1	$\overline{A}\overline{B}\overline{C}\overline{D}$
0	0	0	1	1	$\overline{A}\overline{B}\overline{C}D$
0	0	1	0	1	$\overline{A}\overline{B}C\overline{D}$
0	0	1	1	1	$\overline{A}\overline{B}CD$
0	1	0	0	0	
0	1	0	1	0	
0	1	1	0	0	
0	1	1	1	1	$\overline{A}BCD$
1	0	0	0	0	
1	0	0	1	0	
1	0	1	0	0	
1	0	1	1	0	
1	1	0	0	0	
1	1	0	1	0	
1	1	1	0	0	
1	1	1	1	1	$ABCD$

Next write the S-of-P expression for output Z.

$$Z=\overline{A}\,\overline{B}\,\overline{C}\,\overline{D}+\overline{A}\,\overline{B}\,\overline{C}D+\overline{A}\,\overline{B}C\overline{D}+\overline{A}\,\overline{B}CD+\overline{A}BCD+ABCD$$

Now that we have the S-of-P expression, we can use the Karnaugh Map to simplify the Boolean expression Z.

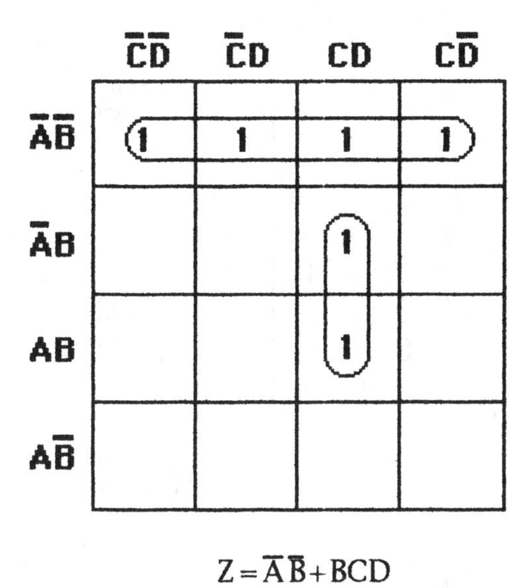

$$Z=\overline{A}\,\overline{B}+BCD$$

Finally implement the Boolean expression $Z=\overline{A}\,\overline{B}+BCD$ (Figure P4.11(a)).

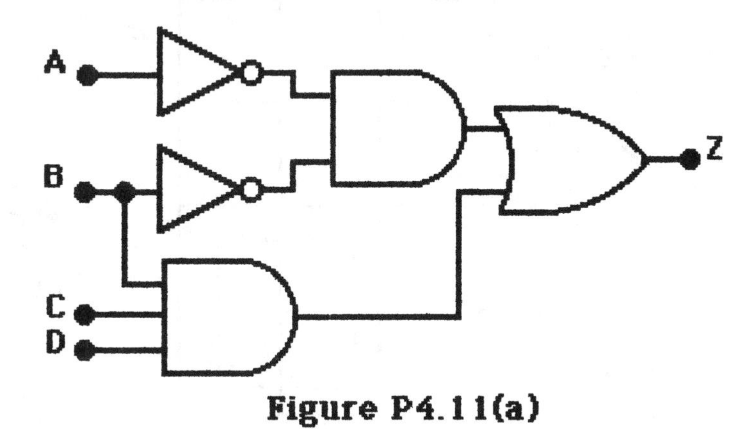

Figure P4.11(a)

(b) Apply the same step-by-step procedure used in question (a).

K	L	M	N	Z	
0	0	0	0	1	$\overline{K}\,\overline{L}\,\overline{M}\,\overline{N}$
0	0	0	1	1	$\overline{K}\,\overline{L}\,\overline{M}N$
0	0	1	0	0	
0	0	1	1	0	
0	1	0	0	1	$\overline{K}L\overline{M}\,\overline{N}$
0	1	0	1	1	$\overline{K}L\overline{M}N$
0	1	1	0	0	
0	1	1	1	0	
1	0	0	0	0	
1	0	0	1	0	
1	0	1	0	1	$K\overline{L}M\overline{N}$
1	0	1	1	1	$K\overline{L}MN$
1	1	0	0	0	
1	1	0	1	0	
1	1	1	0	1	$KLM\overline{N}$
1	1	1	1	1	$KLMN$

$$Z=\overline{K}\,\overline{L}\,\overline{M}\,\overline{N}+\overline{K}\,\overline{L}\,\overline{M}N+\overline{K}L\overline{M}\,\overline{N}+\overline{K}L\overline{M}N+K\overline{L}M\overline{N}+K\overline{L}MN+KLM\overline{N}+KLMN$$

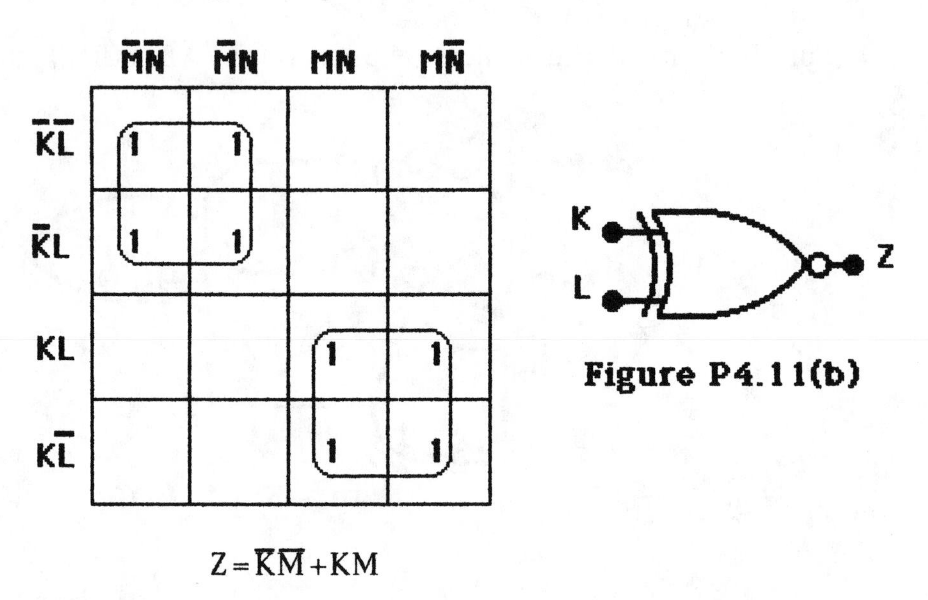

Figure P4.11(b)

$$Z=\overline{K}\,\overline{M}+KM$$

The expression $Z=\overline{K}\,\overline{M}+KM$ is implemented in Figure P4.11 (b)

SECTIONS 4.7 *Parity Generator and Checker*

4.12 The first step is to set up the truth table for the problem. Note that output Y will be the "ODD parity" output, while Z is the "EVEN parity" output. By looking at the truth table it is obvious that output Z is the complement of output Y. Therefore, we can design the circuit for the output Y and then add an INVERTER to the output of that circuit to obtain output Z.

C	B	A	Y	Z
0	0	0	1	0
0	0	1	0	1
0	1	0	0	1
0	1	1	1	0
1	0	0	0	1
1	0	1	1	0
1	1	0	1	0
1	1	1	0	1

$Y = \overline{A}\overline{B}\overline{C} + AB\overline{C} + A\overline{B}C + \overline{A}BC$

$Y = \overline{C}(\overline{A}\overline{B} + AB) + C(A\overline{B} + \overline{A}B)$

Let C=X and $W = A\overline{B} + \overline{A}B$

Hence, $Y = \overline{X}\overline{W} + XW$

$Y = \overline{X \oplus W}$

The expression $Y = \overline{X \oplus W}$ is implemented in Figure P4.12.

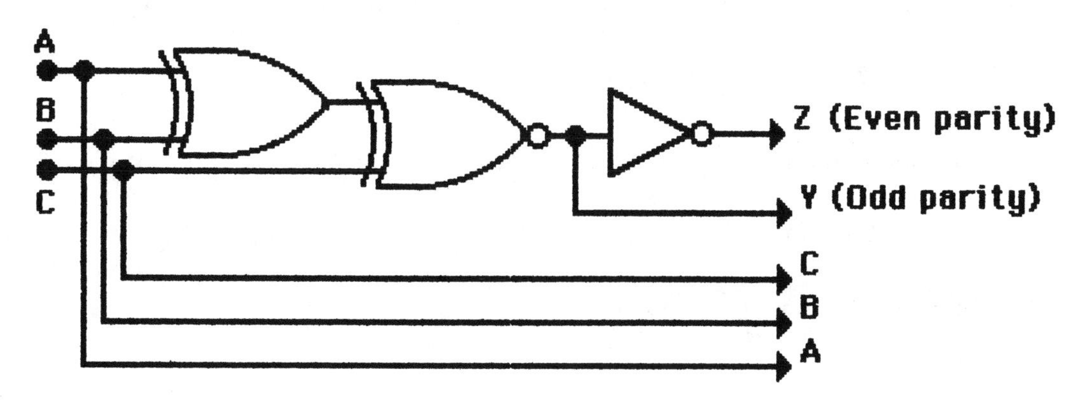

Figure P4.12: 3-bit Even/Odd parity generator.

4.13 The output of the 3-bit Odd/Even Parity Checker (Figure P4.13) will be HIGH if there is an error in the data received from circuit of Figure P4.12.

Y	C	B	A	E	
0	0	0	0	1	$\overline{Y}\overline{C}\overline{B}\overline{A}$
0	0	0	1	0	
0	0	1	0	0	
0	0	1	1	1	$\overline{Y}\overline{C}BA$
0	1	0	0	0	
0	1	0	1	1	$\overline{Y}C\overline{B}A$
0	1	1	0	1	$\overline{Y}CB\overline{A}$
0	1	1	1	0	
1	0	0	0	0	
1	0	0	1	1	$Y\overline{C}\overline{B}A$
1	0	1	0	1	$Y\overline{C}B\overline{A}$
1	0	1	1	0	
1	1	0	0	1	$YC\overline{B}\overline{A}$
1	1	0	1	0	
1	1	1	0	0	
1	1	1	1	1	$YCBA$

$$E=\overline{Y}\overline{C}\overline{B}\overline{A}+\overline{Y}\overline{C}BA+\overline{Y}C\overline{B}A+\overline{Y}CB\overline{A}+Y\overline{C}\overline{B}A+Y\overline{C}B\overline{A}+YC\overline{B}\overline{A}+YCBA$$

$$E=\overline{Y}\overline{C}(\overline{A}\overline{B}+AB)+\overline{Y}C(\overline{A}B+A\overline{B})+Y\overline{C}(\overline{A}B+A\overline{B})+YC(\overline{A}\overline{B}+AB)$$

Let $X=\overline{A}\overline{B}+AB$

$$E=\overline{Y}\overline{C}X+\overline{Y}C\overline{X}+Y\overline{C}\overline{X}+YCX$$

$$E=X(\overline{Y}\overline{C}+YC)+\overline{X}(\overline{Y}C+Y\overline{C})$$

$$E=(\overline{A\oplus B})(\overline{Y\oplus C})+(A\oplus B)(Y\oplus C)$$

Let $K=(Y\oplus C)$ and $Q=(A\oplus B)$

$$E=\overline{Q}\overline{K}+QK=\overline{Q\oplus K}$$

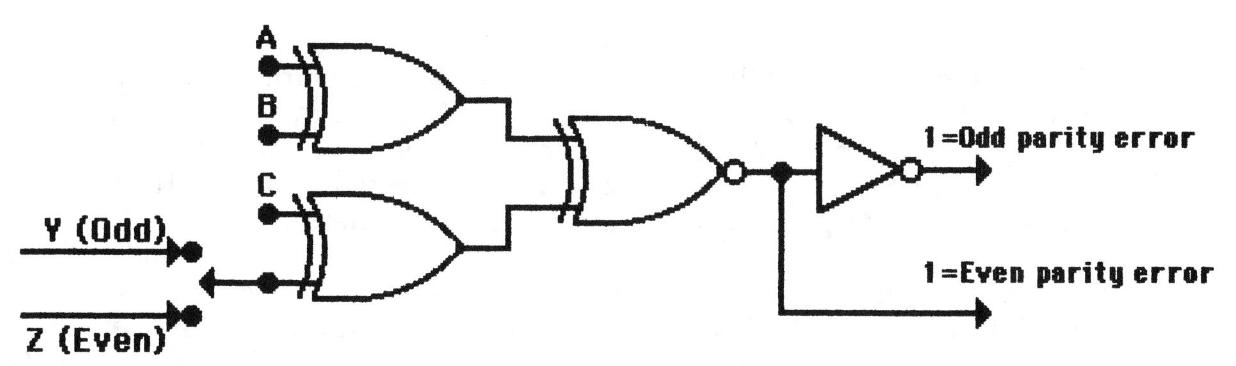

Figure P4.13: 3-bit Odd/Even Parity Checker.

SECTIONS 4.9 *Basic Characteristics of Digital ICs*

4.14 (a) Bipolar transistors are used by the TTL logic family, while the CMOS logic family uses P-channel and N-channel MOSFETs.
(b) If a TTL input is left unconnected, that input acts like a logic HIGH.
(c) If a CMOS input is left floating, the IC may become overheated and eventually destroy itself.
(d) If an IC has fewer than 12 gates built on its substrate, then it's considered Small Scale Integration. However, if it has over 10,000 gates built on its substrate, then the IC falls into the category of Very Large Scale Integration.

SECTIONS 4.10 *Troubleshooting Digital Systems*

4.15 The signal at Z1-6 and Z2-2 should be a logic HIGH. The logic probe indicates a logic LOW at both points. The following are the most probable causes of the malfunction:

1. An internal short to ground at Z2-2 (Bad Z2).
2. An internal short to ground at Z1-6 (Bad Z1).
3. An external short to ground between the two points.

Remote possibility:

4. By checking the specification sheet for a 74HC02, we see that Z1-7 is the ground connection for IC-Z1. It is possible that the technician made the connection from Z2-2 to Z1-7, thereby making the connection to the pin adjacent to pin 6.

4.16 Z2-10 is a logic LOW, while Z2-2, which is the same electrical point, is a logic HIGH. This means that the most probable cause for the malfunction is an open between the connection from Z2-10 to Z2-2. Since this is a TTL logic circuit, an open at input Z2-2 would have acted as a logic HIGH, however, the logic probe would have indicated "indeterminate."

4.17 Clearly, the signal at Z1-4 should be present at Z2-1. However, Z2-1 is indeterminate instead of pulsing. Since the technician used IC sockets, it is possible and very common for one of the pins (pin 1) of the IC socket or IC Z2 itself to get bent during installation, thereby causing one of the IC or IC socket pins (pin 1) not to get connected to the rest of the circuit. In either case an indeterminate state would have been recorded by the logic probe at Z2-1.

4.18 In cases such as this, it helps to know what the circuit looks like. To determine that, we must use the procedure discussed earlier in the chapter to set up the truth table, write the S-of-P expression, simplify and implement it.

1. Truth table and simplification of the S-of-P expression.

D	C	B	A	X	
0	0	0	0	0	
0	0	0	1	0	
0	0	1	0	1	$\overline{D}\overline{C}B\overline{A}$
0	0	1	1	1	$\overline{D}\overline{C}BA$
0	1	0	0	0	
0	1	0	1	0	
0	1	1	0	0	
0	1	1	1	0	
1	0	0	0	0	
1	0	0	1	1	$D\overline{C}\overline{B}A$
1	0	1	0	X	Don't care condition*.
"	"	"	"	"	" " "
1	1	1	1	X	Don't care condition.

* *These are don't care conditions since a BCD code cannot have binary numbers greater than 1001_2.*

$X = \overline{D}\overline{C}B\overline{A} + \overline{D}\overline{C}BA + D\overline{C}\overline{B}A$

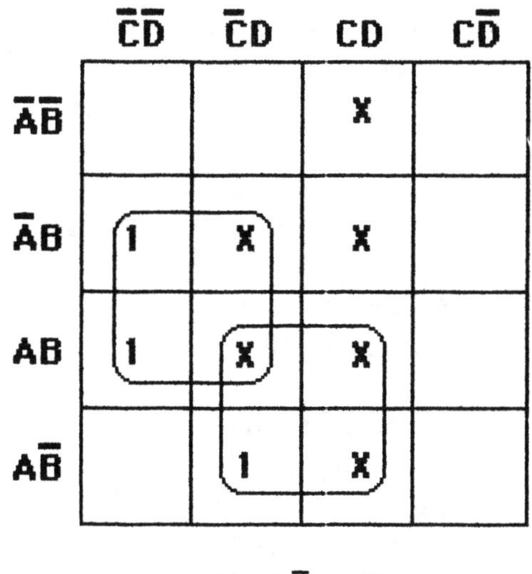

$X = B\bar{C} + AD$

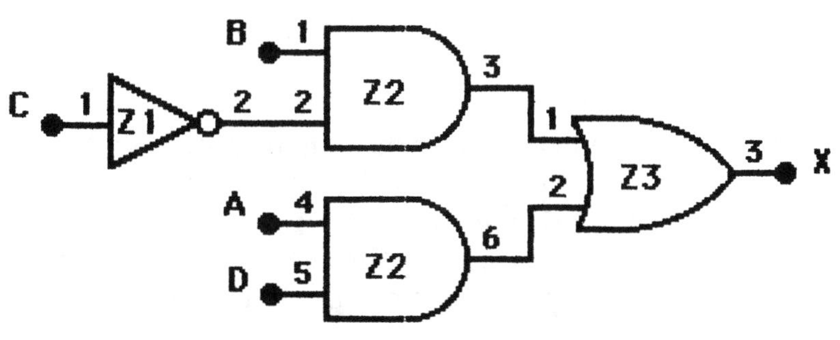

Figure P4.18

Figure P4.18 shows the circuit implementation for the $X = B\bar{C} + AD$ expression. Remember that the technician noticed that output X was always HIGH. The following are 5 possible causes:

1. Output Z3-3 is open or shorted to Vcc.
2. Input Z3-1 is open or shorted to Vcc.
3. Input Z3-2 is open or shorted to Vcc.
4. Output Z2-6 is open or shorted to Vcc.
5. Output Z2-3 is open or shorted to Vcc.

4.19 The problem with this TTL circuit is that points Z2-6, and Z2-10 are indeterminate instead of HIGH. Likewise, points Z2-11 and Z2-9 are indeterminate instead of pulsing.

(a) No. This would cause input Z2-13 to float and thereby assume a logic HIGH. Under normal operating conditions, Z2-13 should be HIGH, therefore, choice (a) could not cause the observed malfunction.

(b) No. This would make Z2-11 and Z2-9 assume a permanent logic HIGH. This does not explain the indeterminate logic levels at Z2-6 and Z2-10.

(c) No. This would cause Z2-9 to float, or assume a logic HIGH, and would not prevent Z2-11 from pulsing.

(d) Yes. If these two points were shorted together, then Z2-6/9/10 would try to follow the pulsing output of Z2-11. However, output Z2-6 under normal conditions is a constant HIGH. Thus, when Z2-11 goes LOW, an indeterminate logic level may occur at Z2-6.
This particular problem occurs when a point in a circuit is forced to be LOW and HIGH at the same time.

(e) Yes. This would cause indeterminate logic levels at the various outputs of Z2.

4.20 The technician needs three 2-input NAND gates, and she uses three 3-input NAND gates. To use them in the circuit, she must connect the unused input from each NAND gate to +5V (unused CMOS inputs must be connected to +5v or ground). If these CMOS inputs are left floating, then the results reported can, and most probably will occur.

5 FLIP-FLOPS AND RELATED DEVICES

SECTIONS 5.1-5.3 *NAND Gate Latch/NOR Gate Latch/ Troubleshooting Case Study*

5.1 (a) The output of a NOR gate latch is invalid, when the SET and the CLEAR inputs are HIGH.
(b) The Q output of a NAND gate latch is LOW, when the SET input is HIGH and the CLEAR input is LOW.
(c) The output of a NAND gate latch will not change, if both inputs are HIGH.

5.2 

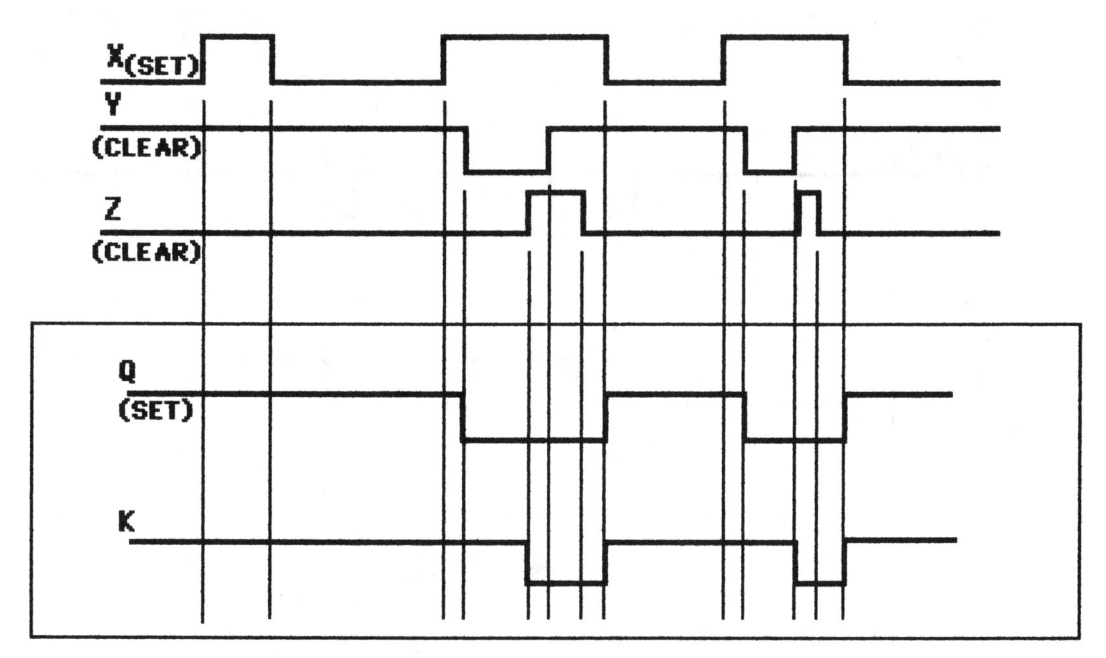

Figure P5.2

5.3 There are two possibilities for the malfunction:

(a) If the K output is permanently HIGH, then it can be said that the K latch is always SET. The conditions necessary for a NOR latch to Set are: Clear=0, SET=1. Since FF K is a TTL NOR latch, if its SET input is open or shorted to Vcc (either case is a TTL HIGH), then the latch will always be Set.
This condition in conjunction with waveform Z will cause, at certain times, FF K to have a logic HIGH at both the SET and CLEAR inputs. This is a violation of the truth table for the NOR latch and may cause erroneous Flip-Flop operation.

(b) Output K is internally or externally connected to Vcc.

SECTIONS 5.4-5.5 *Clock Signals and Clocked Flip-Flops/ Clocked S-C Flip-Flop*

5.4

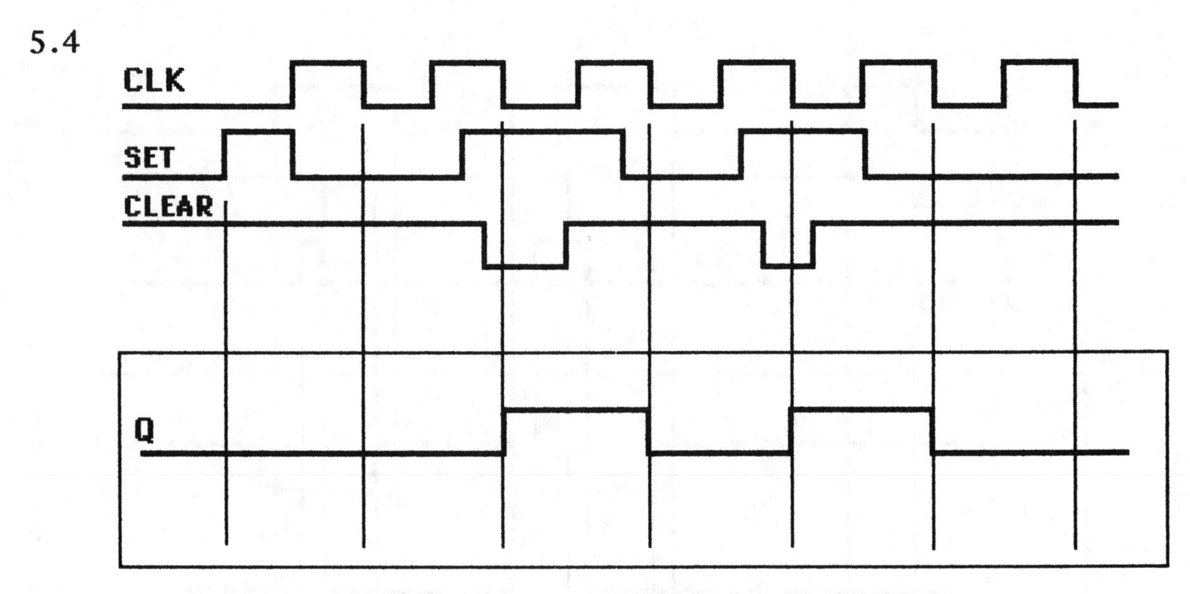

Figure P5.4

5.5 (a) Change the word asynchronous to *Synchronous.*
(b) Change the word Set-up to *Hold Time.*

SECTIONS 5.6-5.8; 5.25 Clocked J-K Flip-Flop / Clocked D Flip-Flop / D Latch (Transparent Latch / Troubleshooting Flip-Flop Circuits.

5.6 (a) No. In order for the output of the flip-flop to be half the frequency of the input signal, the flip-flop would have to be operating in the toggle mode (J=K=1 during the NGT). Figure P5.6 (a) clearly shows that this is not the case.

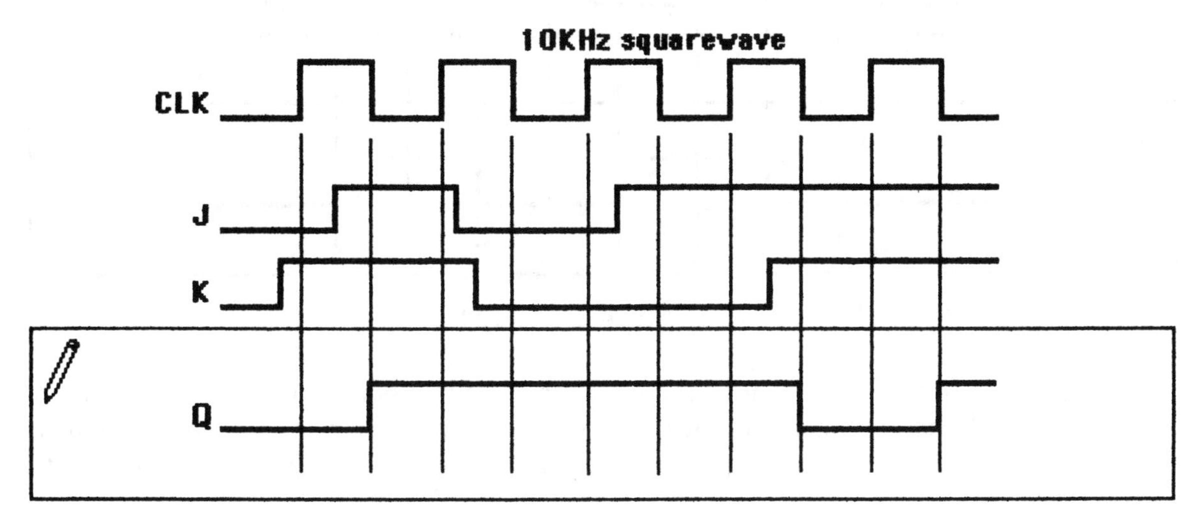

Figure P5.6(a)

(b) No. If the J input was permanently grounded, the flip-flop would never set or be in the toggle mode. Figure P5.6 (b) shows what the Q waveform would look like if this problem existed.

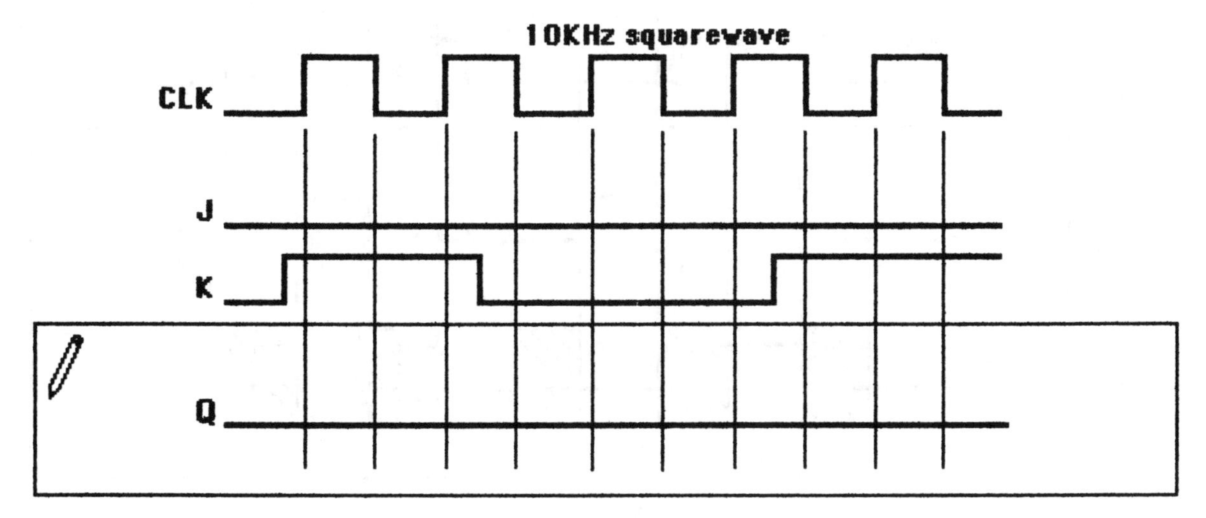

Figure P5.6(b)

(c) Yes. As demonstrated in Figure P5.6 (c), if K had been internally shorted to Vcc the output would have been a 5-KHz squarewave.

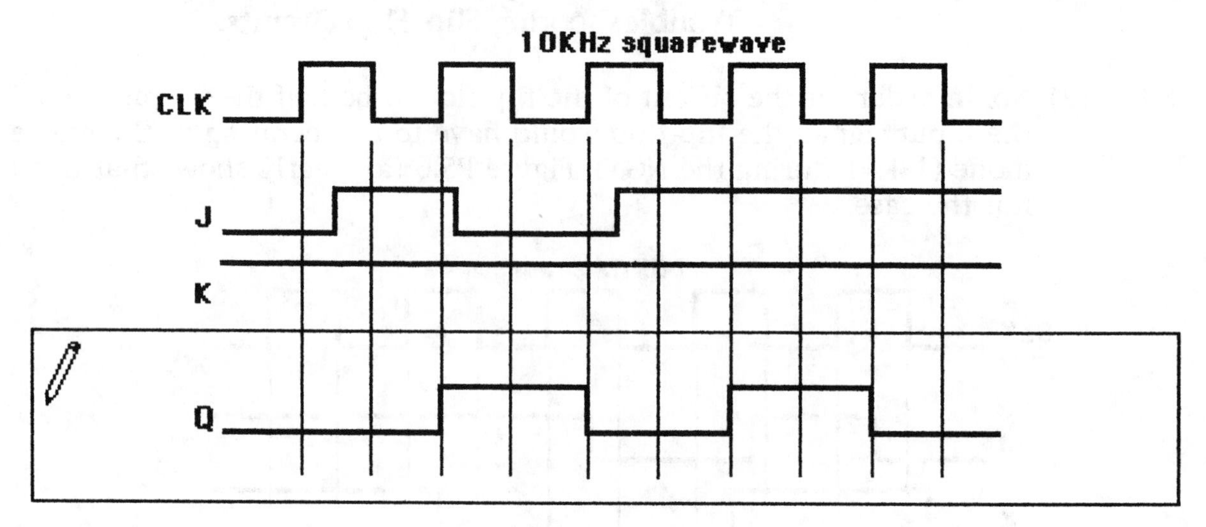

Figure P5.6(c)

(d) If Q and $\overline{Q}$ were connected together, the Q waveform observed on the oscilloscope would have had indeterminate logic levels, since each of the two flip-flop outputs would be trying to change to opposite states.

5.7 Let's assume that the flip-flops are initially LOW. The shaded area of Figure P5.7 shows when the D Latch is in its transparent mode.

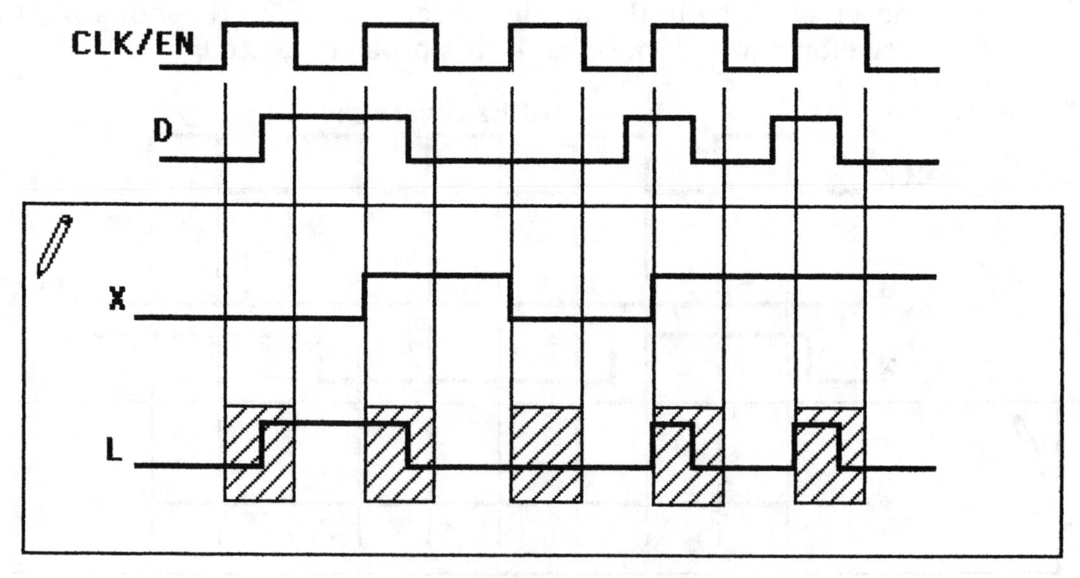

Figure P5.7

5.8 The D Latch is in its 'transparent' mode whenever EN is HIGH. Thus, if EN is either internally or externally connected to Vcc, or open if it is a TTL device, output L will follow the D input.

SECTIONS 5.9-5.10 *Asynchronous Inputs/IEEE-ANSI Symbols*

5.9

C1	1J	1K	R	S	1Q	
NGT	1	1	1	0	1	(Set asynchronously)
NGT	1	0	1	1	1	(Set synchronously)
NGT	0	0	0	1	0	(Clear asynchronously)
NGT	1	1	1	1	1	(Clear)

Figure P5.9

SECTIONS 5.11-5.12 *Flip-Flop Timing Considerations/ Potential Timing Problem in FF Circuits*

5.10 The input clock signal is 1-MHz, 7.5% D.C. Therefore, T=1μs and t_P=75ns. Figure 5.6 shows that $t_{W(H)}$=100ns for the 4013B flip-flop. It is evident that the $t_{W(H)}$ requirement has been violated, and consequently the operation of the flip-flop cannot be guaranteed by the manufacturer.

5.11 The Z waveform cannot be properly sketched, because the signal at the asynchronous input D is changing at the same time as the clock (Positive-Going-Transitions a and b). This becomes a problem because the required hold-time for the 7474 flip-flop is 5ns and in this situation the allowed hold-time is basically 0ns. Therefore, waveform Z can only be predicted with any degree of certainty up to the shaded area. It is anyone's guess what happens during the shaded area. This problem does not exist for output M, because the hold-time requirement for a 4013B flip-flop is 0ns.

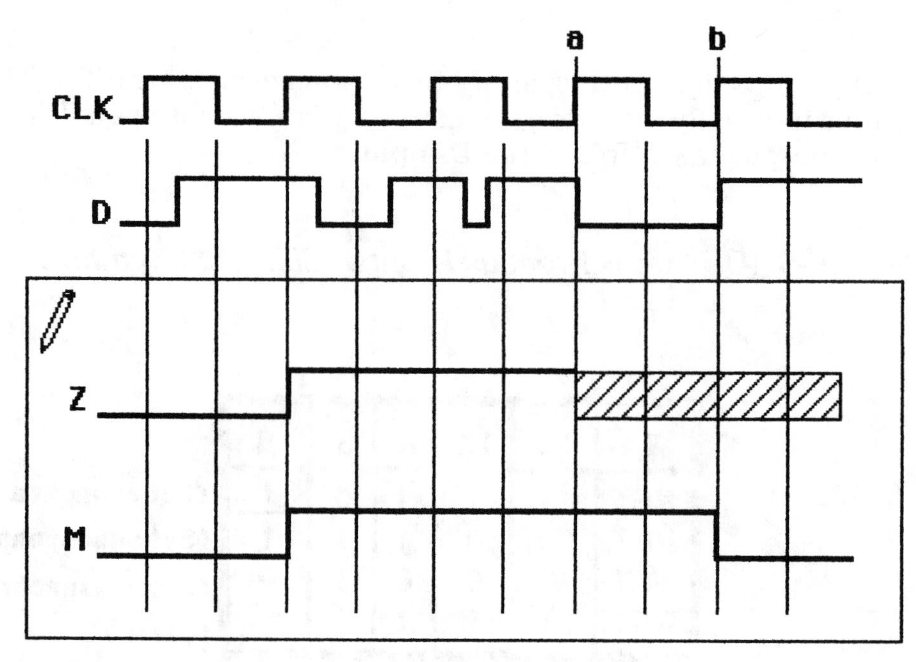

Figure P5.11

<u>SECTION 5.13</u> *Master-Slave Flip-Flops*

5.12 In a M/S flip-flop, the control inputs must be held stable while the CLK is HIGH or unpredictable operation will occur. In an M/S flip-flop with Data Lock-out, the flip-flop can only be affected during a clock transition.

<u>SECTIONS 5.17-5.18</u> *Data Storage and Transfer/ Serial Data Transfer: Shift Registers*

5.13 (a) D flip-flop Y_0 of the Y-Register, should be Set after the fourth shift pulse. Instead, it stays at a logic LOW. Thus, one possible cause for the malfunction is that Y_0 output may be internally or externally shorted to ground. Another possible cause is an open CLK input to flip-flop Y_0.

(b)

X3	X2	X1	X0	Y3	Y2	Y1	Y0	
1	0	1	1	0	0	0	0	<-- Before pulses applied.
0	1	0	1	1	0	0	0	<-- After the first pulse.
0	0	1	0	1	1	0	0	<-- After the second pulse.
0	0	0	1	0	1	1	0	<-- After the third pulse.
0	0	0	0	1	0	1	0	<-- After the fourth pulse.

SECTION 5.19 *Frequency Division and Counting*

5.14

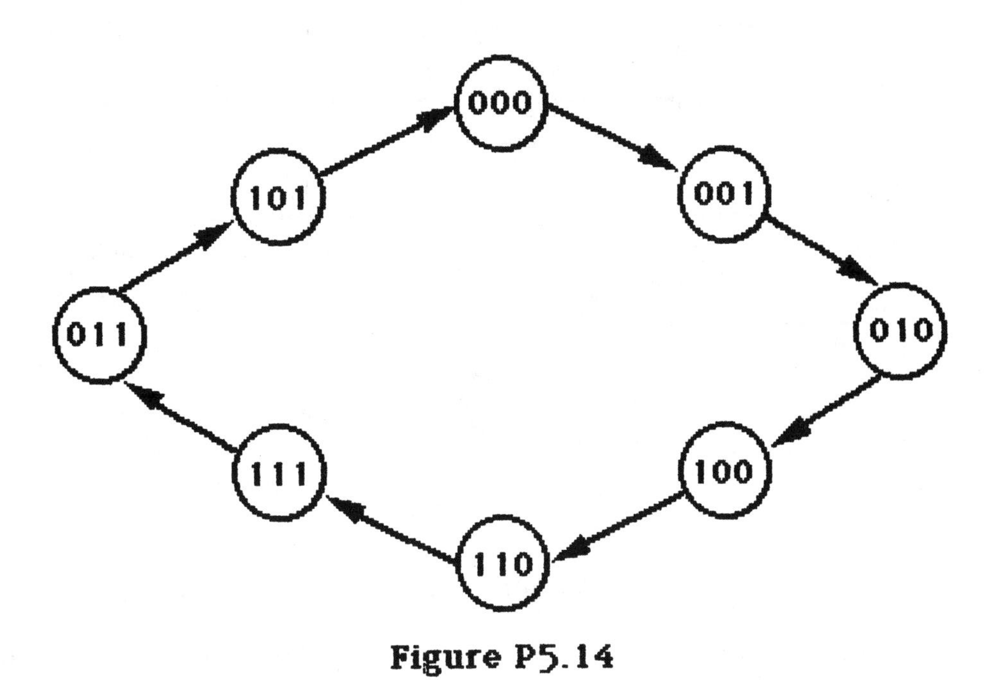

Figure P5.14

5.15 This counter is called a MOD-8 counter because it has 8 different states.

5.16 The counter will recycle every 8 counts. Thus, 3673/8=459.125. In other words, the counter will be at count 101_2 after it recycles 459 times. However, 0.125/8 represents one more count. Therefore, at the end of 3673 pulses the counter will hold count 000_2.

5.17 (a) Let's look at the timing diagram of Figure P5.17(a). This counter is a MOD-8 counter. The input frequency (100 KHz, 10% D.C.) is divided by two at X_0, divided by four at X_1, and divided by eight at X_3. Furthermore, the resulting signals are 50% Duty Cycle waveforms. Thus, squarewaves are present at X_0, X_1 and X_3 outputs, and their frequencies are 50 KHz,, 25 KHz and 12.5 KHz respectively.

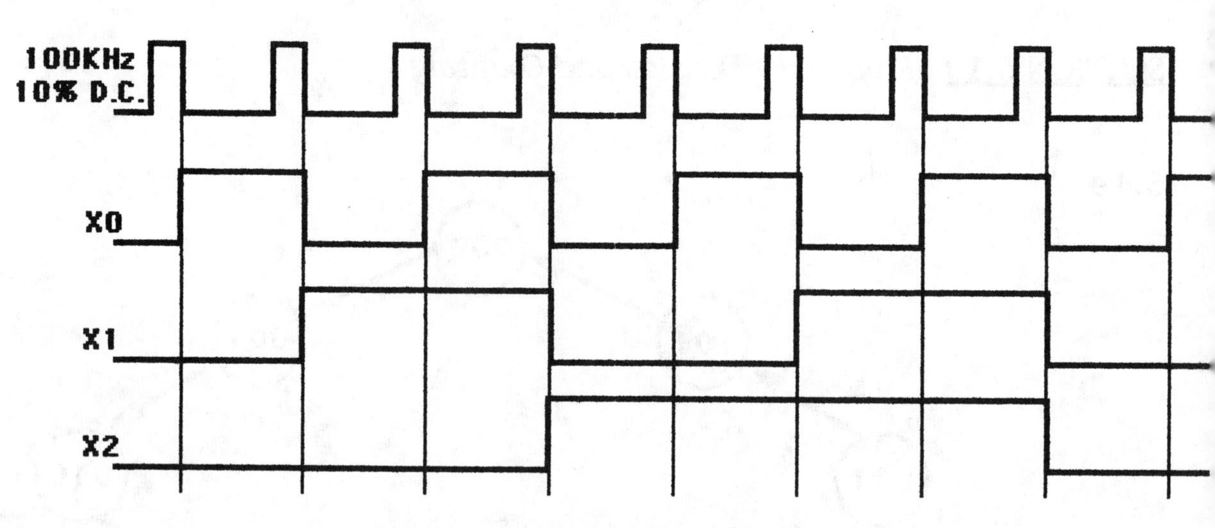

Figure P5.17(a)

(b) Total number of states is equal to 2^N, or 2^8, or 256.
Thus, there are a total of 256 different states (0_{10}-255_{10}).

(c) The maximum binary count that this counter can reach is 11111111_2.

<u>SECTION 5.20</u> *Microcomputer Applications*

5.18 Let's look at the necessary logic levels on the Address Bus lines A_0-A_{15}:

A_{15}	A_{14}	A_{13}	A_{12}	A_{11}	A_{10}	A_9	A_8	A_7-A_0	
0	0	0	0	1	1	1	1	X-X	= $0Fxx_{16}$
0	0	0	0	1	1	1	0	X-X	= $0Exx_{16}$
0	0	0	0	1	1	0	1	X-X	= $0Dxx_{16}$
0	0	0	0	1	1	0	0	X-X	= $0Cxx_{16}$

The circuit of Figure P5.18 shows one possible way of decoding the Address Bus for the design of the "Decoding Logic" circuitry.

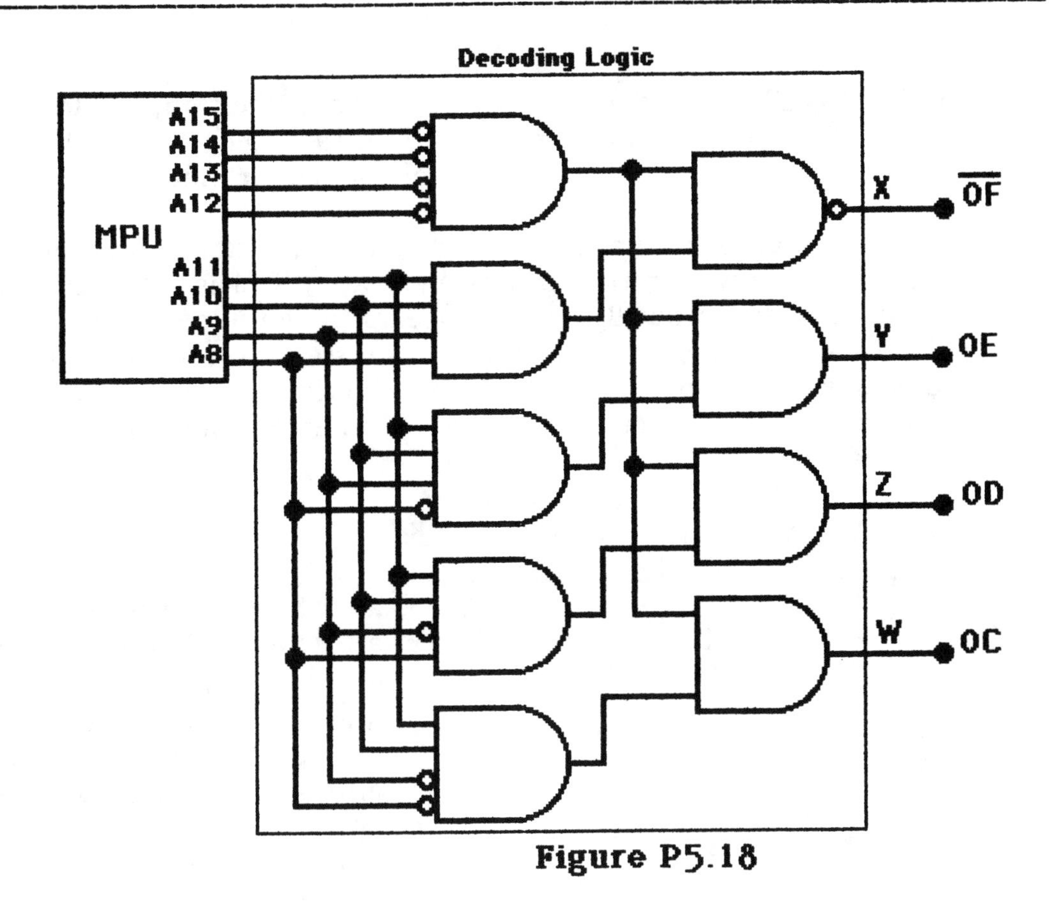

Figure P5.18

5.19 An external open on A_9 has the same affect as a TTL logic HIGH input to the decoding logic. Address line A_9 must be LOW for the decoding of addresses $0Dxx_{16}$ and $0Cxx_{16}$. Therefore, if an external open exists on the Address Bus line A_9, outputs W and Z will always be LOW regardless of the logic levels on the Address Bus.

SECTION 5.22 *One-Shot*

5.20 A retriggerable One-shot can be retriggered while it is in the quasi-stable state, and it will begin a new t_p interval. A Nonretriggerable One-Shot will not react to any signal on its trigger input while it is in its quasi-stable state. A Nonretriggerable One-Shot must time-out before it will react to another trigger pulse.

SECTION 5.25 *Troubleshooting Flip-Flop Circuits*

5.21 (a) The data present at inputs A_3-A_0 will be transferred to outputs B_0-B_3 after SW1 is depressed. Thus, after the transfer, outputs B_3=1, B_2=0, B_1=0, and B_0=1.

(b) The INVERTERS are used between the J and the K inputs, to insure that each J-K flip-flop will never operate in the toggle mode. Thus, upon clocking the flip-flops, they can only get Set or Clear.

5.22 By careful analysis of the table of Figure P5.22, it can be determined that outputs B_0-B_2 always reflect inputs A_0-A_2 after SW1 is activated. The problem is with output B_3, which appears to be faulty at times (*). It should also be noted that, when SW1 is pressed for a second time after a faulty transfer has occurred, the resulted output is correct. Thus, the question is, what particular fault with flip-flop B_3 could cause the observed results?

	A_3	A_2	A_1	A_0	B_3	B_2	B_1	B_0	
	1	0	0	1	0	0	0	0	after SW2 is depressed.
	1	0	0	1	1	0	0	1	after SW1 is depressed.
	1	1	0	0	1	0	0	1	after # A_0-A_3 is changed.
*	1	1	0	0	0	1	0	0	after SW1 is depressed.
	1	1	0	0	1	1	0	0	after SW1 is depressed again.
	1	1	1	1	1	1	0	0	after # A_0-A_3 changed.
*	1	1	1	1	0	1	1	1	after SW1 is depressed.
	1	1	1	1	1	1	1	1	after SW1 is depressed again.
	1	1	1	1	0	0	0	0	after SW2 is depressed.

Figure P5.22

If the K input to flip-flop B_3 is permanently HIGH, then flip-flop B_3 may occasionally get in the toggle mode (when J=K=1), and consequently changes states with each clock transition. Without the usage of some troubleshooting equipment, we can only guess about possible causes:

(a) Either the K input of flip-flop B_3, or the output of the INVERTER is internally or externally connected to Vcc.
(b) An open connection between the output of the INVERTER and the K input of flip-flop B_3, would also cause the same results.

5.23 The following steps describe the sequence of events necessary for a transfer to occur:

1. Binary data must be present and stable at inputs A_3-A_0.
2. A 10 Hz squarewave must be present at Z2-1.

3. SW1 is actuated momentarily, and the Q output of the One-Shot goes to its quasi-stable state for 10ms.
4. The 10ms logic HIGH at Z2-2, allows a Negative-Going-Transition of the 10Hz squarewave to occur at output Z2-3.
5. The Negative-Going-Transition at Z2-3 will clock the data from inputs A_3-A_0 to outputs B_0-B_3.

(a) The problem is with step 4. The designer of this circuit did not make any provisions for the synchronization between output Q of the One-Shot and the 10Hz squarewave. Thus, the pulse from the One-Shot can occur at any time during the 10Hz squarewave. Hence, there is a good chance that the pulse from the One-Shot will occur during the off time of the 10Hz squarewave; one possible scenario is shown in Figure P5.23(a). At such times, flip-flops B_0-B_3 won't be clocked and no transfer of data will take place. As a matter of fact, it may take a few actuations of switch SW1 for data to be successfully transferred to the output.

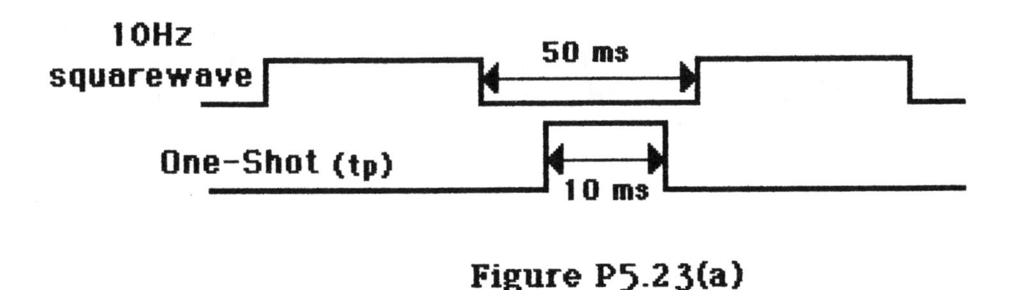

Figure P5.23(a)

(b) There are various ways of fixing this synchronization problem:

1. Increase the squarewave frequency so that its period is much smaller than the pulse width t_P of the One-Shot.
2. Increase t_P of the One-Shot to a value greater than the period of the squarewave.
3. A more complex, but much more elegant way of fixing the problem, is to use an extra D-type flip-flop and the $\overline{Q}$ output of the One-shot. Figure P5.23(b) shows the logic circuit that will synchronize the clocking of flip-flops B_0-B_3 and the 10Hz squarewave.

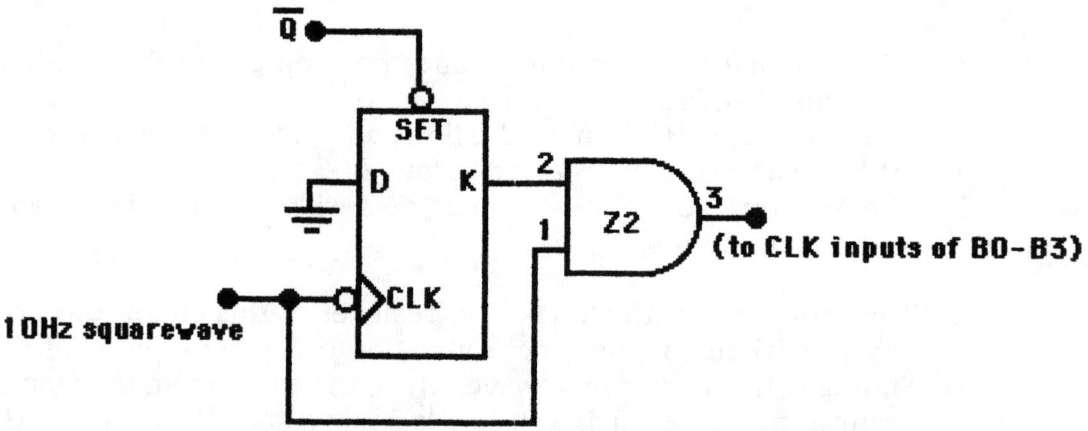

Figure P5.23(b)

5.24 Flip-flops B_0-B_3 are clocked when Z2-3 goes LOW. Z2-3 goes LOW when either or both Z2-2 and Z2-1 go LOW. If the output of the OS goes LOW while Z2-1 is HIGH, a race condition will occur between the synchronous inputs of the flip-flops and the clock signal. In other words, the flip-flops are clocked at the same time that the logic levels at the J and K inputs are changing. This will result in random data being loaded in register B_0-B_3. This problem can be eliminated by using J-K flip-flops with zero hold time requirements (t_H=0).

6 DIGITAL ARITHMETIC: OPERATIONS AND CIRCUITS

SECTION 6.1 *Binary Addition*

6.1 (a)

$$\begin{array}{r} 110110_2 \\ +\ 010010_2 \\ \hline 1001000_2 \end{array}$$

(b)

$$\begin{array}{r} 10101.001_2 \\ +\ 01011.000_2 \\ \hline 100000.001_2 \end{array}$$

(c)

$$\begin{array}{r} 11101_2 \\ +\ 10000_2 \\ \hline 101101_2 \end{array}$$

SECTION 6.2 *Representing Signed Numbers*

6.2 (a) $\underline{0}1110_2 = +14_{10}$ Note: sign bit (MSB) is <u>zero</u> since this is a <u>positive</u> number.

(b) *To express a negative binary number using the 2's Complement System, first write the number as if it was a positive number.*

$0100000_2 = +32_{10}$

Then take the 1's complement of the binary number by inverting each bit.

1011111_2 = 1's Complement

Finally, obtain the 2's complement by adding 1 to the 1's complement.

$$\begin{array}{r} 1011111_2 \\ +\ 0000001_2 \\ \hline \underline{1}100000_2 \end{array}$$

$\underline{1}100000_2$ = 2's Complement = -32_{10}

sign bit (MSB) is <u>one</u> since this is a <u>negative</u> number.

(c) $\underline{0}1000_2 = +8_{10}$

$$
\begin{array}{rl}
 & 10111_2 = \text{1's Complement} \\
+ & 00001_2 \\
\hline
 & \underline{1}1000_2 = \text{2's Complement} = -8_{10}
\end{array}
$$

(d) $\underline{0}110111_2 = +55_{10}$

(e) $0101100011_2 = +355_{10}$

$$
\begin{array}{rl}
 & 1010011100_2 = \text{1's Complement} \\
+ & 0000000001_2 \\
\hline
 & 1010011101_2 = \text{2's Complement} = -355_{10}
\end{array}
$$

(f) $01111011_2 = 123_{10}$

SECTIONS 6.3-6.4 *Addition in the 2's-Complement System/ Subtraction in the 2's-Complement System*

0 <--------(sign bit is 0, therefore number is positive)

6.3 (a) $\underline{0}10110_2=(1x2^4)+(0x2^3)+(1x2^2)+(1x2^1)+(0x2^0)=+22_{10}$

1 <--------(sign bit is 1, therefore number is negative)

(b) $\underline{1}10111_2$

$$
\begin{array}{rl}
 & 011000_2 = \text{1's Complement.} \\
+ & 000001_2 = \text{Add 1 to make a 2's Complement number.} \\
\hline
 & 001001_2 = \text{2's Complement.}
\end{array}
$$

Thus, $001001_2=(0x2^4)+(1x2^3)+(0x2^2)+(0x2^1)+(1x2^0) = +9_{10}$

Since the original binary number 110111_2 represents a negative number, the final answer is -9_{10}

0 <--------(sign bit is 0, therefore number is positive)

(c) $\underline{0}1110011_2=(1x2^6)+(1x2^5)+(1x2^4)+(0x2^3)+(0x2^2)+(1x2^1)+(1x2^0)=+115_{10}$

1 <--------(sign bit is 1, therefore number is negative)
(d) $\underline{1}1000_2$
00111_2 = 1's Complement.
+ 00001_2 = Add 1 to make a 2's Complement number.

01000_2 = 2's Complement.

Thus, $01000_2=(1x2^3)+(0x2^2)+(0x2^1)+(0x2^0)=+8_{10}$

Since the original binary number 11000_2 represents a negative number, the final answer is -8_{10}

1 <--------(sign bit is 1, therefore number is negative)
(e) $\underline{1}11001_2$
000110_2 = 1's Complement.
+ 000001_2 = Add 1 to make a 2's Complement number.

000111_2 = 2's Complement.

Thus, $000111_2=(0x2^4)+(0x2^3)+(1x2^2)+(1x2^1)+(1x2^0)=+7_{10}$

Since the original binary number 111001_2 represents a negative number, the final answer is -7_{10}

6.4 (a) [8+5]-----------> +8 = $\underline{0}1000_2$
+5 = $\underline{0}0101_2$
+
01101_2 = $+13_{10}$

(b) [-12+15]-----------> +12 = $\underline{0}1100_2$
10011_2 1's Complement.
00001_2 Add 1.
+
10100_2 = 2's Complement = -12_{10}

+15 = 01111_2

-12 = 10100_2
+15 = 01111_2
+
1 $\underline{0}0011_2$ = $+3_{10}$
|__ Overflow bit is disregarded.

(c) *To perform a subtraction using the 2's complement system, we must take the 2's complement of the subtrahend and then add it to the minuend.*

[+32-(-8)]-----> $+32 = \underline{0}100000_2$ (minuend)
$+8 = \underline{0}001000_2$
1110111_2 1's Complement.
0000001_2 Add 1.
+ ________
$\underline{1}111000_2 = -8_{10}$ (subtrahend)
0000111_2 1's Complement of the subtrahend.
0000001_2 Add 1.
+ ________
$\underline{0}001000_2$ 2's Complement of the subtrahend.

Now, let's add the minuend with the 2's Complement of the subtrahend.

$\underline{0}100000_2$
$\underline{0}001000_2$
+ ________
$\underline{0}101000_2 = +40_{10}$

(d) [9-13]--------> $+9 = \underline{0}1001_2$ (minuend)
$+13 = \underline{0}1101_2$ (subtrahend)
10010_2 1's Complement of the subtrahend.
00001_2 Add 1.
+ ________
$\underline{1}0011_2$ 2's Complement of the subtrahend.

Now, let's add the minuend with the 2's Complement of the subtrahend.

$\underline{0}1001_2$
$\underline{1}0011_2$
+ ________
$11100_2 = -4_{10}$

When the final answer is negative (Sign bit=1), the result is expressed in the 2's complement form.

SECTIONS 6.5-6.6 *Multiplication of Binary Numbers/Binary Division*

6.5 (a) $10111_2 \rightarrow 23_{10}$
x $00101_2 \rightarrow 5_{10}$

```
  10111
 00000
10111
```

$1110011_2 = 115_{10}$

(b) $1100.1_2 \rightarrow 12.5_{10}$
x $010.01_2 \rightarrow 2.25_{10}$

```
    11001
   00000
  00000
 11001
```

$11100.001_2 = 28.125_{10}$

(c) 0000 0101 | 0010 0011, quotient 111 $=7_{10}$

```
-1 01
  111
 -101
  101
 -101
    0
```

(d) 0000 1111 | 1000 0111, quotient 1001 $=9_{10}$

```
-111 1
  1111
 -1111
     0
```

SECTIONS 6.7-6.8 *BCD Addition/Hexadecimal Arithmetic*

When the addition of two BCD digits results in an illegal BCD code, the correction factor 6_{10} (0110_2) must be added to the result.

6.6 (a) $[25_{10} + 26_{10}]$

$25_{10} = 0010\ 0101_{BCD}$
$26_{10} = 0010\ 0110_{BCD}$
+

0100 1011
0110 --> Correction factor
+

$0101\ 0001_{BCD} = 51_{10}$

(b) $[50_{10} + 32_{10}]$

$$
\begin{array}{rl}
50_{10} = & 0101\ 0000_{BCD} \\
32_{10} = & 0011\ 0010_{BCD} \\
+ & \overline{} \\
& 1000\ 0010_{BCD} = 82_{10}
\end{array}
$$

(c) $[2357_{10} + 1250_{10}]$

$$
\begin{array}{rl}
2357_{10} = & 0010\ 0011\ 0101\ 0111_{BCD} \\
1250_{10} = & 0001\ 0010\ 0101\ 0000_{BCD} \\
+ & \overline{} \\
& 0011\ 0101\ 1010\ 0111 \\
& 0110 \text{ ----------> Correction factor} \\
& \overline{0011\ 0110\ 0000\ 0111}_{BCD} = 3607_{10}
\end{array}
$$

(d) $[9_{10} + 9_{10}]$

$$
\begin{array}{rl}
& 1 \text{ -----------> carry from previous digit} \\
9_{10} = & 0000\ 1001_{BCD} \\
9_{10} = & 0000\ 1001_{BCD} \\
+ & \overline{} \\
& 0001\ 0010 \\
& 0110 \text{ -->*Correction factor} \\
& \overline{0001\ 1000}_{BCD} = 18_{10}
\end{array}
$$

**Note:* *Even though the result of an addition may not produce an illegal BCD code, if a 1 is generated from one digit position to the next, the correction factor (0110_2) must be added.*

6.7 (a) $[75F_{16} + 32D_{16}]$

One way:

$$
\begin{array}{rl}
75F_{16} = & 0111\ 0101\ 1111_2 \\
32D_{16} = & 0011\ 0010\ 1101_2 \\
+ & \overline{} \\
& 1010\ 1000\ 1100_2 = A8C_{16}
\end{array}
$$

When the addition of two hexadecimal digits results in a number equal or greater than 16, then 16 must be subtracted from that number in order to obtain the proper result, in addition, a CARRY is generated into the next hex digit position.

Another way:

$$
\begin{array}{l}
\ \ 7\ 5\ F_{16} \\
\ \ 3\ 2\ D_{16} \\
+\overline{\qquad\quad}
\end{array}
$$

| | | |___ $(15+13)=28;\ 28-16 = 12_{10}$ + Carry $= C_{16}$

| |___ $(5+2+\text{Carry}) = 8_{10} = 8_{16}$

|___ $(7+3)=10_{10} = A_{16}$

Therefore, $[75F_{16}+32D_{16}] = A8C_{16}$

(b) $[12A_{16}-FF_{16}]$

$FF_{16} =$ $0000\ 1111\ 1111_2$ subtrahend
$1111\ 0000\ 0000_2$ 1's Complement
$1111\ 0000\ 0001_2$ 2's Complement of the subtrahend.

$12A_{16} =$ $0001\ 0010\ 1010_2$ minuend
$+$ $1111\ 0000\ 0001_2$ 2's Complement of the subtrahend.

$\underline{1}\ 0000\ 0010\ 1011_2 = 2B_{16}$

|__ Overflow is disregarded in the subtraction.

Another way:

$$
\begin{array}{rl}
1\ 2\ A_{16} & \text{(minuend)} \\
-\ 0\ F\ F_{16} & \text{(subtrahend)} \\
\hline
\end{array}
$$

| | |___ $((10+16)-15) = {}^{*}11_{10} = B_{16}$

| |___ $((1+16)-15) = 2_{10} = 2_{16}$

|___ $(1-1) = 0_{10} = 0_{16}$

Therefore, $[12A_{16}-0FF_{16}] = 02B_{16}$

* When a hexadecimal number is being subtracted from a smaller number ($A_{16}-F_{16}$ in the above example), then a 1_{16} must be borrowed from the next digit (2_{16}) of the minuend and added to the previous digit of the minuend (A_{16}).

This process is repeated with the next digit position. Note that the hexadecimal 1_{16} that is borrowed weights 16_{10} due to its relative position to the digit that is borrowing.

(c) [7834_{16}+$ABCD_{16}$]

$$\begin{array}{rl} 7834_{16} & = 0111\ 1000\ 0011\ 0100_2 \\ ABCD_{16} & = 1010\ 1011\ 1100\ 1101_2 \\ + & \overline{} \\ & 1\ 0010\ 0100\ 0000\ 0001 = 12401_{16} \end{array}$$

|_ NOT disregarded in the addition.

Another way:

$$\begin{array}{r} 7\ 8\ 3\ 4_{16} \\ A\ B\ C\ D_{16} \\ + \overline{} \end{array}$$

(4+13)=17; 17-16 = 1_{16} + Carry
(3+12+Carry)=16; 16-16 = 0_{16} + Carry
(8+11+Carry)=20; 20-16 = 4_{16} + Carry
(7+10+Carry)=18: 18-16 = 2_{16} + Carry

Therefore, [7834_{16}+$ABCD_{16}$] = 12401_{16}

SECTIONS 6.9-6.15 *Arithmetic Circuits/Parallel Binary Adder/ Complete parallel Adder with Registers/ Integrated-Circuit Parallel Adder/ 2's Complement System*

6.8

	[A]	[B]	[S]	
#1	1101	0110	10011	---------> Correct result
#2	1000	0111	10111	---------> Incorrect result
#3	0100	0101	01001	---------> Correct result
#4	0011	0100	01111	---------> Incorrect result

We must try to find a common pattern on the incorrect additions #2 and #4. Additions #2 and #4 should have yielded 01111_2 and 00111_2 respectively. By careful analysis of all four additions, it can be concluded that additions #2 and #4 are the only ones where Carry C_3 is generated into the input C_{in} of the MSB's Full-Adder.

Since this is a TTL logic circuit, if the C_3 connection became open, a logic 1 would always be added to the B_3 and A_3 bits. Thus, an incorrect result would occur whenever C_3 results in a logic LOW. This of course, would not affect additions where C_3 results in a logic HIGH.

6.9

$[A_0\text{-}A_7]$	$[B_0\text{-}B_7]$	SW1	Outputs (C_{out}, S_7-S_0)
01100111	10001011	0 (add)	011110010_2
11000001	10100010	1 (sub.)	000011111_2
10100101	11000010	0 (add)	101100111_2
11100110	11000110	1 (sub.)	000100000_2

SECTION 6.16 *BCD Adder*

6.10

$A_3A_2A_1A_0$	$B_3B_2B_1B_0$	$S_3S_2S_1S_0$	X	$\Sigma_3\Sigma_2\Sigma_1\Sigma_0$
1001	1001	0010	1	1000
1000	0011	1011	1	0001
0111	0101	1100	1	0010
0110	0011	1001	0	1001

6.11 A TTL logic '1' would always be added to the final BCD sum (Σ_3-Σ_0). Thus, the sum would always be greater by a factor of '1', and in some cases the final BCD sum would be an illegal BCD code.

6.12 Whenever output X from the 'correction logic' is HIGH, a correction factor of 0110_2 has to be added to the results of the 4-bit parallel adder 7483 (see Figure 6.3). Figure P6.12 shows how two Half-Adders and a Full-Adder can be connected to implement the 'correction adder'.

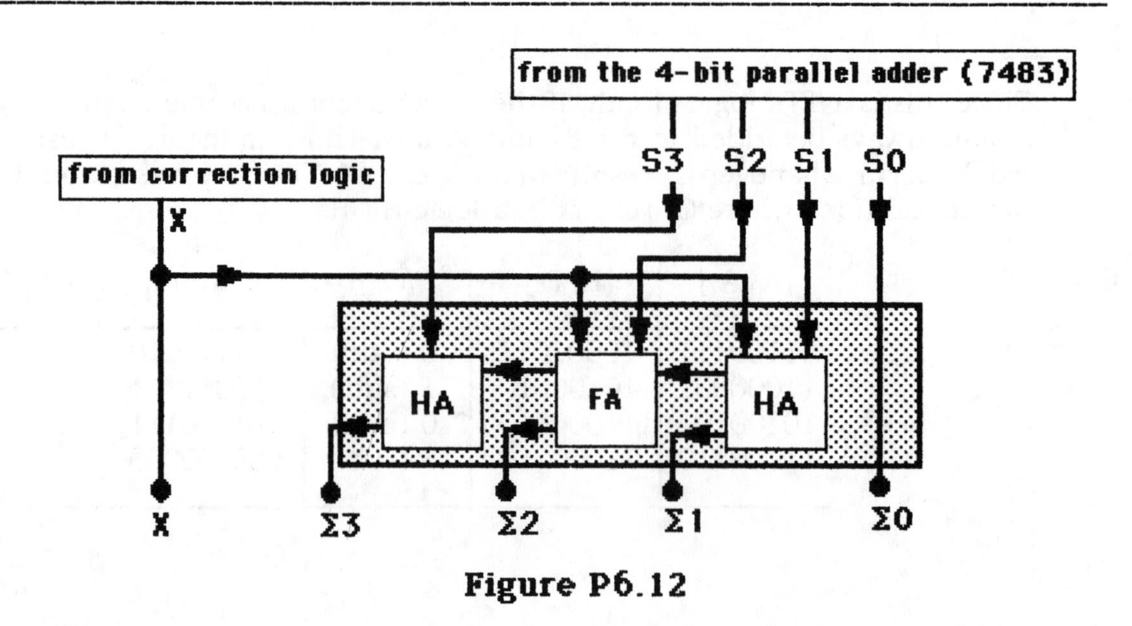

Figure P6.12

<u>SECTIONS 6.17,6.19,6.20</u> *Binary Multipliers/IEEE-ANSI Symbols/ Troubleshooting Case Study*

6.13 By looking at the table, we can see that whenever X=1, the BCD sum is correct. When X=0, the result exceeds the correct answer by four, and in one case causes an illegal BCD code. Thus, we can conclude that the 'Correction Adder' circuit always adds four to the answer, and in cases when X=1, it adds six. The following are the possible problems which could have caused the malfunction:

1) Output X is connected to only A_1 of the 'Correction Adder', while A_2 is externally open (always HIGH).
2) Input A_2 of the 'Correction Adder' is internally open or shorted to Vcc.

6.14 The problem with this circuit is that it occasionally produces the wrong result *. For certain multiplications, depending on the combinations of the multiplier and multiplicand, the result can exceed or be under the correct answer by one. Let's compare the answers obtained with the expected results and try to see if we can detect a common pattern among the wrong answers.

		Correct Answer	Answer Obtained
[11x13]	= 143	10001111_2 = 143	10001111_2 = 143
[10x13]	= 130	10000010_2 = 130	10000001_2 = 129*
[3x3]	= 9	00001001_2 = 9	00001010_2 = 10*
[6x2]	= 12	00001100_2 = 12	00001100_2 = 12
[2x2]	= 4	00000100_2 = 4	00000100_2 = 4
[5x2]	= 10	00001010_2 = 10	00001001_2 = 9*

By looking at the results, it can be observed that whenever the 'Correct Answer' has the two least significant bits equal, the 'Answer Obtained' is correct. When the two least significant bits are different, then the 'Answer Obtained' is incorrect. It appears that if the two least significant bits of the 'Answer Obtained' column had been switched, the correct results would always be obtained.

Thus, it must be concluded that the technician made a mistake when wiring the circuit. The technician probably connected outputs S_0 to the D input of flip-flop A_1 and S_1 to the D input of flip-flop A_0.

6.15 Each one of the digital circuits which make the multiplier of Figure 6.4, has an inherent propagation delay. It takes a certain amount of time for the sum outputs to be transferred from the 8-bit parallel adder to the outputs of the Accumulator, likewise for the B and X shift registers to shift left and right respectively. If the Clock pulses are occurring at a faster rate than these circuits can propagate, then random and unpredictable results are obtained from the multiplier circuit.

6.16 If the D input of Flip-Flop X_3 became open, then after four clock pulses the X shift register would have 1111_2 rather than 0000_2. This would allow transfer pulses at the output of the AND gate to occur continuously, thereby producing incorrect multiplication results.

6.17 This is an 8-bit parallel adder.

7 COUNTERS AND REGISTERS

SECTION 7.1 *Asynchronous (Ripple) Counters*

7.1

	MOD-#	Input freq.	Output freq.	J-K FFs
a)	64	256KHz	4.0KHz	6
b)	32	350KHz	10.93KHz	5
c)	8	800KHz	100.0KHz	3
d)	128	500KHz	3.9KHz	7

(a) $\text{MOD\#} = \dfrac{\text{Input Frequency}}{\text{Output Frequency}} = \dfrac{256\text{KHz}}{4\text{KHz}} = 64$

# of Flip-Flops = N
$2^N = 64$
$N = 6$

(b) $\text{Output Frequency} = \dfrac{\text{Input Frequency}}{\text{MOD\#}} = \dfrac{350\text{KHz}}{32} = 10.93\text{KHz}$

# of Flip-Flops = N
$2^N = 32$
$N = 5$

(c) MOD# $= 2^N$
$= 2^3 = 8$
Input Frequency = (Output Frequency) x (MOD-#)
= 100KHz x 8
= 800KHz

(d) MOD-# $= 2^7$
$= 128$

Output Frequency$= \frac{500KHz}{128} = 3.9KHz$

SECTIONS 7.2-7.3,7.24 ***Counters With MOD Numbers <2^N/ IC Asynchronous Counters/ Troubleshooting.***

7.2 (a) If the inputs to the 3-input NAND gate of Figure 7.2 are A, B and C, then its output will be LOW when the counter reaches the count of seven (0111_2). Therefore, the counter will count from binary 0000_2 to 0110_2 and then clear at the count of 0111_2. Since this counter goes through seven different states before recycling, it is called a MOD-7 counter.

(b) For this combination of inputs, the counter will be cleared at the binary count of thirteen (1101_2). This counter will go through thirteen different states (0000_2-1100_2), and therefore it is called a MOD-13 counter.

(c) Since one of the inputs to the NAND gate is always a logic 1 (+Vcc), its output will be LOW when inputs A and B are HIGH. Thus, this counter will count from binary 0000_2 to 0010_2 and then clear on the next count, binary 0011_2. Therefore, this is a MOD-3 counter.

7.3 When the circuit of Figure 7.3 is working properly, and it reaches the binary count 110_2, the Set/Clear flip-flop Q will be Set and counter A B C will clear. Thus, any condition which may prevent the Set/Clear flip-flop from clearing the counter, will cause the counter to count from binary 000_2 to 111_2 repeatedly, or operate as a MOD-8 counter. The following are some of the possible causes for the malfunction:

1. The connection from the output of the decoding 2-input NAND gate to the SET input of the Set/Clear flip-flop is open.

2. The connection between the Q of the Set/Clear flip-flop and the CLR inputs of the J-K flip-flops is open.

3. The connection from the CLK input to the INVERTER is open.

4. The connection between the output of the INVERTER and the CLEAR input of the Set/Clear flip-flop is shorted to ground.

7.4 Under normal conditions flip-flops Q_1-Q_3 of counter Z1 form a MOD-8 counter. Thus, the frequency at output Q_3 of Z1 is 8.64KHz/8 = 1.08KHz. Counter Z2 is wired as a MOD-9 counter and thus its Q_3 output should be 1.08KHz/9 = 120Hz. This signal is connected to CP_0 of Z1 which divides the frequency in half to produce a 60Hz squarewave. If counter Z2 is wired as a MOD-10 counter, then its Q_3 output is 1.08KHz/10 = 108Hz. This signal is then halved by counter Z1 and a 54Hz squarewave will be present at point X. Thus, in order for counter Z2 to be wired as a MOD-10, its MR1 input would have to be connected to its Q_1 output instead of Q_0.

SECTIONS 7.5,7.24 *Propagation Delay in Ripple Counters/ Troubleshooting*

7.5 The ripple counter of Figure 7.5 is incremented from count 7 (111_2) to count zero (000_2). Therefore, all three J-K flip-flops change from a logic HIGH to a logic LOW (t_{pHL}). From the 74LS112 data sheet in Appendix II in your text book, we obtain a value of 24ns for t_{pHL}. Thus, it will take 72ns (3x24ns) before FF C changes states.

7.6 By analyzing the counting sequence, it should be obvious that the logic states for outputs A and B are correct. Therefore, it can be deduced that the problem is with output C. Output C should be LOW for four counts and HIGH for four counts. When the counter sequences properly, output C changes after output B goes through a negative going transition. Unfortunately, output C is doing just the opposite; it changes when output B goes through a positive going transition. Thus, the most probable cause for this malfunction is a mistake in the wiring of the circuit. If output $\overline{B}$ instead of B is connected to the CLK of flip-flop C, then the counting sequence would be as recorded.

SECTIONS 7.6-7.9,7.24 *Synchronous (Parallel) Counters/ Synchronous Down and Up/Down Counters Presettable Counters/ The 74193 (LS193/HC193) Counter/ Troubleshooting.*

7.7 Asynchronous Counter:

Advantage - Since it requires the fewest amount of components to build, simplicity of construction is the major advantage of this type of counter.

Disadvantage - The speed at which it operates is dependent on the number of flip-flops in the counter. As the number of flip-flops increase, the maximum allowable input clock frequency decreases. This is because of the inherent propagation time delay of each flip-flop. Thus, for example, in a serial counter the Nth flip-flop (MSB) can change only after all of the other flip-flops in the counter have changed.

Synchronous Counter:

Advantage - The speed at which it operates is independent of the number of flip-flops in the counter. The flip-flops which make this type of counter are all clocked simultaneously (in parallel) by the input clock signal. Therefore, no flip-flop in the counter has to wait for the previous flip-flop in the chain to change in order for it to change.

Disadvantage - This type of counter requires more logic circuitry and a greater number of connections than the asynchronous counter.

7.8

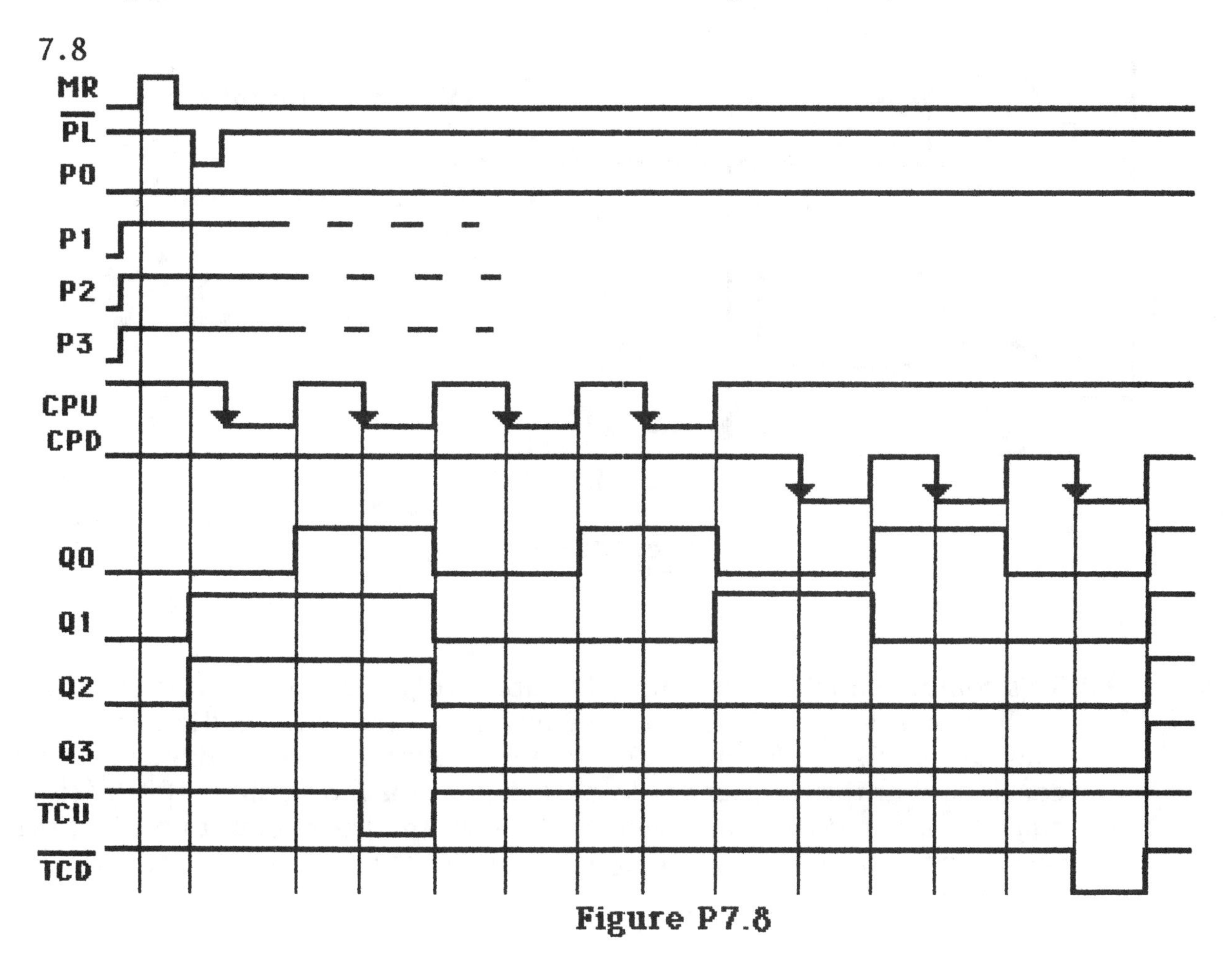

Figure P7.8

7.9

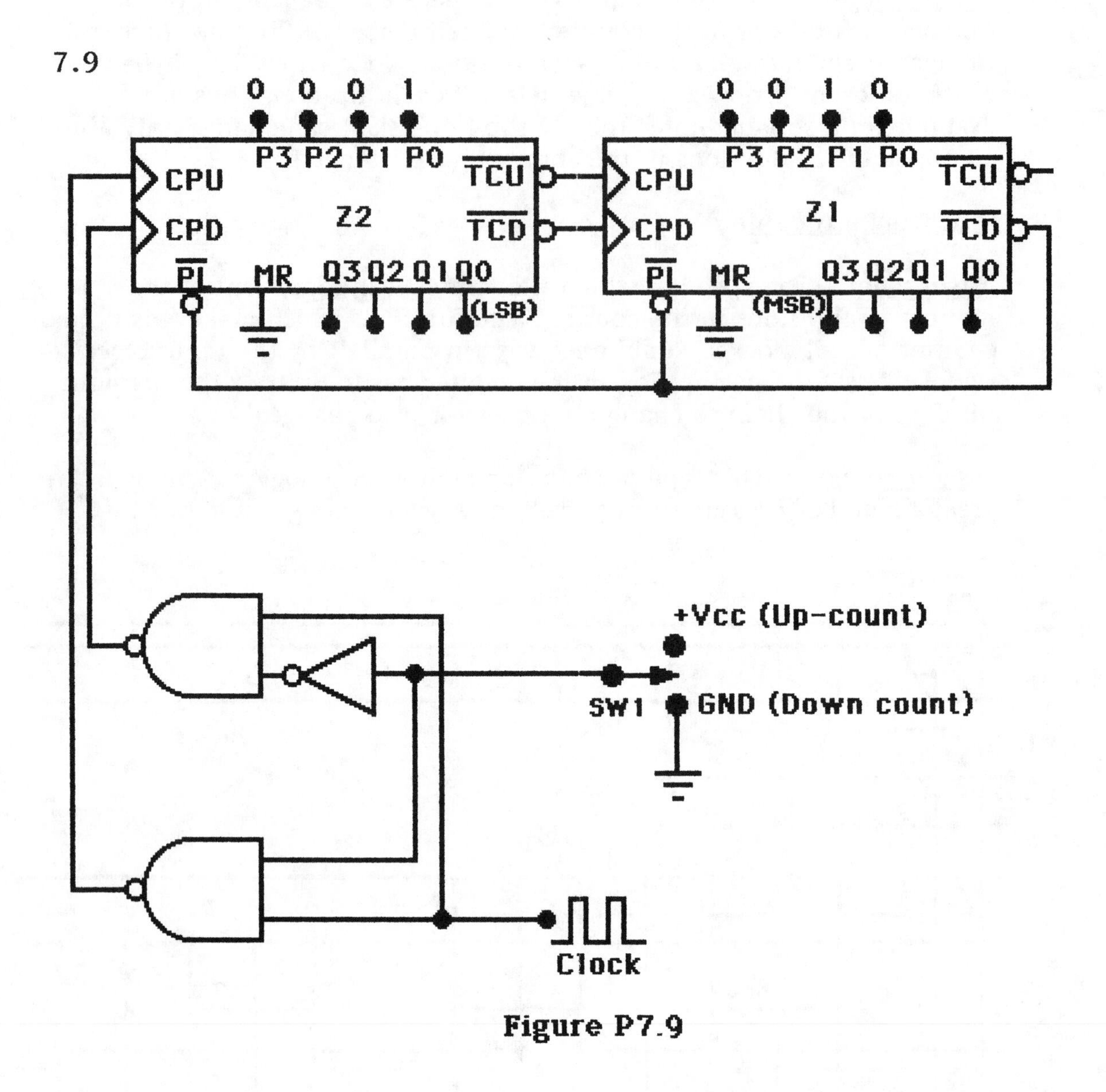

Figure P7.9

7.10 This counter should be dividing the input frequency by 256 when SW1 is in the Count-Up position. Since the technician measured 39Hz at the MSB (output Q_3 of Z1 in Figure P7.9) when the counter was counting up, we can conclude that the counter works properly in this mode, since 10KHz/256 is equal to about 39Hz. When switch SW1 is in the Down-Count position, the frequency at the MSB (output Q_3 of Z1 of Figure P7.9) should be 294.1Hz (10KHz/34).

Since the measured frequency is about 39Hz regardless of the position of switch SW1, it can be concluded that the counter always divides the 10KHz input frequency by 256. In order for the circuit of Figure P7.9 to operate as a MOD-256 counter when SW1 is in the Down-Count position, each of the 74193 counters must be working as a MOD-16 counter. Thus, it can be concluded that counters Z1 and Z2, of Figure P7.9 never get loaded with the binary number 00100001_2 $(33_{10})^*$, but rather are allowed to count down from binary 11111111_2 to 00000000_2 repeatedly.

For the counter (Z1 and Z2) to be a MOD-34 down counter, it must be loaded with binary 00100001_2 (33_{10}) and be allowed to count down to 00000000_2 (0_{10}). In other words, go through 34 different states.

The following are a few of the possible causes for the malfunction:

1. Connection between the $\overline{PL}$ inputs and the output $\overline{TCD}$ of Z1 is open.
2. Parallel inputs to Z1 and Z2 are all floating, or HIGH.
3. The ground connection from switch SW1 is open.

SECTION 7.10 *More on the IEEE-ANSI Dependency Notation*

7.11 (a) This MOD-256 counter (CTRDIV 256) can count from 0-255.
(b) MS1 and MS2 are inputs to an Exclusive-OR gate. Whenever these two inputs are equal, the counter is loaded with the count 150_{10} (CT=150).
(c) See (b).
(d) Label '3D' establishes the dependency of flip-flops A thru H on the common input C3.
(e) At count 255 (1CT=255), output $\overline{CO}$ will go LOW, provided input 'Down' is LOW.
(f) When either MR1 or MR2 are HIGH, this counter will be cleared (CT=0).
(g) When output $\overline{BO}$=0, it means that the counter is cleared or is at binary count 00000000_2, and that input 'Up' is LOW.
(h) The symbol inside of the box whose inputs are MR1 and MR2, would have to be changed to (&).
(i) The '+' symbol at the input UP, indicates that the counter will increment as pulses are applied to this input.

SECTIONS 7.11-7.13, 7.24 *Decoding a Counter/Decoding Glitches/ Cascading BCD Counters/Troubleshooting*

7.12 (a) When the output of a ripple counter is being decoded, temporary states caused by the propagation delays of the individual counter flip-flops are also decoded. Figure P7.12(a) shows the counting sequence of the ripple counter, including the temporary states. The temporary states are marked with an asterisk. The states shaded are decoded (000_2) by the AND gate Z1-A. Flip-flop Z2 is cleared when the counter reaches the count of 011_2, and set when the temporary state 000_2 occurs.

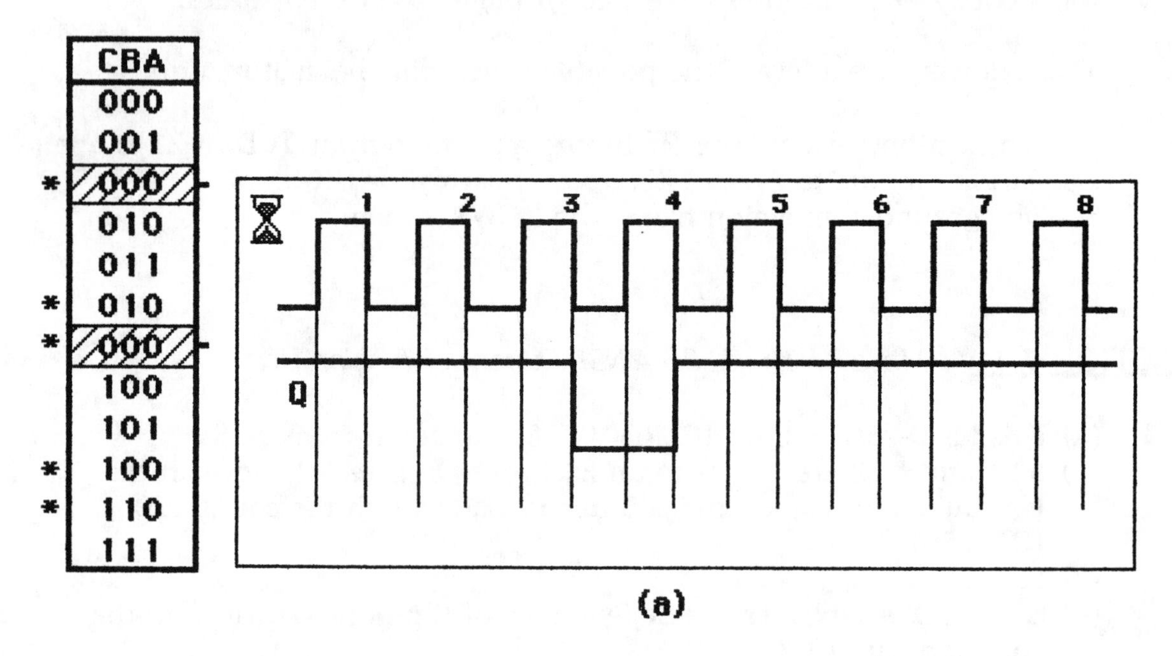

(a)

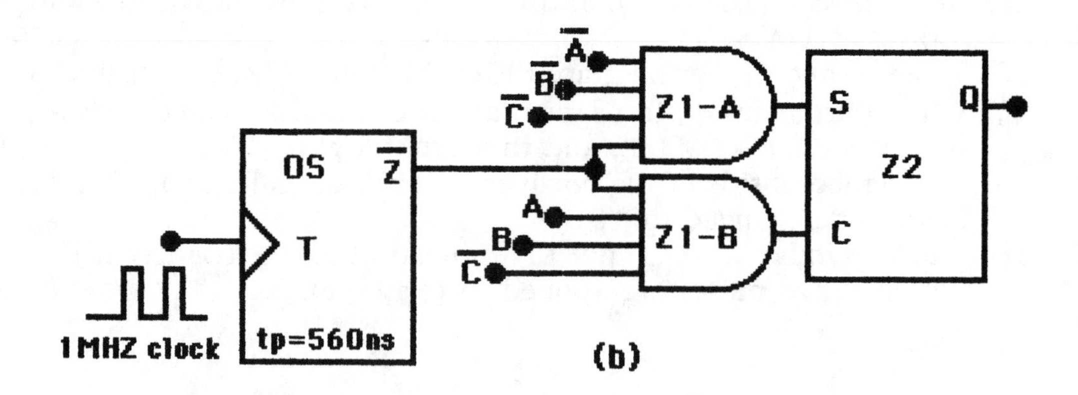

(b)

Figure P7.12

(b) The unused inputs of AND gates Z1-A and Z1-B can be utilized for a strobe input. After determining the maximum propagation delay for the flip-flops used by the ripple counter, a one-shot with the proper t_p value, could be used to eliminate the decoding of the temporary state (000_2) by flip-flop Z2.

Figure P7.12 (b) shows the circuit modification. It is assumed that the ripple counter has an input clock of 1MHz (T=1μs) and that the propagation delay of each of its flip-flops is 20ns maximum.

Circuit Operation:

When a NGT of the 1MHz clock occurs, the ripple counter is incremented. The transitional states of the counter occur shortly after this NGT. The worst case is when the counter is cycling from 111_2 to 000_2 and all three flip-flops are changing from a set to a clear condition. This will cause transitional states to be present at the counter outputs ABC for about 60ns maximum (3x20ns).

When a PGT of the 1MHz clock occurs, the One-shot triggers for 560ns and consequently its $\bar{Z}$ output goes LOW for 560ns. This LOW will keep the outputs of AND gates Z1-A and Z1-B LOW, and thereby preventing any of the transitional states from being decoded during the 560ns interval. Note that the transitional states will be present for 60ns (worst case) after a NGT of the clock occurs, which is 560ns after the One-shot gets triggered by a PGT. After the One-shot goes back to its normal state, the transitional states are no longer present and therefore no transitional state will ever be decoded by the AND gates Z1-a and Z1-B.

7.13 The Decoder/displays show the correct number of input pulses up to 79. When one more input pulse is counted the displays show 180. What has to be determined is why did the "Hundreds" BCD counter incremented at this particular time?

From count 79 to 80, the outputs of the "Tens" BCD counter went from 0111_2 to 1000_2. Since it is at this time that the "Hundreds" BCD counter is incremented, we can probably guess that output C of the "Tens" BCD counter, which just went through its first negative going transition, clocked the "Hundreds" BCD counter.

This most likely wiring mistake would cause the recorded display readings and thus, anytime the "Tens" BCD counter goes from count seven to eight (0111_2-->1000_2), the "Hundreds" BCD counter is incremented.

SECTION 7.14 *Synchronous Counter Design*

7.14 The following six step procedure can be used in the design of any synchronous counter, although the reader may choose to use less than six and still obtain the proper counter design.

(a) *Step 1: Determine the desired number of bits (FFs) and the desired counting sequence.*

In this case the number of bits are three. Let's assume that they are 'ABC' where A is the LSB and D is the MSB.

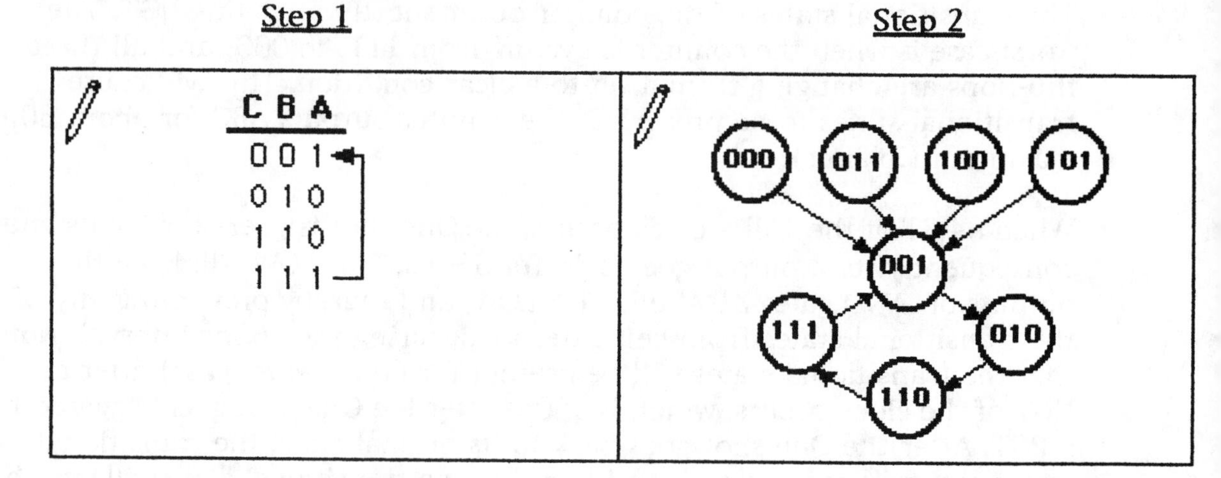

Step 2: Draw the state transition diagram showing all possible states, including those that are not part of the desired counting sequence.

Step 3: Use the state transition diagram to set up a table that lists all present states and their next states.

PRESENT state C B A	NEXT state C B A
0 0 0	0 0 1
0 0 1	0 1 0
0 1 0	1 1 0
0 1 1	0 0 1
1 0 0	0 0 1
1 0 1	0 0 1
1 1 0	1 1 1
1 1 1	0 0 1

Step 4: Add a column to this table for each J and K input. For each present state, indicate the levels required at each J and K input in order to produce the transition to the next state.

PRESENT State CBA	NEXT State CBA	Jc	Kc	Jb	Kb	Ja	Ka
000	001	0	x	0	x	1	x
001	010	0	x	1	x	x	1
010	110	1	x	x	0	0	x
011	001	0	x	x	1	x	0
100	001	x	1	0	x	1	x
101	001	x	1	0	x	x	0
110	111	x	0	x	0	1	x
111	001	x	1	x	1	x	0

Note: The table below shows the levels required at each J and K input in order to produce the transition from the present state (Q) to the NEXT state (Q+1).

Q	Q+1	J	K
0	0	0	x
0	1	1	x
1	0	x	1
1	1	x	0

Step 5: Design the logic circuits to generate the levels required at each J and K input.

Simplify the Sum-of-Products expressions for Ja, Ka, Jb, Kb, Jc, and Kc by using the Karnaugh map method:

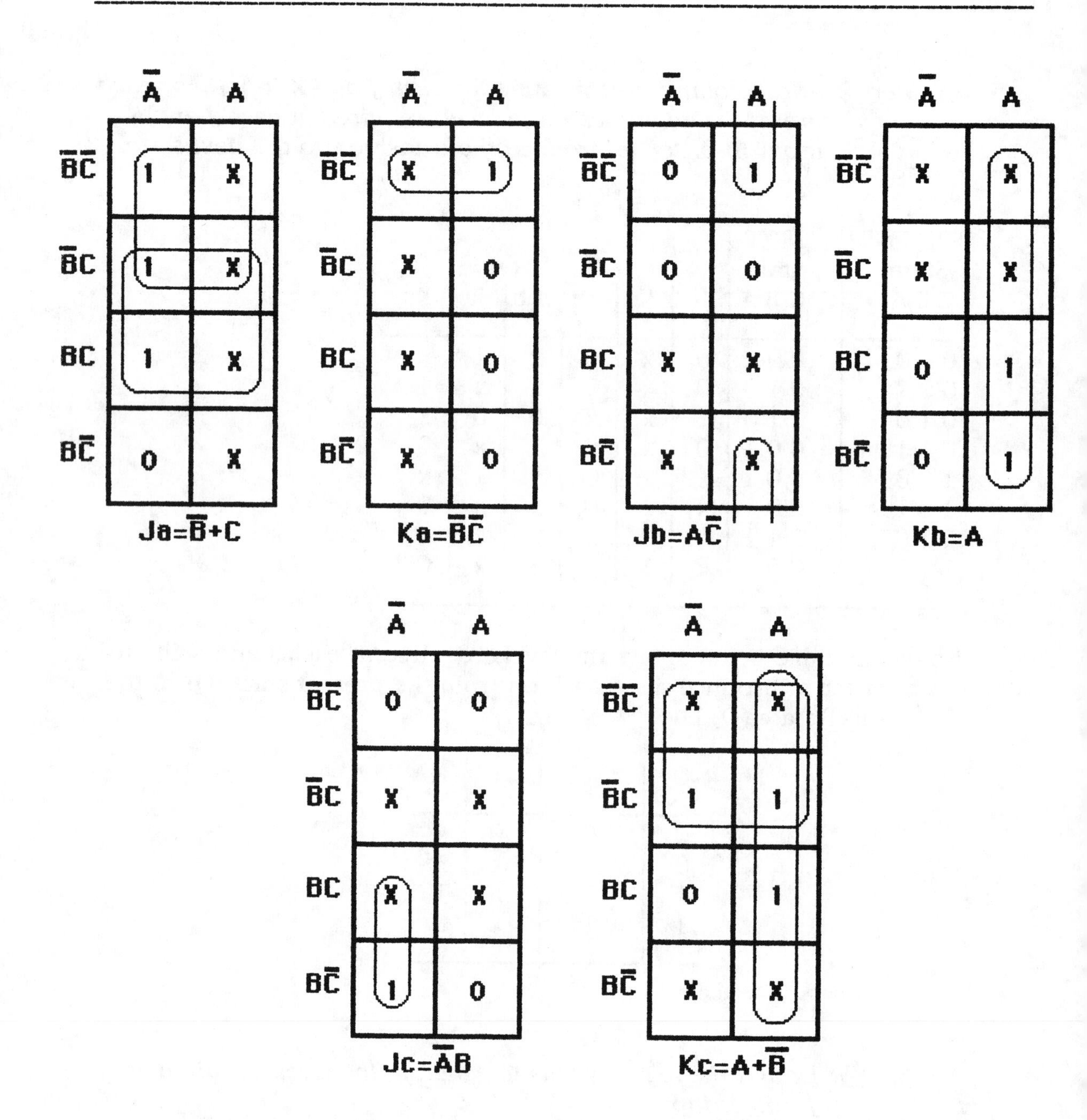

<u>Step 6</u>: *The last and final step is to draw the synchronous counter circuit of Figure P7.14(a).*

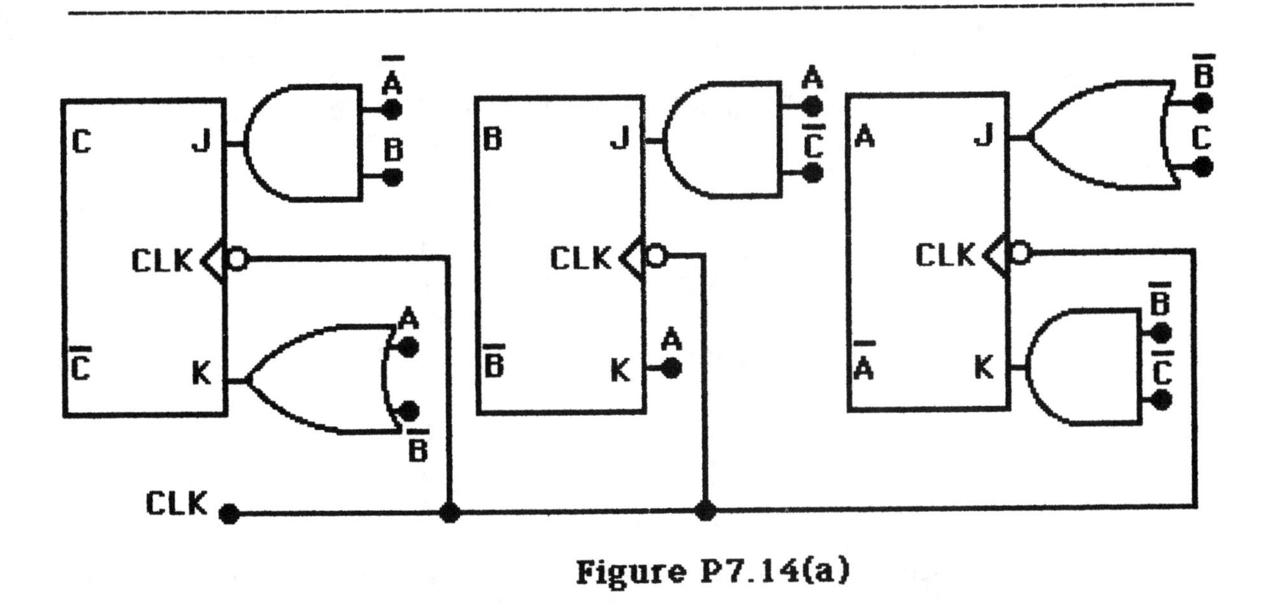

Figure P7.14(a)

(b) Use the same step-by-step procedure established in problem 7.14(a).

<u>Step 1</u>: *Determine the desired number of bits (FFs) and the desired counting sequence. (Same number of bits and counting sequence as the previous problem.)*

<u>Step 2</u>: *Draw the state transition diagram showing all possible states.*

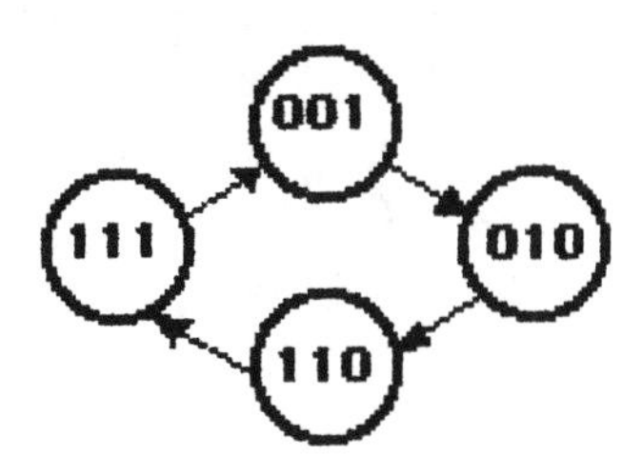

<u>Step 3</u>: *Use the state transition diagram to set up a table that lists all present states and their next states.*

PRESENT state CBA	NEXT state CBA
000	XXX
001	010
010	110
011	XXX
100	XXX
101	XXX
110	111
111	001

Step 4: Add a column to this table for each J and K input. For each present state, indicate the levels required at each J and K input in order to produce the transition to the next state.

PRESENT State CBA	NEXT State CBA	Jc	Kc	Jb	Kb	Ja	Ka
000	XXX	x	x	x	x	x	x
001	010	0	x	1	x	x	1
010	110	1	x	x	0	0	x
011	XXX	x	x	x	x	x	x
100	XXX	x	x	x	x	x	x
101	XXX	x	x	x	x	x	x
110	111	x	0	x	0	1	x
111	001	x	1	x	1	x	0

Step 5: Design the logic circuits to generate the levels required at each J and K input.

Simplify the S-of-P expressions for Ja, Ka, Jb, Kb, Jc, and Kc by using the Karnaugh map method:

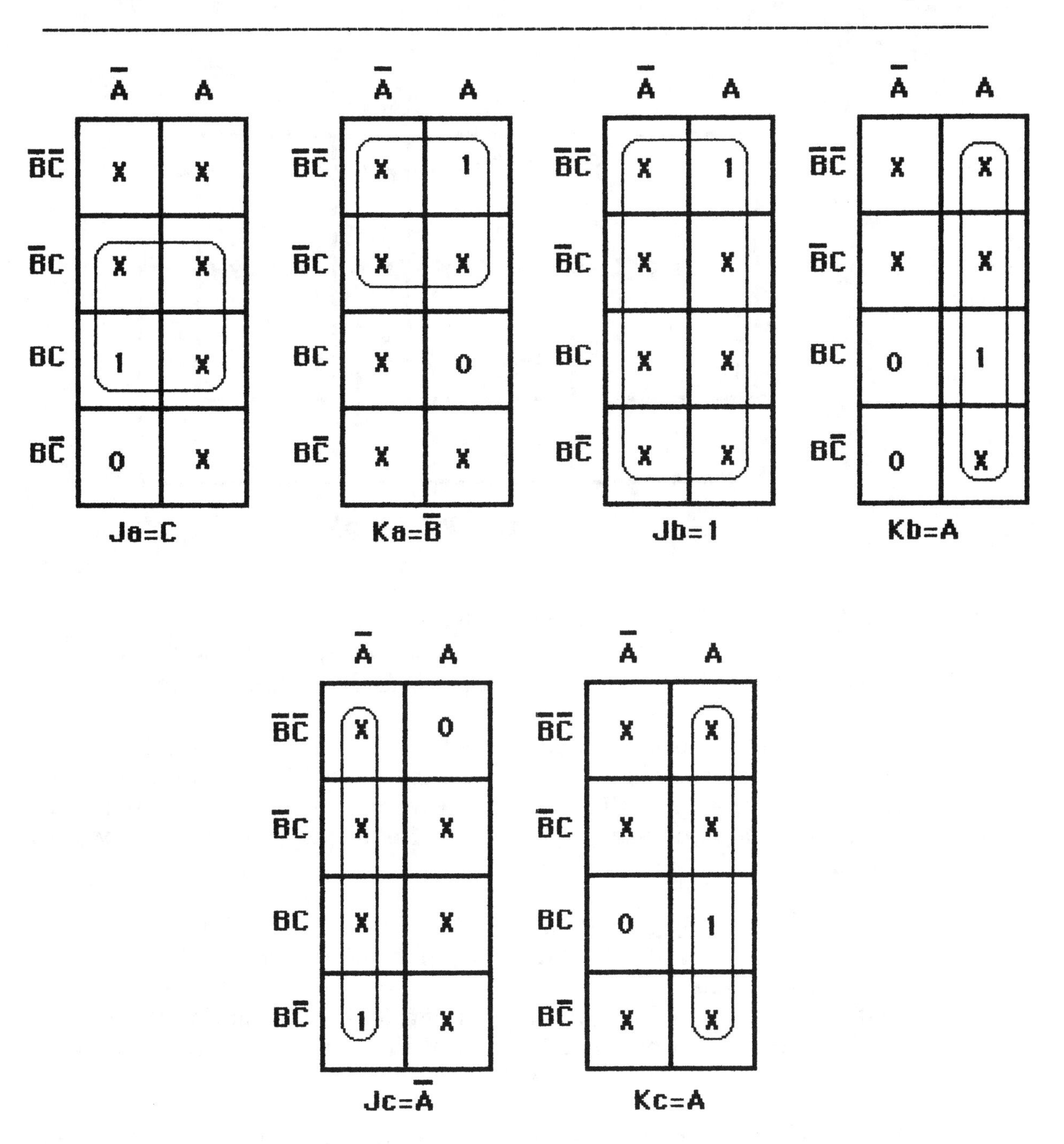

<u>Step 6</u>: *The last and final step is to draw the synchronous counter circuit of Figure P7.14(b)*

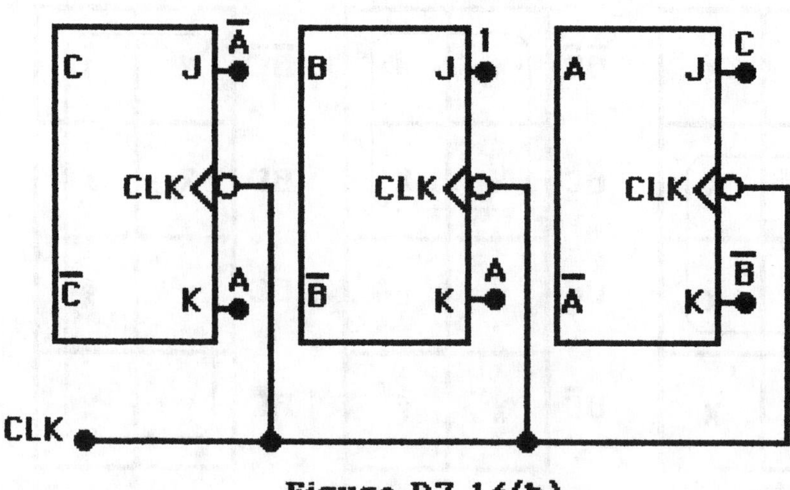

Figure P7.14(b)

SECTIONS 7.15-7.17, 7.24 — *Shift-Register Counters/ Counter Applications: Frequency Counter and Digital Clock/Troubleshooting*

7.15 Waveform Q_0 of Figure 7.9 repeats itself every six clock pulses. Since Q_0 is one of the outputs of this Ring counter, there must be six flip-flops in the counter. Note that each flip-flop output waveform of this counter has a frequency equal to one-sixth of the clock frequency, since this is a MOD-6 counter.

7.16 (a) To build a MOD-23 Ring Counter, it requires 23 flip-flops.
(b) To decode any state of a MOD-14 Johnson Counter, it requires a TWO input AND gate.
(c) If the frequency of the input clock of a 6-bit Johnson Counter is 60 KHz, then the output signal at any of the flip-flops in the counter is equal to 5 KHz, 50% Duty Cycle.

7.17 The counter is incremented by the "Unknown frequency" pulses only when signals "SAMPLE pulses" and "X" are HIGH (see Figure 7.10). Note that signal "X" stays HIGH for two "SAMPLE pulses" (FF X is a MOD-2 counter). Therefore, if the counter was incremented during the time interval which signal "X" is HIGH, the resulting reading on the Decoder/Display would be a value that is larger by a factor of two. Thus, a broken connection between the "SAMPLE pulses" signal and the input of the AND gate could cause the circuit to behave as it does. Note that this essentially creates a constant HIGH at that input of the AND gate.

7.18 In the circuit of Figure 7.11, the output of the NAND gate is active only when flip-flop X is set and the BCD counter reaches the count of three (0011_2). When these conditions are present, flip-flop X is cleared and the BCD counter is parallel loaded with binary 0001_2, thereby causing this section of the digital clock to cycle from 12:59 to 01:00. If flip-flop X fails to be cleared at this time, then the BCD counter is parallel loaded with binary 0001_2 every time it reaches the binary count of 0011_2. This of course causes the "Hours" section of the digital clock to oscillate between the hours of 11 and 12. Therefore, a possible cause for the malfunction is the nonexistent or broken connection between the CLR input of the X flip-flop and the output of the 3-input NAND gate.

SECTION 7.21, 7.24 *Parallel In/Serial Out-(The 74165/74LS165/74HC165/)/ Troubleshooting.*

7.19 Since input Ds changes while data is being clocked into the register, it helps to visualize the flow of data within the 74165 IC, (Figure P7.19(a)) before we draw the waveforms (Figure P7.19(b)).

Ds	Q_0	Q_1	Q_2	Q_3	Q_4	Q_5	Q_6	Q_7	
0	1	0	1	1	1	0	0	1	After loading.
0	0	1	0	1	1	1	0	0	After 1st shift pulse.
1	1	0	1	0	1	1	1	0	After 2nd shift pulse.
0	0	1	0	1	0	1	1	1	After 3rd shift pulse.
0	0	0	1	0	1	0	1	1	After 4th shift pulse.
0	0	0	0	1	0	1	0	1	After 5th shift pulse.
0	0	0	0	0	1	0	1	0	After 6th shift pulse.
0	0	0	0	0	0	1	0	1	After 7th shift pulse.
0	0	0	0	0	0	0	1	0	After 8th shift pulse.
0	0	0	0	0	0	0	0	1	After 9th shift pulse.
0	0	0	0	0	0	0	0	0	After 10th shift pulse.
0	0	0	0	0	0	0	0	0	After 11th shift pulse.

Figure P7.19(a)

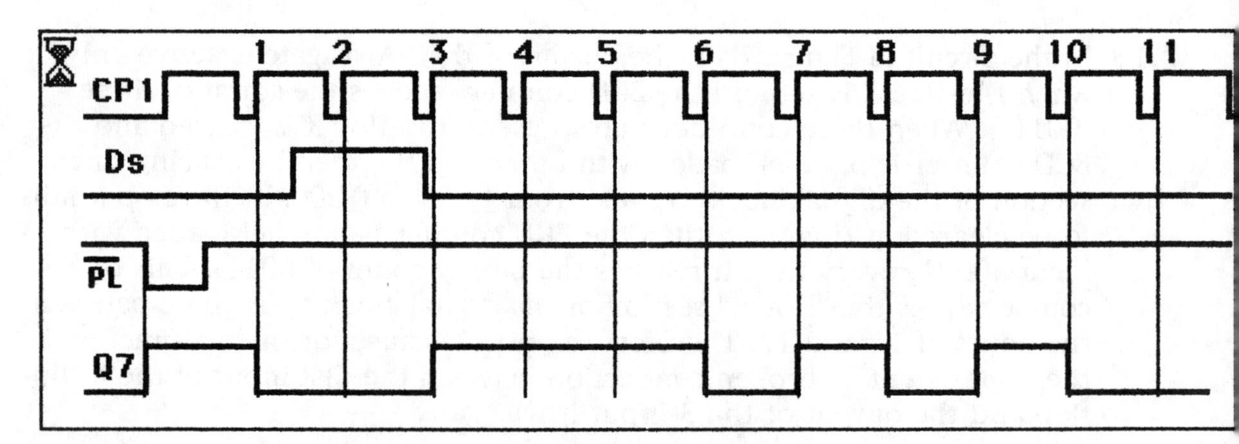

Figure P7.19(b)

SECTION 7.23 *IEEE-ANSI Register Symbols.*

7.20 The best way of determining the function of this, or any, dependency notation diagram, is to decipher and understand each bit of information within the IEEE/ANSI diagram.

1. When the CLR signal is HIGH the outputs are cleared (LOW).
2. Outputs are enabled when inputs M and N are both LOW.
3. When both gated data enable inputs $\overline{G}1$ and $\overline{G}2$ are LOW, data at the D inputs are loaded into their respective flip-flops on the next PGT of the clock.
4. Data inputs (1D) in the control block, show dependency on the circuit C1 (shaded area 3).
5. Four registers with tri-state outputs.

8 INTEGRATED-CIRCUIT LOGIC FAMILIES

SECTIONS 8.1-8.3 *Digital IC Terminology/The TTL Logic Family/ Standard TTL Series Characteristics.*

8.1 (a) V_{OL}, is the voltage level at the output of a TTL logic circuit in the LOW state.
(b) I_{IL}, is the current that flows into a TTL input when a specified LOW-level voltage is applied to that input.
(c) t_{pLH}, is the delay time from a logic 0 to a logic 1.
(d) Fan-out, is the maximum number of standard logic inputs that an output can reliably drive.

8.2 (a) The average current of any TTL IC is equal to: $I_{CC(Ave.)} = \frac{I_{CCH} + I_{CCL}}{2}$.
Using the TTL data sheets in Appendix II of your text book:

7400 IC: I_{CCH}=8mA, I_{CCL}=22mA, $I_{CC(Ave.)}$=15mA

7404 IC: I_{CCH}=12mA, I_{CCL}=33mA, $I_{CC(Ave.)}$=22.5mA

7420 IC: I_{CCH}=4mA, I_{CCL}=11mA, $I_{CC(Ave.)}$=7.5mA

7432 IC: I_{CCH}=22mA, I_{CCL}=38mA, $I_{CC(Ave.)}$=30mA

The total $I_{CC(Ave.)}$ for the circuit is 75mA

The average power dissipation is equal to: P_D=$I_{CC(avg)}$ x Vcc.

Thus, for the circuit of Figure 8.1, the maximum average power dissipation is equal to 75mA x 5.25V or 393.75mW.

(b) The average propagation delay of a TTL circuit is: $t_{p(Ave.)} = \frac{t_{PLH} + t_{PHL}}{2}$.

From Appendix II and for the 7404 Inverter, we obtain t_{PLH}=22ns and t_{PHL}=15ns. Therefore, the maximum average propagation delay of one 7404 Inverter is 18.5ns.

(c) The speed-power product of any TTL gate is equal to: (gate power dissipation x propagation delay).

Thus, maximum power dissipation of one 7404 IC is P_D=$I_{CC(avg)}$xVcc. The average current I_{CC}(Ave.) = 22.5mA. The average current of one Inverter is 3.75mA (22.5/6). Therefore, the maximum power dissipation of one 7404 Inverter is 19.68mW (3.75mA x 5.25V). Therefore, the maximum speed-power product of a single 7404 Inverter is 18.5ns x 19.68mW or 364.1pJ .

SECTIONS 8.4-8.6 *Improved TTL series/TTL Loading and Fan-Out/ Other TTL Characteristics.*

8.3 (a) Since the input to IC1 is tied to Vcc, its output is a constant logic LOW. When a TTL's Totem-Pole output is LOW, it sinks current from the driving inputs. In this situation, IC1 has to sink the cumulative I_{IL} currents from the inputs of IC2, IC3, and IC4. The following table shows how much I_{IL} flows out of each input:

IC	I_{IL}
7404	-1.6mA (1 U.L.)
7420	-1.6mA (1 U.L.)*
7432	-1.6mA (1 U.L.)

* *When the inputs of a TTL NAND gate are tied together, and they are at a logic LOW, the total current that flows out of the inputs is equal to I_{IL} or 1 U.L. in the LOW state. Furthermore, current I_{IL} is leaving that input and therefore it is preceded by a negative sign.*

Thus, the output of IC1 must sink current I_X, which is 4.8mA (1.6mA x 3).

(b) If the input of IC1 is LOW, then its output is HIGH. When a Totem-Pole output is HIGH, that output sources current to the inputs which are being driven.

In this situation, the output of IC1 has to supply the needed cumulative I_{IH} currents to the inputs of IC2, IC3, and IC4. The following table shows how much current I_{IH} flows into each input:

IC	I_{IH}
7404	40μA (1 U.L.)
7420	160μA (4 U.L.)*
7432	40μA (1 U.L.)

* When the inputs of a TTL NAND gate are tied together, the total current that flows into the inputs is equal to (I_{IH} x # of inputs tied together). Thus, current Ix is 240μA (40μA x 6 inputs), and flows from the output of IC1 into the inputs of ICs 2, 3 and 4.

(c) If IC1 is changed to a 54LS04 and its input is HIGH, the current Ix will not change and will continue to be 4.8mA. Nevertheless, a problem arises because the fan-out for the 54LS04 is exceeded. For question (a), IC1 (7400) could sink as much as 16mA (I_{OL}). Since it had to sink an Ix current of 4.8mA, its rating for the maximum sinking current was far from being exceeded. However, this is not the case if IC1 is a 54LS04. From the data sheet for a 54LS04 in Appendix II, we obtain an I_{OL} rating of 4mA (2.5 U.L.), which would clearly be exceeded by Ix. This will cause the voltage at point Z (V_{OL}) to be greater than what is normally expected, and perhaps even exceed the maximum allowable value for an input TTL voltage ($V_{IL(max)}$). Furthermore, it will reduce the dc noise margins of the logic circuits being driven by IC1, thereby making them more susceptible to noise.

8.4 When the output of IC1 is LOW, it must sink 1 U.L. from IC2, 1 U.L. from IC3, and 1 U.L. from IC4. Thus, IC1 must sink a total of 3 standard Unit Loads or 4.8mA (3x1.6mA).

When the output of IC1 is HIGH, it must supply 1 U.L. to IC2, 4 U.L. to IC3, and 1 U.L. to IC4. Thus, IC1 a total of 6 standard Unit Loads in the HIGH state or 240μA (6x40μA).

8.5 LOW-state dc noise margin for any TTL logic family: $V_{NL} = V_{IL(max)} - V_{OL(max)}$
HIGH-state dc noise margin for any TTL logic family: $V_{NH} = V_{OH(min)} - V_{IH(min)}$

<u>For the 74F04 IC:</u> V_{NL}=0.8V-0.5V; V_{NL}=<u>400mV</u> ; V_{NH}=2.5V-2.0V; V_{NH}=<u>500mV</u>

<u>For the 74LS04 IC:</u> V_{NL}=0.8V-0.5V; V_{NL}=<u>400mV</u> ; V_{NH}=2.7V-2.0V; V_{NH}=<u>700mV</u>

The 74LS04 IC has the same LOW-state dc noise margin (400mV) as the 74F04 IC. However, its HIGH-state dc noise margin (700mV) is a little bit better than that of the 74F04 IC (500mV).

SECTIONS 8.7-8.8 *Connecting TTL Outputs Together/ Tristate (3-State) TTL.*

8.6 To calculate the minimum value for resistor Ra, we must determine the loading on the output of the 7405 Inverter when it is in the LOW state. The clock input of the 74LS112 flip-flop has a rating of 0.5 U.L. (I_{IL}=0.8mA), and the input of the 7401 has a rating of 1 U.L. (I_{IL}=1.6mA). Thus, by using Kirchhoff's current law, the current that the output of the 7405 must sink is 2.4mA (0.8mA+1.6mA). The maximum current that the 7405 Inverter can sink is 16mA (I_{OL}). Therefore, I_{Ra} is 13.6 mA (16mA - 2.4mA). The maximum allowable LOW voltage at the output of the 7405 is 0.4V ($V_{OL(max)}$).

Thus, by using ohm's law we can determine that:

$$R_{a(min)} = \frac{(V_{CC} - V_{OL(max)})}{I_{Ra}} = \frac{(5V - 0.4V)}{13.6mA} = 338.2\Omega$$

The process used to calculate the minimum value of Ra can also be used to find the minimum value of Rb. The CLR input of each 74S112 flip-flop has a rating of 4.375 U.L. (7mA) and the input of the 7404 has a rating of 1 U.L. (1.6mA). The maximum allowable sink current for the 7401 is 16mA ($I_{OL(max)}$). By using KVL, we obtain an I_{Rb} current of 0.4mA (16mA-1.6mA-14mA). The maximum allowable LOW voltage at the output of the 7401 is 0.4V ($V_{OL(max)}$.

Thus, by using ohm's law we can determine that:

$$R_{b(min)} = \frac{(V_{CC} - V_{OL(max)})}{I_{Rb}} = \frac{(5V - 0.4V)}{0.4mA} = 11.5K\Omega$$

8.7 The 220Ω Ra resistor, will allow the voltage at the output of the 7405 Inverter to be >0.4V ($V_{OL(max)}$). This will reduce the dc noise margin at the clock input of the 74LS112 to less than 400mV. Hence, the reason why the 74LS112 flip-flop is much more susceptible to ambient noise.

8.8

Ring Counter W Y X	Z1 A_1 B_1 C_1	Z2 A_2 B_2 C_2	Z3 A_3 B_3 C_3	Data Bus A B C
0 0 1	1 1 0	0 1 0	**0 1 1**	**0 1 1**
0 1 0	1 1 1	**0 0 1**	1 0 0	**0 0 1**
1 0 0	**1 1 0**	1 1 0	0 1 0	**1 1 0**
0 0 1	1 1 0	0 0 0	*1 0 1*	*1 1 1*
0 0 1	0 1 1	1 1 1	*0 0 0*	*0 1 0*

When output X from the Ring counter is HIGH the contents of counter Z3 are placed on data bus ABC, via the tristate buffers Z7. Likewise, the contents of counters Z2 and Z1 are placed on the data bus when the outputs of the Ring counter Y or W are HIGH respectively. As can be seen from the above table, the data underlined in bold type are placed on the data bus at the right time and without getting changed. A problem exists however with the data expressed in italic type from counter Z3. As stated before, when output X from the Ring counter is HIGH, the outputs of counter Z3 should be allowed to be placed on the data bus. This does not always happen, particularly when B3 is LOW. It appears that whenever counter Z3 is selected by the Ring counter, output B3 always ends up as a logic HIGH on the data bus. The circuit will behave according to the table, if the connection from output B3 to the input of the tristate buffer B of Z7 was open or shorted to Vcc.

8.9 The tristate buffers at the output of counter Z1 are enabled when output W from the Ring counter is HIGH. Since the circuit of Figure 8.3 is a TTL circuit, if the connection from output W was open the buffers from Z5 would always be enabled, thereby allowing the outputs from counter Z1 to always be present on the data bus. This open line would make it possible for two MOD-8 counters to place their data on the data bus at the same time. This particular situation is often referred to as "Bus contention."

SECTIONS 8.10-8.15 *MOS Digital Integrated Circuits/ The MOSFET/Digital MOSFET Circuits/ Characteristics of MOS Logic/CMOS Logic/ CMOS Series Characteristics.*

8.10 (a) TTL logic uses Bipolar transistors, while CMOS uses Field Effect transistors.
(b) In a circuit which utilizes CMOS logic, the power dissipation is directly proportional to the frequency at which it operates.

(c) CMOS should be chosen over TTL, if the main consideration is packing density.
(d) Among all the MOSFET circuits, N-MOS has the lowest packing density.
(e) The usage of negative voltages is a disadvantage of P-channel MOSFETs.
(f) Current spikes are drawn from the biasing power supply, each time a CMOS output switches from LOW to HIGH.
(g) In both the HIGH and LOW states, the dc noise margin of CMOS logic can be determined by multiplying V_{DD} by 30%.
(h) Damage can occur to a CMOS circuit, if its inputs are left floating.
(i) High-Speed CMOS is an improved version of the 74C series which has a tenfold increase in switching speed.
(j) BiCMOS logic combines the best of features of bipolar and CMOS logic.
(k) Because of parasitic PNP and NPN transistors embedded in the substrate of CMOS ICs latch-up can occur under certain circumstances.

SECTIONS 8.16-8.21 *CMOS Open Drain and Tristate Outputs/ CMOS Transmission Gate (Bilateral Switch)/ IC Interfacing/TTL Driving CMOS/ CMOS Driving TTL/Low-Voltage Technology - LVT*

8.11 By careful analysis of the tabulated results of Figure 8.6, it can be observed that whenever inputs A and B are different, an indeterminate logic level results at points C and D respectively. Furthermore, if either or both of the 4016 transmission gates are enabled, indeterminate logic levels will be present at outputs X and/or Y. The most probable cause for the malfunction is a short between points C and D in the circuit of Figure 8.5. This would cause the voltage levels at points C and D to be about 2.5V (1/2 Vcc). This situation occurs only when CMOS outputs are tied together and forced to go to different logic levels.

8.12 The problem with the circuit of Figure 8.7 is that different logic families were interfaced without any special design consideration. First, the 7404 TTL Inverter cannot drive the 4000B CMOS Inverter directly. A 10KΩ pull-up resistor should be used between the output of IC1 and the input of IC2. This allows the TTL output voltages to be comparable with the CMOS input voltages. Another way of achieving the same results is to use either a 74HCT04 IC or a 74ACT04 IC instead of the CMOS 4000B IC.

The next problem that should be addressed is that of the outputs of the 4000B CMOS Inverter driving the inputs of a standard TTL gate directly. When the output of a 4000B Inverter is LOW it can only sink 0.4mA.

Since the 7410 NAND gate is a standard TTL circuit, its inputs present 1 TTL U.L. in the LOW state (I_{IL}=1.6mA) to the output of the 4000B Inverter. Clearly the CMOS Inverter cannot sink the 1.6mA current coming from the 7410 inputs. Since a 4000B output can reliably drive one 74LS input, the replacement of the 7410 IC with a 74LS10 would take care of the problem.

8.13 The problem with this circuit is the different biasing voltages between the 4000B Inverter and the 7404 Inverter. The voltage level at the output of the 4000B Inverter will be either 0V or 10V. Understandably the 10V level is too high as a logic HIGH input for the TTL 7404 Inverter. A Voltage-Level translator that converts the high voltage (+10V) to a +5V output has to be interfaced between the output of IC1 and the input of IC2. A 4050B IC is one possible voltage-level translator that could be used to solve the problem.

8.14 Fan-out in the <u>LOW</u> state = $\frac{I_{OL}}{I_{IL}} = \frac{64mA}{1.6mA} = 40U.L.$ Therefore, in the LOW state a typical 74LVT device can drive 40 U.L.

Fan-out in the <u>HIGH</u> state = $\frac{I_{OH}}{I_{IH}} = \frac{32mA}{0.4mA} = 80U.L.$ Therefore, in the HIGH state a typical 74LVT device can drive 80 U.L.

<u>SECTION 8.22</u> *Troubleshooting.*

8.15 <u>One way to isolate the problem</u> :

While monitoring point C with the logic probe, inject pulses with the logic pulser into input A. The logic probe should indicate the presence of pulses by blinking its indicator light. Move the logic probe to point D and inject pulses again with the logic pulser into input A. Again, the logic probe should indicate the presence of pulses. We can than conclude that from the results obtained with the logic probe, a short must exist between points C and D.

8.16 <u>One way to isolate the short</u> :

Place the logic pulser at the output of tristate buffer B of Z5. Then place the current tracer just to the right of the logic pulser. While actuating the logic pulser, move the current tracer slowly away from the logic pulser towards node K on the data bus. The current tracer should cease indicating the presence of current pulses just after crossing node K, which is where the short to ground exists.

8.17 The following are the most probable circuit faults which could make output Z to be always HIGH.

1. Input to IC1 is shorted to ground.
2. The output of IC1 or the input of IC2 is shorted to Vcc.
3. The output of IC2 is shorted to ground.
4. The node at the input of IC3 is shorted to ground.
5. Output Z is shorted to Vcc.

Procedure to determine faults [1] or [2] :

Inject a pulse into the input of IC1 using the logic pulser, while monitoring the output of IC1 with a logic probe. If the logic probe indicates the presence of a pulse, then faults [1] and [2] do not exist. If the logic probe doesn't detect the pulse, then the current tracer along with the logic pulser may be used to determine where the input of IC1 is grounded or the input of IC2 is shorted to Vcc.

Procedure to determine faults [3] or [4] :

Inject a pulse into the input of IC2 using the logic pulser, while checking the output of IC2 with a logic probe. If the logic probe indicates the presence of a pulse at the output, then we can eliminate possible faults [3] and [4]. However, if the logic probe does not detect a pulse between the output of IC2 and inputs of IC3, then the current tracer along with the logic pulser may be used to determine where the short to Vcc exists.

If the previous procedures do not reveal the fault, than the most probable fault is [5]. This can be verified by pulsing the node at the input of IC3 with the logic pulser, while monitoring its output with the logic probe. If output Z is indeed shorted to Vcc, then the logic probe indicator will remain lit, thereby indicating a permanent logic HIGH at the output.

9 MSI LOGIC CIRCUITS

SECTIONS 9.1-9.2 *Decoders/BCD-to-7-Segment Decoder-Drivers.*

9.1 (a) A 1-of-16 decoder is a circuit which accepts four binary inputs and has 16 outputs.
(b) A BCD-to-Decimal decoder can also be called a 1-of-10 decoder.

9.2 For output $\overline{O_{13}}$ of Figure 9.1 to be active, the 74LS138 with outputs $\overline{O_8}$-$\overline{O_{15}}$ must be selected. In order to select this decoder, signal A_3 must be HIGH and signal A_4 must be LOW. Once this decoder is selected, the binary combination at its inputs A_2, A_1, and A_0 must be such that output $\overline{O_{13}}$ is active. Thus, for output $\overline{O_{13}}$ to be LOW, inputs A_0-A_4 must be at the following logic levels: A_0=1, A_1=0, A_2=1, A_3=1, and $A_4$4=0.

9.3 The selection of the first and last decoders ($\overline{O_0}$-$\overline{O_7}$, $\overline{O_{24}}$-$\overline{O_{31}}$) are not affected by the wiring mistake, since both logic levels at inputs A_3 and A_4 are equal when these particular decoders are active. However, a problem arises when signals A_3 and A_4 are at different logic levels. Whenever the logic conditions at inputs A_0-A_4 dictate the selection of the decoder with outputs $\overline{O_8}$-$\overline{O_{15}}$, the decoder with outputs $\overline{O_{16}}$-$\overline{O_{23}}$ is selected instead, and vice-versa.

SECTIONS 9.3-9.4 *Liquid Crystal Displays/Encoders.*

9.4 (a) LCDs and LEDs have different operational requirements and therefore cannot be used interchangeably. For example, to light up a LCD segment, requires an AC signal applied between the segment and the backplane. LEDs don't have backplanes at all. LEDs require much more power to operate than LCDs. For that reason LCDs are generally found in designs which utilize CMOS devices.

(b) Reflective LCDs use ambient light, while back-lit LCDs use a light source.

9.5 The following are some of the reasons as to why segment 'a' of Figure 9.2 would be lit intermittently:

1. Output 'a' of the 4511B is open or floating.
2. Input 'a' to the LCD is open or floating.
3. The 40-Hz connection to the input of the exclusive OR gate which controls segment 'a' is open or floating.

9.6 (a) When switch SW8 is depressed, the outputs $\overline{O}_3$-$\overline{O}_0$ of the 74147 encoder are 0111_2 respectively.

(b) Since the 74147 encoder is a priority encoder, whenever two of its inputs are active simultaneously, the output will respond to the highest-numbered input. Therefore, if switches SW3 and SW8 are depressed at the same time, the output will be 0111_2, which corresponds to SW8.

9.7 The major difference between priority and non-priority encoders is that a non-priority encoder may have an ambiguous output when two of its inputs are active at the same time, while a priority encoder will not. When two or more inputs are activated simultaneously, the priority encoder will respond to the highest-numbered input.

<u>SECTIONS 9.5-9.6</u> *IEEE-ANSI Symbols/Troubleshooting.*

9.8

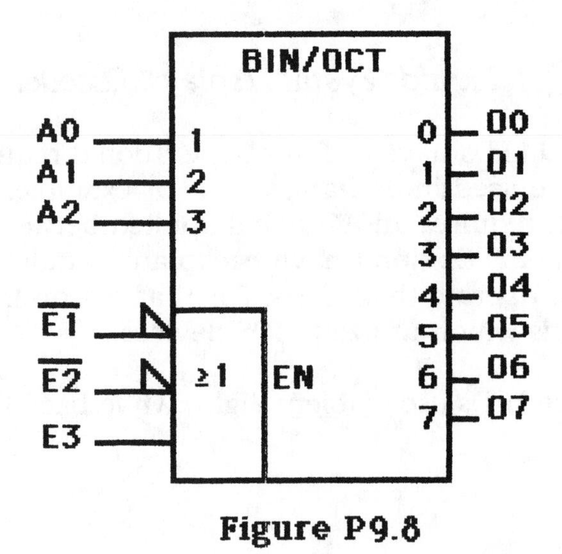

Figure P9.8

9.9 It can be concluded from the sequence of events that, as the switches are actuated, the corresponding BCD code is entered into the registers and displayed on the displays. However, some entries get displayed in the wrong LED display. By careful analysis of the recorded results, we can conclude that displays K3 and K2 always have the wrong data. These results would be observed if the connections from flip-flops Y and Z to modules K2 and K3 were switched with each other.

SECTIONS 9.7-9.11 *Multiplexers/Multiplexers Applications/ Demultiplexers/More IEEE-ANSI Symbology/ More Troubleshooting.*

9.10 (a) When inputs I_3 and I_4 are multiplexed, output Z stays at a voltage level which is halfway between a logic LOW and a logic HIGH. Two CMOS outputs will behave this way when they are tied together and each is trying to be at different logic levels at the same time. Thus, it can be concluded that multiplexer inputs I_3 and I_4 are shorted.

(b) The contents of the Storage register can be determined by looking at waveform Z, with the exception of bits X_3 and X_4. Since it was determined in the previous problem that these two inputs were shorted together and simultaneously trying to be at different logic levels, storage registers X_3 and X_4 may have either combination of bits shown by the shaded areas in Figure P9.10.

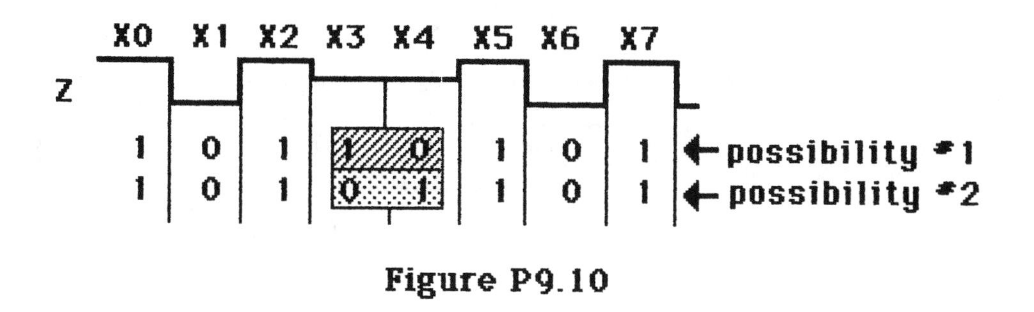

Figure P9.10

9.11 Let's set up the truth table since the expression $Z=\overline{C}\overline{B}A+C\overline{B}\overline{A}+CBA$ cannot be simplified any further either algebraically or by Karnaugh mapping:

C	B	A	Z	
0	0	0	0	
0	0	1	1	$\overline{C}\overline{B}A = 001_2 = 1_{10}$
0	1	0	0	
0	1	1	0	
1	0	0	1	$C\overline{B}\overline{A} = 101_2 = 4_{10}$
1	0	1	0	
1	1	0	0	
1	1	1	1	$CBA = 111_2 = 7_{10}$

Now, let's connect input variables ABC to S_0, S_1, and S_2 respectivelly. The levels on the inputs S_0, S_1, S_2 will determine which data at the input of the Multiplexer will be present at output Z. Clearly, output Z will be a logic HIGH only when inputs CBA are 001_2, 101_2, or 111_2. Any other combination of CBA will result in a logic LOW at output Z. Thus, the circuit of Figure P9.11 will implement the Boolean expression $Z = \overline{C}\overline{B}A + C\overline{B}\overline{A} + CBA$.

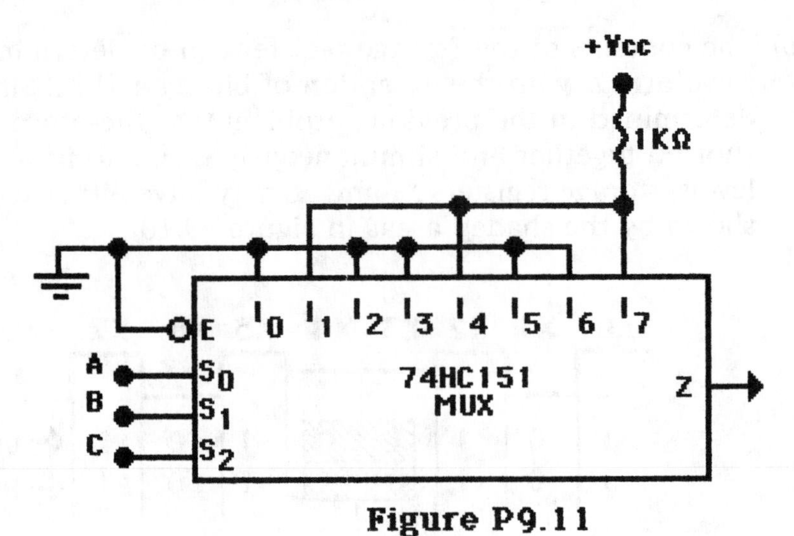

Figure P9.11

9.12 It can be readily seen that the wrong LED flashes when doors 1, 2, 5, or 6 are open. What has to be determined is if there is any commonality among these four situations.

Upon closer examination of the circuit, it can be concluded that during these four situations the MOD-8 counter has its outputs Q_0 and Q_1 with logic levels that are different from each other. Furthermore, any of these four conditions causes LED 1 to flash when LED 2 should, and vice-versa. Similar behavior is observed for LEDs 5 and 6.

On the other hand, whenever logic levels of Q_0 and Q_1 are equal, the proper LED flashes. . These facts take us to the conclusion that either of the following faults may exist:

(1) Q_0 is connected to S_1, and Q_1 is connected to S_0.
(2) Q_0 is connected to A_1, and Q_1 is connected to A_0.

9.13 The $G\frac{0}{7}$ inside an IEEE/ANSI MUX symbol, denotes the AND dependency between the select inputs (S_0 S_1 S_2) and each of the 7 data inputs (I_0 - I_7).

9.14 (a) No. This would cause the counter to stop at count three (011_2).

(b) No. This situation would cause I_6 to be always HIGH and thus allow the counter to go through the regular counting sequence. However, the time that would take the counter to increment from count 110_2 to 111_2 would be equal to the internal propagation delay of the MUX.

(c) Yes. If this was the fault, at the time when the counter reached binary 101_2, the proper input signal to actuator #5 of the Physical process, would not have been active. This would cause Sensor 5 to remain LOW and thereby disabling output $\bar{Z}$ of the MUX from clocking the counter.

(d) Yes. Same as (c).

SECTIONS 9.12-9.16 *Magnitude Comparator/Code Converters/ Data Busing/The 74173-LS173-HC173 Tristate Register/Data Bus Operation.*

9.15 The Truth Table of the 74HC85 comparator in your text book, or TTL data manual shows that, when inputs A and B are $11001101_2(205_{10})$, and 11001110_2 (206_{10}) respectivelly then the outputs are: $I_{A>B}=0$, $I_{A<B}=1$, $I_{A=B}=0$.

9.16 The circuit of P9.16 shows three 4-bit magnitude comparators arranged to compare two 9-bit binary numbers.

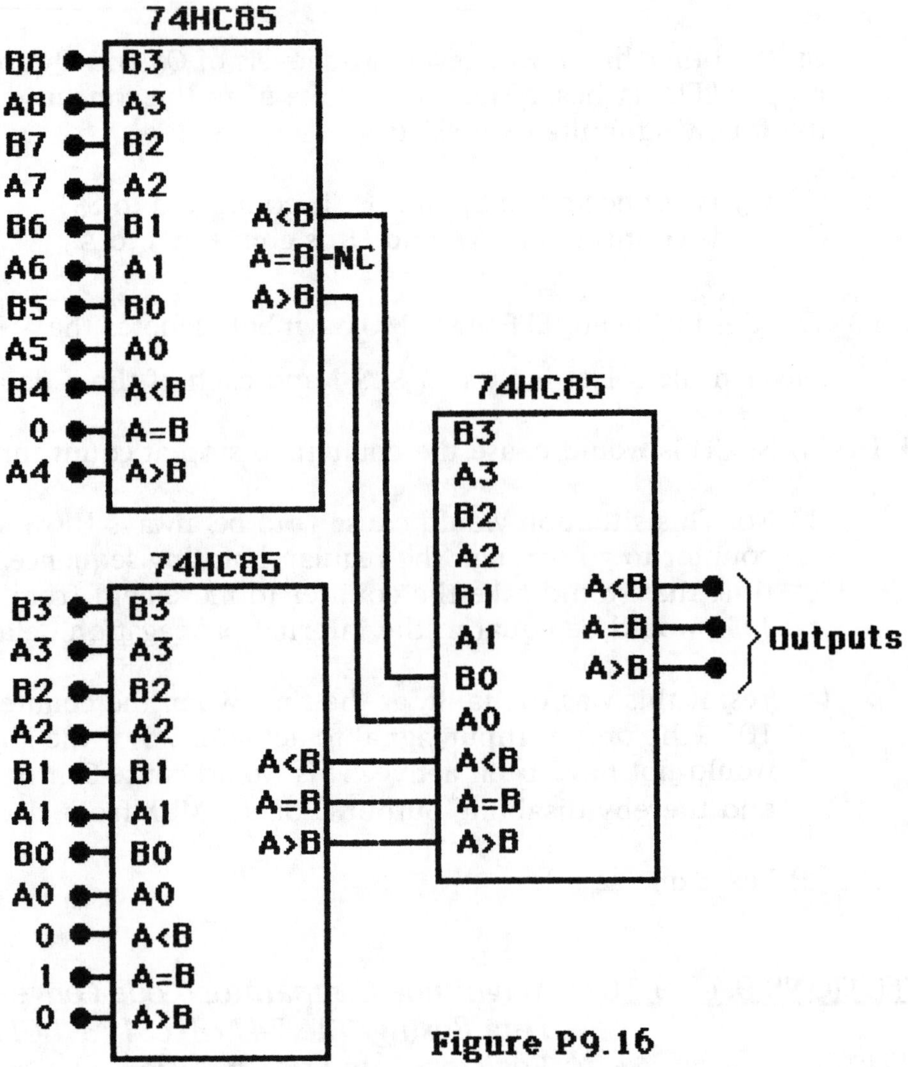

Figure P9.16

9.17 (a) After close examination of the equivalent four bit regular binary and the four bit gray code tables, it can be concluded that:

1. Gray bit G_0 is LOW only when binary bits B_0 and B_1 are equal.
2. Gray bit G_1 is LOW only when binary bits B_1 and B_2 are equal.
3. Gray bit G_2 is LOW only when binary bits B_2 and B_3 are equal.
4. Gray bit G_3 is is always equal to B_3.

Thus, a 4-bit Binary-to-Gray code converter can be implemented by using three exclusive-OR gates (Figure P9.17(a)).

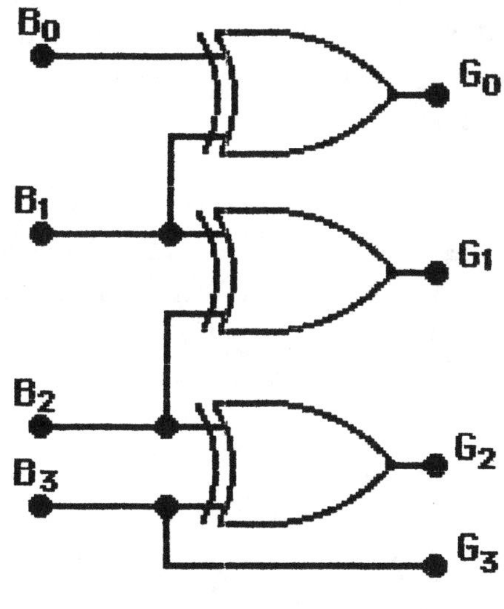

Figure P9.17(a)

(b) Again, after close examination of the equivalent four bit regular binary and the four bit gray code tables, it can be concluded that:

1. Binary bit B_0 is LOW only when gray bit G_0 and binary bit B_1 are equal.
2. Binary bit B_1 is LOW only when gray bit G_1 and binary bit B_2 are equal.
3. Binary bit B_2 is LOW only when gray bit G_2 and binary bit B_3 are equal.
4. Binary bit B_3 is always equal to G_3.

Thus, a 4-bit Gray-to-Binary code converter can be implemented by using three exclusive-OR gates (Figure P9.17(b)).

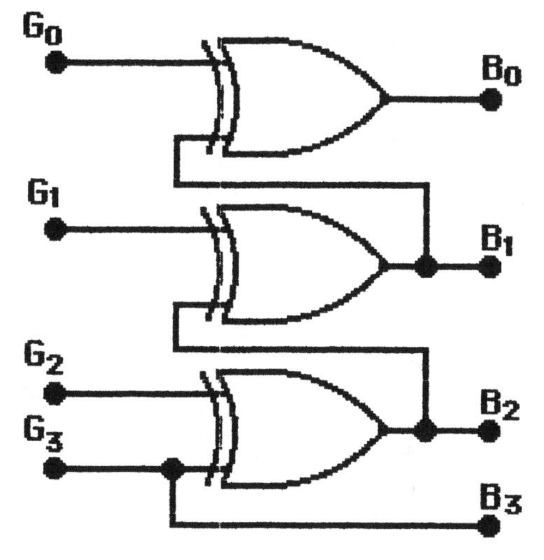

Figure P9.17(b)

9.18 (a) If input $\overline{IE}_B$ is always LOW, then any time a positive going transition occurs on the clock, the data present on the data bus will be loaded into tristate register B.

(b) If input $\overline{OE}_B$ is always LOW, then tristate register B is always placing its contents on the data bus. If either of the tristate registers A or C have their outputs enabled, then 'bus contention' will occur. 'Bus contention' occurs any time two different devices place data on the same data bus simultaneously.

10 INTERFACING WITH THE ANALOG WORLD

SECTIONS 10.1-10.2 ***Interfacing With the Analog World/ Digital-to-Analog Conversion.***

10.1 With 10 bits, there will be 1023 steps ($2^{10} - 1$). Since the full-scale value is 15V, the step-size is 14.66mV (15V/1023). Therefore, when the binary input is 1010110111_2 the output is 10.19V (695 x 14.66mV).

10.2 There are two different ways of determining the % Resolution of a D/A converter:

Method #1: $\% \text{ Resolution} = \frac{\text{Step-Size}}{\text{Full-Scale}} \times 100\% = \frac{14.66\text{mV}}{15\text{V}} \times 100\% = 0.097\%$

Method #2*: $\% \text{ Resolution} = \frac{1}{2^n - 1} \times 100\% = \frac{1}{2^{10} - 1} \times 100\% = 0.097\%$

**Note that % Resolution is only dependent on the number of bits of the Digital-to-Analog Converter.*

10.3 (a) The problem with the DAC's output staircase waveform is that it is missing a step. The missing step is the one corresponding to the input binary code 0110_2.

(b) The effect of this missing binary code is that the motor will not be able to be controlled over a certain range of RPMs. The full-scale value of the DAC is 2mA, and its step size is (2mA/15) = 133.3μA .

Therefore, when the input binary code is 101_2 the motor will be rotating at $\left(\frac{133.3\mu\text{A} \times 5 \times 1000\text{RPM}}{2\text{mA}}\right)$ = 333.3 RPMs. The next time the DAC's output increments, it will be at (133.3μA x 7) = 933.1μA and the motor will be rotating at $\left(\frac{133.3\mu\text{A} \times 7 \times 1000\text{RPM}}{2\text{mA}}\right)$ = 466.6 RPMs.

Thus, the computer is not able to set the motor to any RPM value between 333.3 and 466.6 RPMs.

10.4 (a) If the input D_1 (MSB) weight is 2.24V, then the LSB weight is (2.24V/80) = 28mV, which represents the resolution of the DAC.

(b) The output of this DAC can have a maximum of 99 steps. Thus, the full-scale value is (99 x 28mV) = 2.772V. The % Resolution of the BCD Digital-to Analog converter is $(\frac{28mV}{2.772V} \times 100\%) = 1.01\%$.

(c) If the output is 2.156V, then the DAC input must be $\frac{2.156\,V}{28mV} = 77_{10}$.

Thus, the BCD input is $0111\ 0111_{BCD}$.

SECTIONS 10.3-10.4,10.6 *D/A-Converter Circuitry/DAC Specifications DAC Applications*

10.5 In order for this DAC to have a step-size equal to -78.13mV, its R_f resistor must be equal to $\frac{-78.13mV \times 8K\Omega}{5V} = 125\Omega$.

10.6 The step-size for this DAC is $(\frac{1.590V}{2^{10}-1}) = 1.554mV$. The binary input 1100110110_2 is equivalent to 822_{10}. Therefore, in this situation, the ideal output voltage is (822x1.554mV) = 1.278V. The error that can exist on any conversion performed by this DAC is ±(0.25% x 1.59V) = ±3.975mV. Thus, for this particular binary input the output voltage of 1.283V does not fall within the acceptable accuracy range of 1.274V-1.282V.

10.7 The settling time is the time that takes the output of a DAC to settle within ±1/2 of its step size. From Figure 10.4 it can be determined that it takes about 70ns before the output oscillations stay within 1/2 the step-size (1/2 LSB).

SECTIONS 10.7-10.9 *Troubleshooting DACs/ Analog-to-Digital Conversion/Digital-Ramp ADC*

10.8 Since this is a digital-ramp ADC, its conversion time is directly proportional to the size of the analog signal being converted. The maximum conversion time for this ADC will be when the counter has to increment to its maximum count of (2^8-1) clock cycles.

The minimum conversion time is when the counter has to increment from zero to the first step. Thus, for an 8-bit digital-ramp DAC with a clock frequency of 1MHz we obtain:

The time that takes to increment the DAC by one-step is (1/1MHz) = 1μs
Thus, $t_{C(max)}$=255μs, $t_{C(min)}$=1μs and $t_{C(avg)}$=128μs.

10.9 Step-size for the Digital-to-Analog converter is (12.35V/255) = 48.43mV. Since binary 10110001_2 is equivalent to decimal 177_{10}, output $V_{A'}$ will be (48.43mV x 177) = 8.572V.

10.10 From the previous problem it can be seen that when the counter output is binary 10110001_2, $V_{A'}$ = 8.572V. If V_A = 8.572V, then the comparator will not switch since $V_{A'}$ has not exceeded V_A by V_T (0.1mV). Therefore, signal $\overline{EOC}$ (End Of Conversion) will remain HIGH. This will allow one more clock pulse to increment the counter, thereby causing the final binary count to be 10110010_2.

10.11 Several circuit conditions could cause the counter to continuously count. The following are the most probable causes:

(a) V_A is greater than the DAC's full-scale value.
(b) The output of comparator IC1 is open or shorted to Vcc.
(c) The output of the DAC ($V_{A'}$) is open.

10.12 (a) Yes, but only up to a certain point. First, there is a maximum allowable operating frequency for the counter as well as the AND gate being used. However, it is highly probable that before the counter is affected by the high input frequency, the op-amps' slew-rate within this ADC circuit will be exceeded and the ADC will start behaving erratically or stop working altogether.
(b) No. This will result in a more accurate conversion but it will take longer to perform it.
(c) See answer (a).
(d) Yes, but at the cost of lower accuracy.

SECTIONS 10.10-10.12 *Data Acquisition/Successive-Approximation Analog-to-Digital Converter/Flash ADCs.*

10.13 The step-size of the ADC is about 48.4mV (from problem 10.9). Another way which the step-size can be expressed for this ADC is 4.8mV/100μs.

The digitized waveform is obtained by joining points a thru p of Figure P10.13(a).

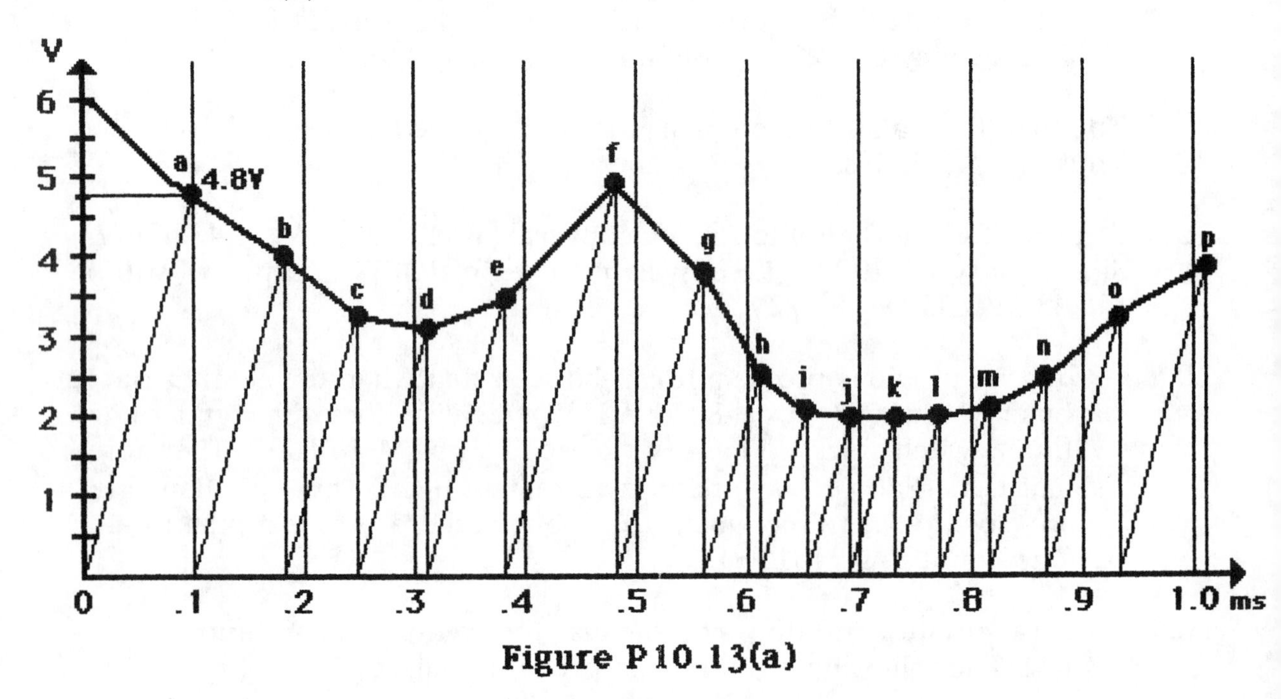

Figure P10.13(a)

(b) If a highly accurate reproduction of the analog signal of Figure 10.6 is desired, a much shorter conversion time has to be obtained. One possible way is to use an 8-bit successive-approximation ADC instead of the 8-bit digital-ramp ADC used in the circuit of Figure 10.5. Note that the average conversion for the 8-bit digital-ramp ADC of Figure 10.5 was 128μs (problem 10.8). If we used the same frequency of 1MHz and an 8-bit successive-approximation ADC, the conversion time would be (1μs x 8bits) = 8μs. A flash ADC is another possibility, since it is among the fastest ADCs available.

10.14 <u>Sample calculation for the VA'</u> : The step size is equal to 10mV. Therefore, when D_7 is HIGH (128_{10}) VA' is equal to (10mV x 128) = 1.28V.

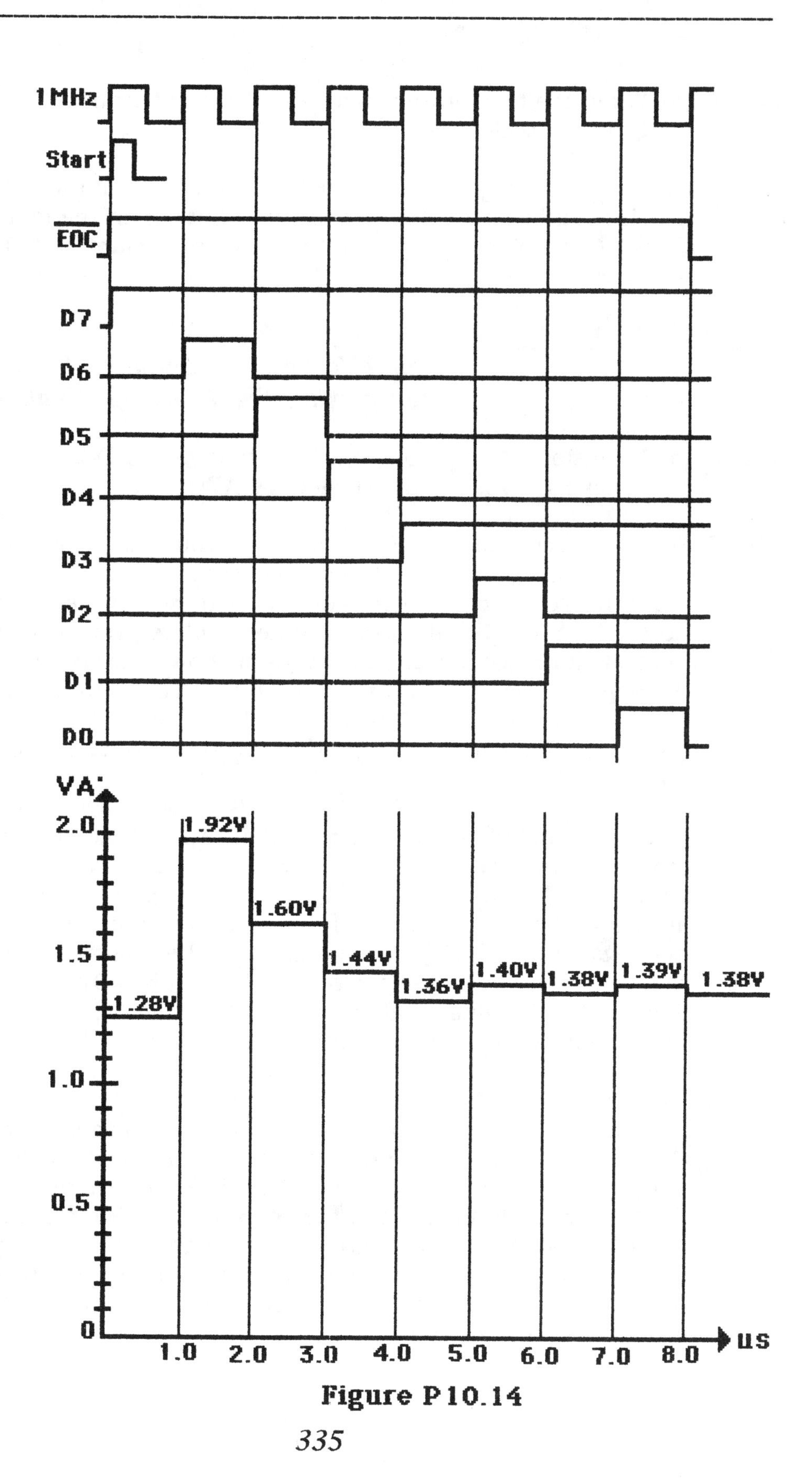

Figure P10.14

10.15 (a) When the temperature reaches 60°C the output of the ADC0804 will be at full-scale or 11111111_2.

(b) The Vref/2 input must be held at +1.5V. Then, the ADC0804's step-size will be (3.0V/255) = 11.8mV. Thus, when the temperature is at 30°C the binary output will be at its maximum or 11111111_2.

SECTIONS 10.13-10.17 *Other A/D Conversion Methods/ Digital Voltmeter/Sample-and-Hold Circuits/ Multiplexing/Digital Storage Oscilloscope.*

10.16 (a) Dual-slope ADC. (b) Voltage-to-frequency ADC.
(c) Tracking ADC. (d) Dual-slope ADC.

10.17 From the recorded results it can be determined that for input signals which are less than 90mV the display shows 000_2. This means that the counters are not being incremented for these particular analog values. If COMP signal is LOW, then the counters will be prevented from counting. Thus, it can be concluded that $V_{A'}$ somehow is always greater than 90mV, thereby keeping the output of the op-amp comparator LOW. The next step is to determine the reason why $V_{A'}$ is always greater than 90mV and how that could cause the rest of the recorded data to appear the way it does.

It can be concluded from the recorded results that the connection from the LSB of the second MSD counter to the BCD-to-Analog Converter is permanently HIGH. This would cause the display to show a count that, at certain times, would exceed the actual value by 100mV. The false readings on the display become evident whenever V_A values are such that will cause the LSB of the second MSD to be LOW (i.e. 1.0V, 8.0V, 9.0V, 9.2V). V_A values that make the LSB of the MSD of the counter to be HIGH would show as good readings by the display (i.e. 1.1V, 8.5V, 9.5V).

10.18 First the computer must make signals S1, S2, and S3 LOW, thereby placing the outputs of the transmission gates in their HI-Z state. Next, the binary data which controls the *positioning controller A* has to be sent into the DAC from the computer. After allowing sufficient time for the conversion to take place, the computer will enable transmission gate IC1-A by making S1 HIGH. IC1 must be enabled long enough for capacitor C1 to charge up to the analog voltage present at the output of the DAC. Next the computer must make S1 LOW, and then output the proper data for *positioning controller B*.

Again, after allowing sufficient time for the conversion to occur, signal S2 is made HIGH, thereby allowing C2 to charge up to the voltage at the output of the DAC. The same procedure is finally used for the last set of data, that correspond to *positioning controller C.* The timing for the sequence described above will be achieved via a computer program (software).

10.19 The maximum rate at which data can be sent to each positioning controller is limited by the conversion time of the DAC, and the time required for the capacitors of the Sample-and-Hold circuits to charge.

10.20 An increase in operational speed would be the major advantage of using three DACs instead of the multiplexing scheme of Figure 10.9. The disadvantage however would be a substantial increase in the cost of the circuit.

10.21 The following are the basic sequence of operations performed by a DSO: Data acquisition / Digitizing / Storage / Data outputting.

11 MEMORY DEVICES

SECTIONS 11.1-11.2 *Memory Terminology/General Memory Operation.*

11.1 a) The capacity of a certain memory device is 1Kx8.
b) The term used to describe an 8-bit word is a byte.
c) Any device which is capable of storing a single bit can be called a memory cell.
d) A Volatile memory is a memory that requires the application of electrical power in order to store information.
e) The amount of time required to perform a read operation is called the access time.
f) A static memory is a semiconductor memory in which the stored data will remain permanently stored as long as power is applied. Data stored in a dynamic memory, on the other hand, does not remain stored even with power applied, unless it is periodically refreshed.
g) Internal memory stores instructions and data the CPU is currently working on.

11.2 The following is the procedure used to Read the contents of memory location 11001_2 of Figure 11.1: First apply address 11001_2 to the Address inputs (A_4-A_0) of the 32x4 memory. Then make the Memory Enable input HIGH. And finally, make the $R/\overline{W}$ input HIGH. Data will now be ready at the Data outputs.

11.3 2^{10}=1024 (memory locations).

SECTION 11.3 *CPU-Memory Connections*

11.4 a) The bus which carries data between the CPU and the memory ICs is *Bidirectional.*
b) During a *READ* operation data may flow from a memory IC into the CPU via the data bus.
c) During a *WRITE* operation data flows out of the CPU via the data bus.

SECTIONS 11.4-11.6 *Read-Only Memories/ROM Architecture/ ROM Timing.*

11.5 When A_0=1, A_1=1, A_2=1, and A_3=0, Row 3 and Column 1 are active. This combination of Column and Row will enable Register 7 to send its contents to the Output buffers D_7-D_0.

11.6 For Register 14 to be selected, its Enable inputs must be HIGH simultaneously. Thus, any anomaly that would cause an open on either enable input will prevent register 14 from being accessed.

11.7 a) Since there are ten address lines going into the memory, it can be determined that there are 2^{10} memory locations or 1024 locations. This memory has four output lines, therefore it can be concluded that each memory location can store any combination of four 1s and 0s. Thus, the capacity of this memory is 1024x4 or 1Kx4.

b) Since this memory has four data inputs, four data outputs and a $R/\overline{W}$ input, it can be deduced that it is a RAM. Remember that, data can be stored (written) only once into ROMs. Thus, there is no need for either data or $R/\overline{W}$ input lines in a ROM.

11.8 The data which is loaded into Register A, cannot be determined. This unpredictability stems from the fact that the Enable input of the memory must be LOW for a minimum of 150ns (t_{OE}) in order for stable data to be present at the output of the memory. Since we are allowing the Enable pulse to be LOW for only 100ns, ambiguous and unpredictable data will be loaded into Register A.

SECTIONS 11.7-11.10 *Types of ROMs/Flash Memory/ROM Applications/ Programmable Logic Devices.*

11.9 $O_0 = \overline{D}\overline{C}\overline{B}A + \overline{D}\overline{C}B\overline{A} + \overline{D}\overline{C}BA + \overline{D}C\overline{B}A + \overline{D}CB\overline{A} + \overline{D}CBA + D\overline{C}\overline{B}A + D\overline{C}B\overline{A} + D\overline{C}BA + DC\overline{B}A + DCB\overline{A} + DCBA$

$O_1 = \overline{D}\overline{C}B\overline{A} + \overline{D}\overline{C}BA + \overline{D}C\overline{B}\overline{A} + \overline{D}C\overline{B}A + D\overline{C}B\overline{A} + D\overline{C}BA + DC\overline{B}\overline{A} + DC\overline{B}A$

$O_2 = \overline{D}C\overline{B}\overline{A} + \overline{D}C\overline{B}A + \overline{D}CB\overline{A} + \overline{D}CBA + D\overline{C}\overline{B}\overline{A} + D\overline{C}\overline{B}A + D\overline{C}B\overline{A} + D\overline{C}BA$

$O_3 = D\overline{C}B\overline{A} + D\overline{C}BA + DCB\overline{A} + DCBA$

Let's use Karnaugh maps to simplify the Sum-of-Products expressions (O_0-O_3).

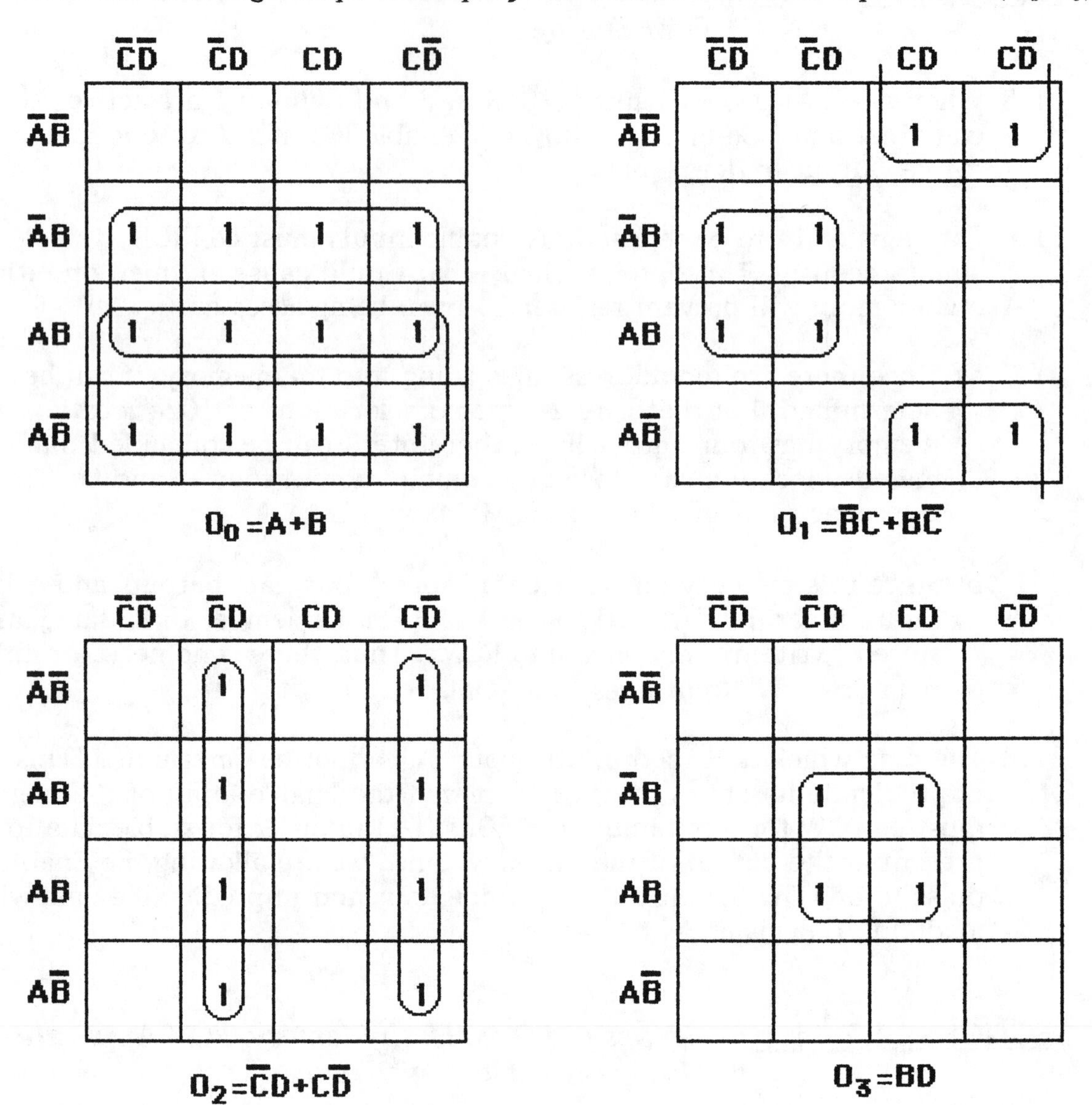

1.10 Figure P11.10 shows the fuse pattern of a PAL whose outputs O0-O3 determined in the previous problem, are equivalent to those of programmed PROM of Figure 11.4.

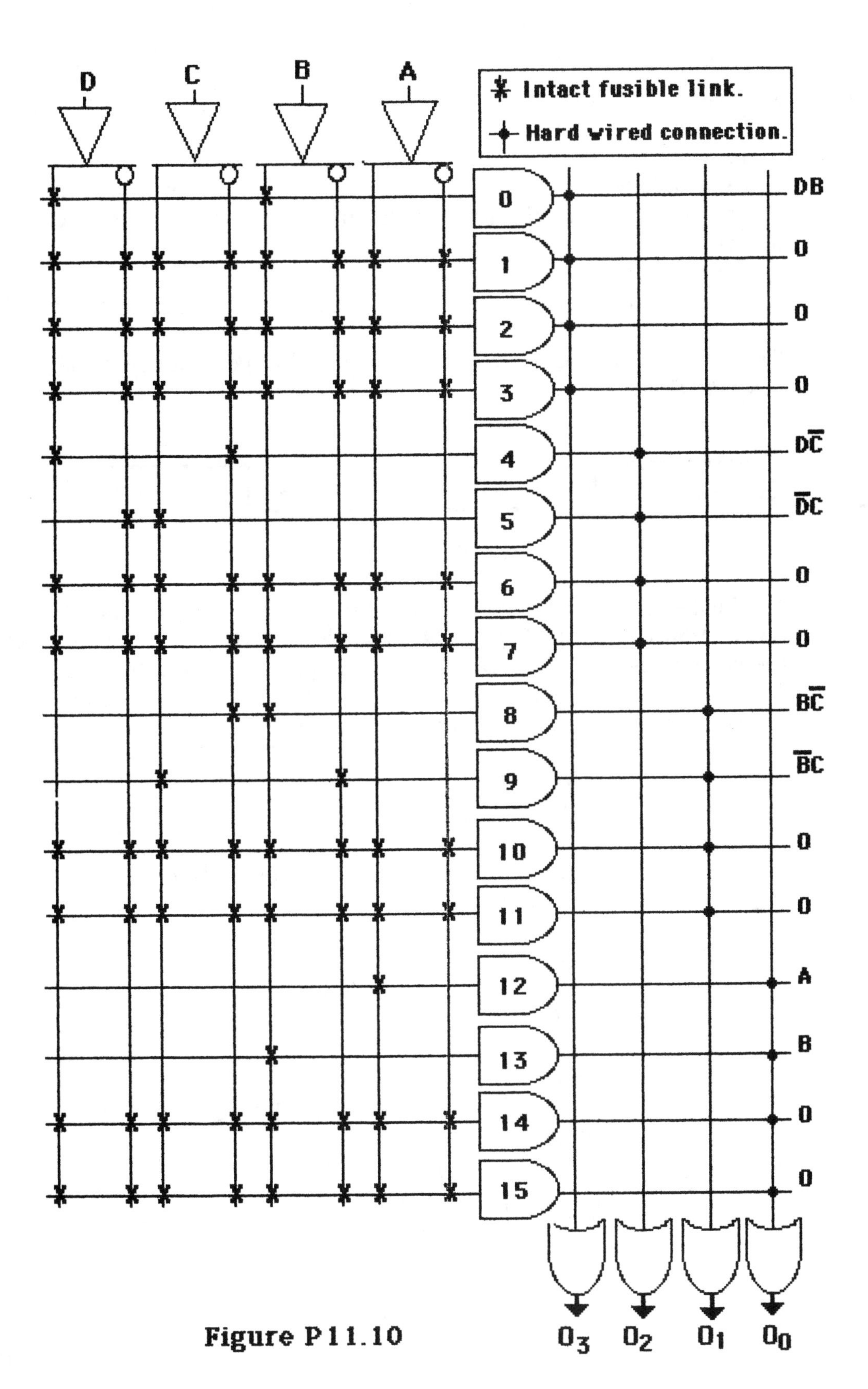

Figure P11.10

11.11 a) False. EPROMs do not need power applied to them in order for them to keep data stored.
b) True.
c) False. MROMs are very expensive to build and therefore should only be used in circuit designs where large quantities are needed.
d) True.
e) True.
f) False. PLAs are often referred to as field-programmable arrays (FPLAs).

11.12 Figure P11.12 is the waveform that should appear at the output X of the 16x1 PROM.

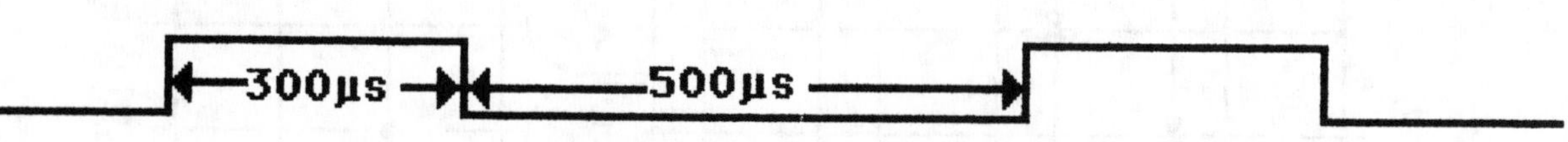

Figure P11.12

The MOD-16 counter is incremented once every 100μs (ClK=10KHz). This means that the 16x1 PROM is being accessed every 100μs. Consequently, the content of each memory location is present at the output for only 100μs. The following is the data that must be stored in the ROM for such the waveform of Figure P11.12 to be generated at output X:

Address	Data	
0000	1	}
0001	1	} ON for 3-bits (300μs).
0010	1	}
0011	0	]
0100	0	]
0101	0	] OFF for 5-bits (500μs).
0110	0	]
0111	0	]
1000	1	}
1001	1	} ON for 3-bits (300μs).
1010	1	}
1011	0	]
1100	0	]
1101	0	] OFF for 5-bits (500μs).
1110	0	]
1111	0	]

11.13 (a) O_3 would be the ORed result of the functions present at the outputs of AND gates 1,2,3.

(b) O_3 would always be a logic 1.

11.14 (a) This two-step sequence erases all cells in the 28F256A CMOS array.

(b) After all memory cells have been erased, all bytes = 11111111_2=FF_{16}.

SECTIONS 11.13-11.17 *Static RAM (SRAM)/Dynamic RAM (DRAM)/ Dynamic RAM Structure and Operation/ DRAM Read/Write Cycles/ DRAM Refreshing*

11.15 a) SRAM memory cells are essentially *flip-flops* that will stay in a given state indefinitely, provided that power to the circuit is not interrupted.

b) If the t_{RC} of a SRAM is 50ns, the CPU can read *20* million words per second.

c) Dynamic RAMs require periodic recharging of the memory cells; this is called *refreshing* the memory.

d) In order to reduce the number of pins on high-capacity DRAM chips, manufacturers utilize address *multiplexing.*

e) During a read cycle the $\overline{RAS}$ signal is activated *before* the $\overline{CAS}$ signal.

f) $\overline{RAS}$-only refresh method is performed by strobing in a row address with $\overline{RAS}$ while $\overline{CAS}$ and $R/\overline{W}$ remain HIGH.

SECTIONS 11.18-11.19 *Expanding Word Size and Capacity/ Special Memory Functions*

11.16 Figure P11.16 shows a 4Kx8 memory. Note that, the shaded area is a 1Kx8 arrangement and that we need three more identical arrangements to have a total of 4Kx8. All eight $R/\overline{W}$ lines are connected together as well as all of the eight $\overline{CS}$ lines.

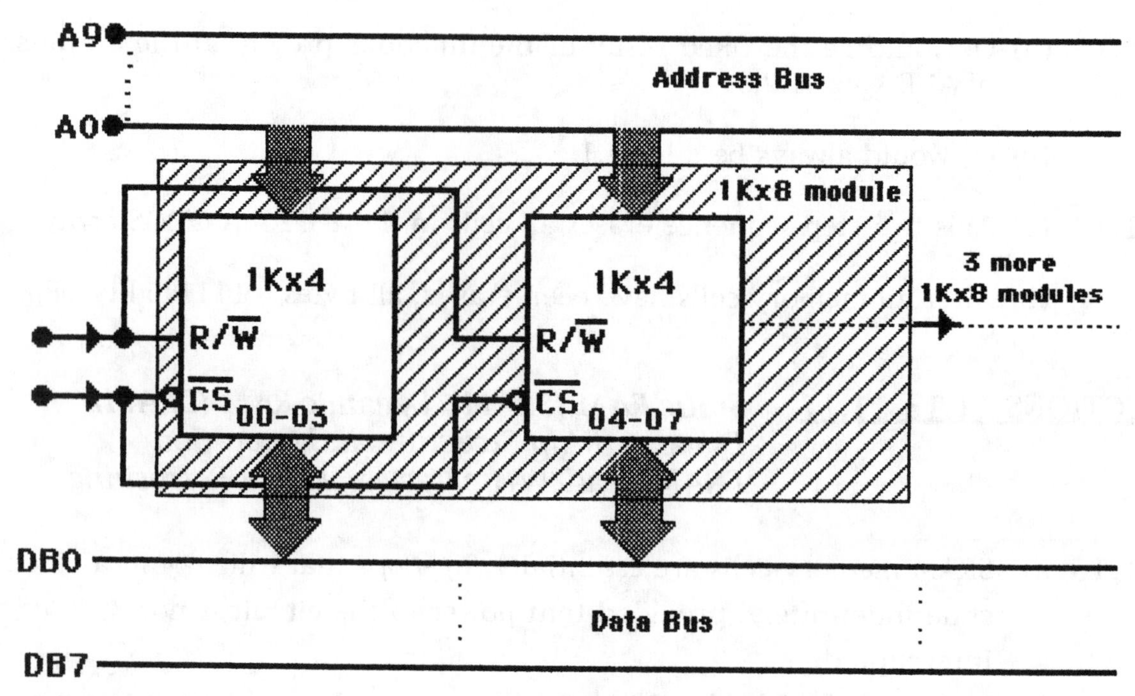

Figure P11.16

11.17 The 2125A RAM is a 1Kx1 memory and therefore it requires 8 ICs to built a 1Kx8 memory block. Thus, it will take 32 of the 2125A RAM ICs (1Kx8s) in order to build a 4Kx8 memory module.

11.18 Since it's stated that regardless of the address on the Address bus the data on the Data bus is always the same, we can conclude that outputs 0 thru 3 of the decoder are always HIGH. Thus, there are only two possible causes for this malfunction. One is an inoperative 3-to-8 line decoder. The other possibility, and most probable, is input C to the decoder has become open. This would allow only outputs 4 thru 7 of the decoder to be active, which of course would not enable any of the PROMs in the circuit.

11.19 One method of preventing the loss of memory data during a system power failure is to store in RAM critical data during a normal system operation. The RAM is powered from backup batteries whenever power is lost. There are special types of CMOS RAMs that have a small lithium battery built right on the chip for this very purpose.

A second method of preventing the loss of memory data during a system power failure is to store critical system data in nonvolatile flash memory. This method has the advantage of not requiring any backup battery power. However, it is more difficult to change the data stored in a flash memory than in a static RAM.

A third method of preventing the loss of memory data during a system power failure is to have the CPU store all data in high-speed, volatile RAM during normal system operation. During a power interruption the CPU executes a short power-down program that resides in ROM which transfers critical data from the system RAM into either battery-backup CMOS RAM or flash memory. Upon resumption of power, the CPU executes a power-up program from ROM that transfers the critical data from the backup storage memory to the system RAM.

SECTIONS 11.20-11.21 *Troubleshooting RAM Systems/Testing ROM.*

11.20

Module	A_{15}	A_{14}	A_{13}	A_{12}	A_{11}	A_{10}	A_9	A_8	A_7	A_6	A_5	A_4	A_3	A_2	A_1	A_0
0	1	0	0	0	0	0	X	X	X	X	X	X	X	X	X	X
1	1	0	0	0	0	1	X	X	X	X	X	X	X	X	X	X
2	1	0	0	0	1	0	X	X	X	X	X	X	X	X	X	X
3	1	0	0	0	1	1	X	X	X	X	X	X	X	X	X	X

Module 0: 8000_{16}-$83FF_{16}$ Module 1: 8400_{16}-$87FF_{16}$
Module 2: 8800_{16}-$8BFF_{16}$ Module 3: $8C00_{16}$-$8FFF_{16}$

11.21 If the decoder is enabled during the time when the address bus is still changing, then it is possible that erroneous data are written into various memory locations. Note that when the address bus is changing from the old to the new address the data bus is in the HI-Z state (floating). If at this time, memory is allowed to be written into, then whatever is floating on the data bus will be stored in memory. Thus, one can deduce that probably the RAM is enabled during the time when the address bus is changing. One of the functions of the ϕ_2 clock is not allowing the decoder outputs to change during the time when the address bus is changing.

Thus, if during this critical time, input $\overline{EI}$ of the decoder is LOW, the results would be as the technician witnessed. This suspicion can be verified by using a dual trace oscilloscope.

The oscilloscope should be triggered on the positive slope of the $ø_2$ clock signal. The outputs of the decoder should then be monitored one-by-one on the other channel of the oscilloscope. Decoder outputs 0 thru 3 should go LOW only during the time when the $ø_2$ clock is HIGH.

11.22

Address	Binary Data	Hex Data
0000	00110110	36
0001	11001100	CC
0010	11000001	C1
0011	00111001	39
0100	11111111	FF
0101	00000000	00
0110	01110011	73
0111	10011001	99
1000	00010001	11
1001	01011111	5F
1010	11000011	C3
1011	10010010	92
1100	00111101	3D
1101	11100000	E0
1110	10000000	80
1111	*01101001*	*69*

The checksum is obtained by adding the fifteen hex data numbers, ignoring carries from the MSD. This can also be done by adding the fifteen 8-bit data words ignoring the carries from the MSB. The following is the procedure which was used to arrive at the checksum of 69:

36+CC=02 ----> 02+C1=C3 ----> C3+39=FC-----> FC+FF=FB----> FB+00=FB
FB+73=6E -----> 6E+99=07 -----> 07+11=18 ----> 18+5F=77----> 77+C3=3A
3A+92=CC ----> CC+3D=09 ----> 09+E0=E9 ----> E9+80=[*69*]

12 APPLICATIONS OF A PROGRAMMABLE LOGIC DEVICE

SECTIONS 12.1-12.2 *The GAL16V8A/Programming PLDs*

12.1 (a) The major components of the GAL devices are the input term *matrix*, the *AND* gates, and the *Output Logic Macro Cells* *(OLMC)*.
(b) Within each OLMC the products are *ORed* together to generate the Sum-of-Products expression.
(c) The GAL16V8A has three different modes of operation. They are the *Simple* mode, the *Complexed* mode, and the *Registered* mode.

12.2 In order for the GAL16V8A to be programmed in the complex mode, the signals *SYN, AC0,* and *AC1* must all be equal to 1.

12.3 $Z = ABC + \overline{A}\overline{B}C + A\overline{B}\overline{C}$

12.4

Mode	*Configuration*	*SYN*	*AC0*	*AC1*
Simple mode	*IN*	*1*	*0*	*1*
	OUT	*1*	*0*	*0*
Complex mode		*1*	*1*	*1*
Registered mode	*Registered*	*0*	*1*	*0*
	Combinational	*0*	*1*	*1*

12.5

1. A personal computer.
2. PLD development software.
3. A programming fixture.
4. Software to drive the programming fixture.
5. A Programming Logic device.

12.6 (a) A *fuse plot* is a file that is like a map that shows which fuses in a programmable device are to be fused open and which ones are to remain intact.

(b) *Downloading* is a term used to describe the action of sending a programming file over a cable to the programming device.

(c) A *JEDEC* is standardized file which is loaded into any *JEDEC*-compatible PLD programmer that is capable of programming the desired type of PLD.

SECTIONS 12.3-12.4 *Development Software/Universal Compiler for Programmable Logic (CUPL).*

12.7 1. *Low-level development systems.*
2. *High-level logic Compilers.*

12.8 (a) False. *High-level* development systems will accept the Boolean equations in an ASCII input file.
(b) True.
(c) True.

12.9 (a) AB ------> A & B
(b) A+B ------> A # B
(c) $\overline{A}$ ------> !A
(d) $A \oplus B$ ------> A $ B
(e) $\overline{A \oplus B}$ ------> !A & !B # A & B

12.10 X = !A # !C & D # A $ C # !B & D

$$X = \overline{A} + \overline{C} \cdot D + A \cdot \overline{B} + \overline{A} \cdot B + \overline{B} \cdot D = \overline{A} + \overline{C} \cdot D + (A \oplus B) + \overline{B} \cdot D$$

12.11

B A	$\overline{O_0}$	$\overline{O_1}$	$\overline{O_2}$	$\overline{O_3}$
0 0	0	1	1	1
0 1	1	0	1	1
1 0	1	1	0	1
1 1	1	1	1	0

Write equations for LOW outputs:

!O0 = !B & !A
!O1 = !B & A
!O2 = B & !A
!O3 = B & A

Source File

```
Name            ;
Partno          ;
Date            ;
Revision        ;
Designer        ;
Company         ;
Assembly        ;
Location    Problem 12.11;
Device          G16V8A;
/*****************************************************************************/
/*                                                                           */
/*  7442 (1-OF-10 DECODER)                                                   */
/*                                                                           */
/*****************************************************************************/
/*              Target Device GAL 16V8A                                      */
/*****************************************************************************/
/*          Inputs    */
pin 1       = A              ;      /* LSB select input                      */
pin 2       = B              ;      /* MSB select input                      */

/*          Outputs   */
pin 19      = !O0            ;      /* Goes LOW for BA = 00                  */
pin 18      = !O1            ;      /* Goes LOW for BA = 01                  */
pin 17      = !O2            ;      /* Goes LOW for BA = 10                  */
pin 16      = !O3            ;      /* Goes LOW for BA = 11                  */
/*****************************************************************************/
/*              Equations                                                    */
/*****************************************************************************/
O0.OE  = 'b'1;
O1.OE  = 'b'1;
O2.OE  = 'b'1;
O3.OE  = 'b'1;

O0 = !B & !A;
O1 = !B &  A;
O2 = B & !A;
O3 = B &  A;
```

12.12

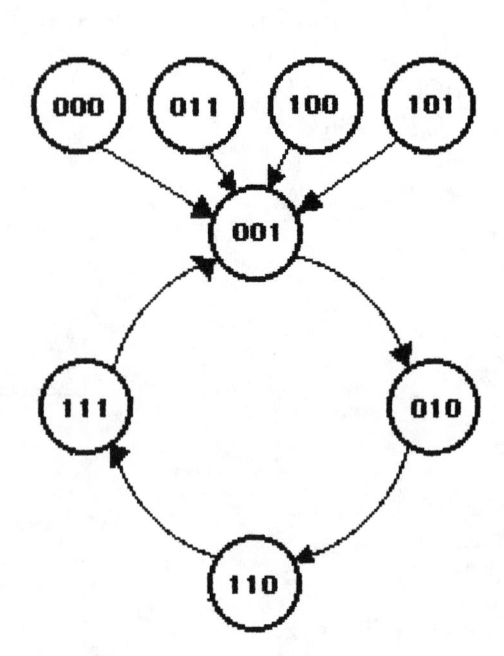

Present			Next		
QC	QB	QA	QC	QB	QA
0	0	0	0	0	1
0	0	1	0	1	0
0	1	0	1	1	0
0	1	1	0	0	1
1	0	0	0	0	1
1	0	1	0	0	1
1	1	0	1	1	1
1	1	1	0	0	1

Next QC

	$\overline{C}$	C
$\overline{A}\,\overline{B}$		
$\overline{A}\,B$	1	1
$A\,B$		
$A\,\overline{B}$		

Next QB

	$\overline{C}$	C
$\overline{A}\,\overline{B}$		
$\overline{A}\,B$	1	1
$A\,B$		
$A\,\overline{B}$	1	

Next QA

	$\overline{C}$	C
$\overline{A}\,\overline{B}$	1	1
$\overline{A}\,B$		1
$A\,B$	1	1
$A\,\overline{B}$		1

$QA = C + \overline{A}\cdot\overline{B} + A\cdot B$

$QB = \overline{A}\cdot B + A\cdot\overline{B}\cdot\overline{C}$

$QC = \overline{A}\cdot B$

Source File

```
Name            ;
Partno          ;
Date            ;
Revision        ;
Designer        ;
Company         ;
Assembly        ;
Location    CHAP.12 problem 12.12;
Device      G16V8A;
/*******************************************************************************************/
/*                                                                                         */
/*  SYNCHRONOUS COUNTER DESCRIBED ON PROBLEM 7.14(a) ON PAGE 125                           */
/*                                                                                         */
/*******************************************************************************************/
/*              Target Device GAL 16V8A                                                    */
/*******************************************************************************************/
/*          Inputs      */
pin 1       = CLOCK             ;   /* COUNTER CLOCK                         */

/*          Outputs     */
pin 11      = OE                ;   /* OUTPUT ENABLE (ACTIVE LOW)            */
pin 12      = QA                ;   /* OUTPUT QA                             */
pin 13      = QB                ;   /* OUTPUT QB                             */
pin 14      = QC                ;   /* OUTPUT QC                             */

/*******************************************************************************************/
/*              Equations                                                                  */
/*******************************************************************************************/
QA.D    =   C # !A & !B # A & B;
QB.D    =   !A & B # A & !B & !C;
QC.D    =   !A & B;
```

13 INTRODUCTION TO THE MICROPROCESSOR AND MICROCOMPUTER

SECTIONS 13.4-13.6 *Basic Computer System Organization/ Basic μC Elements/Computer Words.*

13.1 (a) [7-ALU] (b) [8-Memory Unit] (c) [9-Magnetic-Strip]
(d) [4-Output Unit] (e) [6-Control Unit] (f) [1-CPU]
(g) [5-Byte] (h) [2-Peripherals] (i) [11-Word]

SECTIONS 13.7-13.8 *Instruction Words/Executing a Machine-Language Program.*

13.2 One byte, the op-code.

13.3 The Op-Code is the information contained in a single-byte instruction.

13.4 The last two bytes represent the Operand Address.

13.5 Machine language (1s and 0s) is the only language a computer (machine) understands.

13.6 a) LDA is the mnemonic which instructs the computer to *load the accumulator.*
b) STA $0300 instructs the computer to take the contents of the *accumulator* and store them in memory location whose hexadecimal ($) address is *0300.*
c) The Program Counter is a counter within the *control* unit.
d) The computer is always in one of two kinds of operating cycles: *fetch cycle* or *execute cycle.*

13.7 *The first instruction is LDA $0500.*

This causes the computer to Load the accumulator with the contents of memory location $0500. Since memory location $0500 has FF_{16} stored in it, the accumulator will be loaded with FF_{16}.
Note that $ simply means that the number that follows is a hexadecimal number.

The second instruction is JMP $0300.

This instruction sends the computer directly to address $0300 since it represents an unconditional jump. At address $0300 the computer just continues with the execution of the program.

The third instruction is STA $0400.

This instruction tells the computer to store the contents of the accumulator in memory location $0400. Since the accumulator at this time has FF_{16} stored in it, data FF_{16} will be placed in memory location $0400.

The fourth instruction is HLT.

Upon execution of this instruction the computer will Halt from executing any more instructions. Therefore, at the end of the program, memory location $0400 has [$FF_{16}$] stored in it.

SECTIONS 13.9 *Typical μC Structure*

13.8 The direction of data flow on the data bus is determined by the instruction which the computer is executing. If the instruction requires the computer to perform a READ operation, then data flows into the CPU. If the instruction requires the computer to perform a WRITE operation, then data flows out of the CPU.

13.9 The 8085μP will take a total of four machine cycles and a total of 13 T states to execute the instruction STA $0400.

13.10 *LDA $0500*

READ op-code $3A_{16}$.
READ lower-byte of the operand address 00_{16}.
READ higher-byte of the operand address 05_{16}.
READ the contents of memory location $0500 into the accumulator.

To execute the instruction LDA $0500, it requires four READ operations.

JMP $0300

READ op-code $C3_{16}$.
READ lower-byte of the operand address 00_{16}.
READ higher-byte of the operand address 03_{16}.

To execute the instruction JMP $0300, it requires three READ operations.

STA $0400

READ op-code 32_{16}.
READ lower-byte of the operand address 00_{16}.
READ higher-byte of the operand address 04_{16}.
WRITE the contents of the accumulator into memory location $0400.

To execute the instruction STA $0400, it requires three READ operations and one WRITE operation.

HLT

READ op-code $3F_{16}$.

To execute the instruction HLT, it requires one READ operation.

Thus, during the execution of the program of problem 12.7, the computer performs 11 READ operations and 1 WRITE operation.

TEST 1

1.....(d)
2.....(c)
3.....(d)
4.....(c)
5.....(d)
6.....(b)
7.....(a)
8.....(b)
9.....(c)
10...(d)

TEST 2

1.....(b)
2.....(c)
3.....(b)
4.....(b)
5.....(d)
6.....(d)
7.....(b)
8.....(d)
9.....(a)
10...(a)

TEST 3

1.....(b)
2.....(b)
3.....(d)
4.....(b)
5.....(c)
6.....(d)
7.....(c)
8.....(b)
9.....(a)
10...(b)

TEST 4

1.... (a)
2.... (a)
3.... (c)
4.... (b)
5.... (b)
6.... (a)
7.... (d)
8.... (c)
9.... (a)
10.. (a)

TEST 5

1.... (e)
2.... (d)
3.... (b)
4.... (c)
5.... (e)
6.... (d)
7.... (d)
8.... (e)
9.... (c)
10.. (b)

TEST 6

1.... (b)
2.... (d)
3.... (d)
4.... (c)
5.... (a)
6.... (b)
7.... (c)
8.... (b)
9.... (a)
10.. (b)

TEST 7

1....(c)
2....(a)
3....(d)
4....(b)
5....(a)
6....(c)
7....(d)
8....(b)
9....(a)
10..(b)

TEST 8

1....(b)
2....(d)
3....(c)
4....(d)
5....(b)
6....(b)
7....(d)
8....(b)
9....(a)
10..(a)

TEST 9

1....(d)
2....(c)
3....(b)
4....(c)
5....(a)
6....(a)
7....(b)
8....(d)
9....(b)
10..(a)

TEST 10

1.....(a)
2.....(b)
3.....(a)
4.....(a)
5.....(a)
6.....(a)
7.....(b)
8.....(a)
9.....(a)
10...(b)

TEST 11

1.....(c)
2.....(d)
3.....(a)
4.....(c)
5.....(b)
6.....(d)
7.....(a)
8.....(b)
9.....(b)
10...(b)

TEST 12

1.....(b)
2.....(d)
3.....(c)
4.....(b)
5.....(d)
6.....(b)
7.....(a)
8.....(d)
9.....(b)
10...(b)

TEST 13

1.... (c)
2.... (d)
3.... (c)
4.... (c)
5.... (a)
6.... (d)
7.... (d)
8.... (a)
9.... (b)
10.. (b)